FLIGHT IN THE CHINA AIR SPACE

1910-1950

by

MALCOLM ROSHOLT

Malcolm Rosholt

ROSHOLT HOUSE
Box 104
Rosholt, Wi 54473

Library of Congress Catalog Card No.: 84-061079

ISBN: 0-910417-04-0

Printed in the United States of America by
Palmer Publications, Inc., Amherst, WI 54406

FOREWORD

Flight in the China Air Space presents an overview of aviation of China covering single engine and multi-engine propeller-driven aircraft from 1910 to 1950. No attempt is made to describe every Chinese, American or Japanese fighter or bomber squadron which flew in the China Air Space before or during World War II. I have, however, included several unit citations awarded American squadrons, or groups.

My own interest in China goes back many years. I worked from 1931-1937 as a reporter and later as an editor on the *China Press,* an English-language daily published in Shanghai. I helped cover the first Japanese attack on the city in the so called "Shanghai Incident" in 1932, and I helped cover the second "Shanghai Incident" in 1937 that went on to become a Seven-Year-War.

I returned to the United States in 1938. In 1942 I was commissioned a 1st lieutenant in the Army Air Corps, largely on the merits of my background in the Chinese language which I had studied in Shanghai. After boot training at Miami Beach, I went on to Harrisburg for training in combat intelligence. I returned to China in November 1943, this time flying over the Burma mountains, better known then as "the Hump." After a month in the war room of the 14th Air Force headquarters in Kunming, I was sent out to join Chinese armies in the field to act as a liaison between our squadrons and Chinese ground forces. I was at Changsha with the Chinese 9th War Area headquarters when the Japanese launched *Ichi-Go* (One-Five), code name for the biggest campaign of the Seven-Year war in China. I retreated 200 miles into south China with the Chinese headquarters battalion to avoid encirclement. I was discharged in Washington, D.C., 13 December 1945.

In the spring of 1949, I joined Civil Air Transport, an airline in China organized by Claire L. Chennault and Whiting Willauer. I worked in public relations and continued with the firm until 1952 when I returned to the Zone of Interior. Since 1953, I have been living in the village of my nativity, free lancing and publishing books, mostly on state and local history. Since 1979, I have also served as editor of the annual *Pictorial Magazine* published by veterans of the 14th Air Force Association, Inc., a non-profit organization.

In retrospect, I can say with Virgil's Aeneas: "Many of these things I saw, and some of them I was."

The photographs in the book have come from widely scattered sources, private and public, some from the National Air & Space Museum and some from the San Diego Aero-Space Museum. Air Corps and Signal Corps pictures were also purchased from the Air Force archives which, in 1982, were located at Anacostia Naval Air Station.

I am especially grateful to Leo Braun who served with the 21st Photo Reconnaissance Squadron in China and managed to preserve a prize collection of air views of fields used by American squadrons. Some of these fields have since reverted to rice paddies, and others have been modified for civilian use and are scarcely recognizable to returning veterans on tour.

Most of the pictures of early aircraft flown by Chinese-American pilots are from the Glenn H. Curtiss Museum of Local History, Hammondsport, New York. Merrill Stickler, curator of the museum, also furnished information on the pictures.

William M. Leary, Jr. professor of history at the University of Georgia, provided editorial guidance and assistance on documentary sources and microfilms.

Ray Wagner, author of *American Combat Planes,* provided both assistance and editorial matter. He was in Moscow, USSR, a few years ago and returned with an article from *Red Star* which described the Soviet Air Force in China 1937-39. He generously allowed me to use the translation here.

Jack Schweitzer, Reed Knight, and W.A. Spencer, the only three survivors of the Jouett Mission to China, provided unpublished pictures and information. Sebie Biggs Smith, closely associated with Claire L. Chennault in 1937-38, also helped with both his time and pictures.

For other favors and pictures, I am indebted to Edward Leiser and Santiago Flores Ruiz of the San Diego Aero-Space Museum; to John Underwood for his pictures of Katherine Stinson; to Susi Sato of Arizona State University at Tempe; to James V. Edmundson and Alva Harvey, former group commanders of the 58th Bomb Wing; to Denny Pidhayne, secretary of the 58th Bomb Wing Association; to Ralph F. Mitchell for the Stinson aircraft pictures; and to Ivaloo Watson for permission to use the diary kept by her late husband, Rolfe, covering his years in China, and for her own personal account of a trip over the Burma Road.

A letter to the editors of *Time* Magazine about my friend, John Birch, a fellow intelligence officer in China, brought me to the attention of Harry Kjus, Mayville, Illinois. He sent me pictures and letters written home from China by an older brother, Carl Kjus, and from these materials I pieced together the chapter on "The Flight of the Tin Goose."

To my daughter Mei-fei (Mrs. David Elrick, Guelph, Ontario) and to Margaret, my wife and companion of fifty years, my love and affection for their help and editorial assistance.

In the beginning it was my intention to follow the Wade-Giles system of romanization in the spellings of Chinese place names as found in the CIA *Communist China Administrative Atlas* (1968), but now a new system, *pin-yin* (phonetics) has been introduced by the Peking government, but a system, in my opinion, which is inferior to Wade-Giles. I use the traditional spellings of place names found on English maps of China-Burma from the 1930s and 1940s. The spellings of personal names follow either the spellings found in the original document, or a limited use of Wade-Giles, since no completely accurate rendering of a name can be made under any system without access to the original *tze* or "characters."

Malcolm Rosholt
Box 104, Rosholt, WI 54473

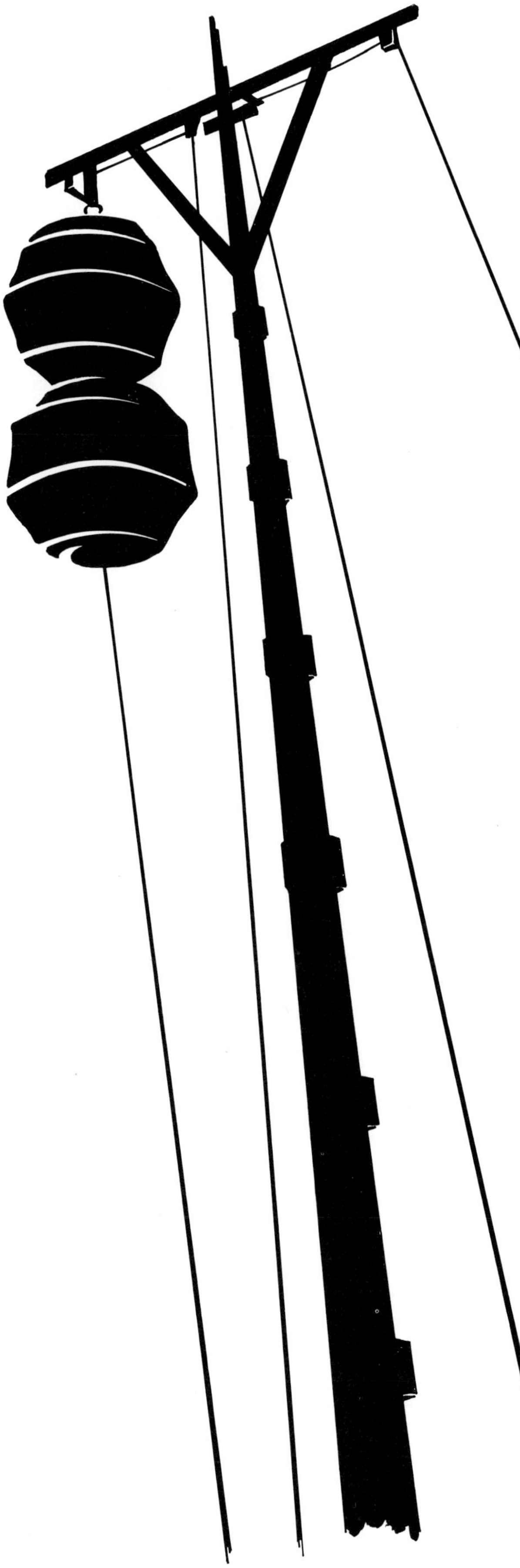

Contents

I. First Into the China Air Space 6
II. The Chinese Air Force 24
III. Aviation in North China 30
IV. Central Aviation School 33
V. Central Aircraft Manufacturing Company 52
VI. Aircraft Sales and Salesmen 58
VII. Commercial Aviation 59
VIII. Flight of the Tin Goose 74
IX. Air Force Day in Shanghai 78
X. Mission Accomplished 85
XI. Soviet Aid to China 87
XII. Combat Over Wu-Han 93
XIII. Bombing of Chungking 99
XIV. The 14th Volunteer Bombardment Squadron 102
XV. The First American Volunteer Group 107
XVI. The China Air Task Force 117
XVII. The Doolittle Raiders 134
XVIII. The 14th Air Force 138
XIX. Mission to Shinchiku 163
XX. The Chinese-American Composite Wing 166
XXI. The 308th Bomb Group (H) 174
XXII. The 58th Bomb Wing (VH) 187
XXIII. The Japanese Air War in China 203
XXIV. Ready for Wheels Down 217

Bibliography 230
Articles 230
Notes 231
Picture Credits 234
Index 235

I

First In The China Air Space

The introduction of aviation to China coincided with the beginning of the Chinese Revolution in 1911 to overthrow the Manchu Dynasty which had ruled China since 1644. Political events that followed in the wake of the Revolution both spurred and delayed the development of aviation. It was spurred by the eagerness of the new leaders to create a modern state, but it was delayed by the failure of these leaders to agree on a common purpose.

The leader of the movement to overthrow the Manchus was Dr. Sun Yat-sen, a Cantonese who had gone abroad in his youth to study medicine. But instead of practicing medicine, he devoted the rest of his life to the Revolution and to making China an equal partner in the family of nations. The Manchu authorities put a price on his head and even managed to kidnap him in London at one time, but he was released by British intervention, and spent the rest of his life traveling among overseas Chinese communities, pleading for funds to send to his revolutionary friends in China. Finally, when the Manchu garrison at Wuchang, across the Yangtze River from Hankow, was occupied by Revolutionary forces, 10 October 1911, Dr. Sun was not in China. But he hurried back and arrived in Hongkong on 22 December 1911 where he was met by a reporter for *South China Morning Post* who described him as a "gritty-looking man, beaming with smiles and wearing a grey coat and black bowler hat."[1]

A Chinese in a black bowler was a rarity, but Dr. Sun had lived abroad so long in the pursuit of his dream that he had become Western in his personal habits and dress but never for a moment anything but a Chinese in his devotion to his country. After pausing briefly in Hongkong, he went on to Canton and here, at a newly-elected assembly of delegates, he was named First President of the Republic of China.

Overseas Chinese in the United States and Southeast Asia were his chief supporters and they not only donated money to him but gifts in kind. Without a doubt, one of the most remarkable donations were two Etrick monoplanes which were shipped to Shanghai by some overseas group in the hope that the birds could be used in the cause of the Revolution.

The Etrick, also called *Taube* (dove) was built in Austria in two models, one with a 100 hp Daimler engine, and one with a 120 hp Daimler. Seen from above, it looked like an enlarged partridge lying breast down with fan tail. It was an ungainly looking airplane no matter how it was viewed, but the two "machines," as airplanes were called at this time, arrived in China too late to be of any help. The revolutionary party was already in power, the Manchus out of power and off the throne.

There was probably only one qualified Chinese pilot in China at the time, aside from the American, Fung Joe Guey of San Francisco, namely Li Ju-jen (also as Y.C. Lee).[2] Li is mentioned in a news story in the *South China Morning Post* (hereafter as SCMP) shortly after he returned from abroad where he had learned to fly, probably in England.[3] The new government did not know what to do with the Etricks and placed Pilot Li in charge of them. He had them shipped to Nanking, and then, presumably, hearing about the new flight school about to be launched at Nanyuan, south of Peking, he took the two Etricks north by rail to the Old Capital.

Meanwhile, Dr. Sun Yat-sen was unable to hold his new government together with slogans of democracy and equality *(T'ien hsia wei-kung)*. He had no private army to back him up; he could only suggest, not command, and in China, one who cannot command, cannot govern. A so-called National Assembly had moved from Canton to Nanking, and Sun, who was in Shanghai at the moment, decided to go to Nanking to hand in his resignation, 1 April 1912. His place was taken by General Yuan Shih-kai, a man who bore all the trappings and habits of the recently-deposed Manchus. In fact, at one point in his four-year rule, he sought to have himself proclaimed *Huang-ti* (emperor) but this backfired and he had to back down.

General Li Yuan-hung, a former Manchu officer, who had allied himself with the revolutionary party in Wuchang, was named vice-president of the new Republic, and when the two Etricks arrived at Nanyuan, he went out to see them, quite impressed.

Later, learning of General Li's interest in aviation, the Caudron Brothers of France sailed for China early in 1913 with a Caudron G-3 trainer in the hold of the ship. At Nanyuan the Caudron made several flights which so impressed Li Yuan-Hung that he made arrangements to place an order for twelve Caudron trainers, six of which were 80 hp and six 50 hp, all biplanes. Two French flight instructors, remembered only as Lieutenants Obre and Bon, and two mechanics, were later engaged to open a flight school at the new field.

In August 1914, France went to war against the Central Powers and the two lieutenants, if they were still in China, were probably recalled to active duty. The training program was continued under Chinese instructors, and since Switzerland was neutral in the Great War, Swiss pilots probably served as instructors after the French left.

But the first man in the China Air Space was actually an American, Thomas Scott Baldwin of Mineola, New York. A stunt man since his youth, he toured the world in 1889-90, accompanied by William Ivy, another stunt man, exhibiting balloon ascents and parachute jumping. They performed in both Shanghai and Hongkong.

The type of balloon used by Baldwin and Ivy was a spherical hot air balloon enclosed in a net, with rings and/or trapeze suspended underneath. There was no basket. After inflating his balloon, Baldwin held on to the rings with his feet on either a spreader-type trapeze bar or rope, his parachute hanging on one side, and harness buckled on. Once he reached altitude, he let go of the rings, his weight breaking the attachment cord holding his chute to the balloon and he floated down. Just before that, he also pulled a rip panel which brought the balloon down too.

Baldwin and Ivy charged their sponsors $1 per foot, that is the higher the jump, the more it cost. In practice, however, it was understood that they would not jump from higher than 1000 feet.

The second attempt to enter the China Air Space in a balloon was not as successful as the first. A Chinese youth, referred to in the press as "Hu Tsung," had built his own balloon and was seeking permission from the local authorities in Canton to take it up.[4] On 19 July it was learned that permission had been given and Hu Tsung was cleared for lift-off. This is how the Canton correspondent for the SCMP described the event:

> When I arrived on the scene, I found thousands of people were attracted by the rarest exhibition ever seen in Canton. One of the long poles erected in the ground for lifting the balloon, was suddenly snapped, and it took some time to replace it with another pole. Before the balloon had scarcely risen to the top of the poles, to the great surprise of the whole crowd, it exploded and fell to the ground. You may well imagine what an uproar was raised by the paying customers. However, it is hoped that tomorrow will be a successful one, if a trial is made again.[5]

But no "trial" was made again.

The first airplane in the China Air Space, according to Major De Sean Chung, an early writer on aviation, was flown over Peking by a Russian pilot (remembered only as "Kouzminski") who arrived in the city in October 1910 with a French-made Bleriot XI, an air-cooled monoplane.[6] After assembling his aircraft, he probably took off from the esplanade that once ran east and west between the Legation Quarter on the south side of the street and the Grand Hotel de Pekin on the north side, not far from Tien An Men.

It seems astonishing, in retrospect, that only seven years or so after the Wright Brothers took off from Kitty Hawk that a Russian pilot was demonstrating a French-made airplane in Peking, China.

The second aviator to fly in China was a French pilot, Louis Vallon. Some aviation historians place Vallon in Shanghai in 1909 but correspondence in the SCMP clearly

Captain Thomas Scott Baldwin, stunt man, balloonist and pioneer in lighter-than-air machines, grasps wheel of a pusher he built called the "Red Devil." This had a tricycle landing gear up front and skid in rear. This may have been the same one he brought to Hongkong in 1911.

Facing page: An artist's conception of airplanes flown in China from 1910 to 1921. Upper and lower right, variations of the Sommer biplane, one of which was flown in Shanghai in 1911 by Louis Vallon of France. Center, an English Avro trainer used at Nanyuan (Peking) flight school after 1919; upper left, a Bleriot XI probably similar to one flown by first Russian pilot over Peking, October 1910; center left, an Etreck (Taube), and lower left, a Caudron-3 trainer used at Nanyuan flight school c. 1913 to 1917. Nanyuan flight school had own runway, but in other cities, pilots used local race course.

shows that he was not in China until 1911. He made at least two exhibition flights over Shanghai in March and April, in a Sommer biplane, but he did not live to exploit his popularity. The SCMP carried this story on 8 May 1911:

> Mr. Vallon, the French aviator who has been exhibiting for some weeks in Shanghai, has met with a terrible death. The 'Echo de Chine' succeeded in raising a fund of $1000 [yuan] to be given to Mr. Vallon for another flight over Shanghai, the condition being that Mr. Vallon would fly from Kiangwan race course. Accepting this offer, Mr. Vallon biplaned from the Kiangwan race course on Saturday afternoon. Flying over the [foreign] settlements, he circled the Shanghai race course during an interval in the [pony] races.
>
> The descent was begun from a height of 1,000 feet. At 600 feet high the aviator checked his downward course and made a horizontal sweep, following which he attempted to descend the remaining 400 feet with a graceful turn. Unfortunately, he made the error of turning too sharply, with the result that his biplane turned turtle, and fell with a sickening splash . . . . thousands of spectators witnessed the fall, including Madame Vallon . . .
>
> The races were continued.[7]

A few weeks before the races were continued, a dispatch from Canton reported that a Mr. Fung U was building an airplane "of sufficient carrying capacity to be of commercial service in aerial navigation."[8]

Actually, he was not building it; he was erecting one he brought with him from California, for Fung U was actually a San Francisco Chinese better known as Fung Joe Guey, the Joe here probably representing a Cantonese corruption for the mandarin *ju,* and the Guey a Cantonese corruption for *kuei.*

Fung left China for the United States when he was twelve years old. After the Wright Brothers flight in 1903, he became devoted to aviation, and four years later, he built his own airplane. Hearing of the young aviator's success, relatives and friends in Canton urged him to return to China and help conquer the China Air Space. He probably left San Francisco by ship in February 1911. On 27 March a story in the SCMP reported that Fung, accompanied by three friends, had the honor of an audience with the Viceroy of Kwantung Province, His Excellency Chang Ming-chi. Fung went to see him to explain that he would be able to manufacture an airplane from his own design that "could carry more than 400 tons, rise to a height of over 900 feet and fly with great speed."

But Fung was misquoted. There was no airplane in 1911 that could carry 400 tons or even four tons. Nothing further is heard about Fung Joe Guey until a year later when this brief story appeared in the SCMP.

> There was a very large gathering and all the soldiers turned out from the barracks to witness his [Fung's] flight. He lectured on his machine in detail before the troops that surrounded him, little dreaming of any fatal accident. He then mounted his machine in high spirits and made a rise too quickly above the top of trees. The aeroplane made three attempts to ascend high above the trees, and during the last attempt it dashed into a bamboo grove and was smashed to pieces.[9]

Fung was taken to a hospital and died, his body punctured by bamboo splinters. He was the first Chinese to die in an airplane.

In January 1911, Thomas Baldwin, former parachute jumper, was back in Hongkong, this time with airplanes and at least two pilots. He is referred to in the SCMP as the "doyen of aviators, being the inventor of the parachute and the owner of the first dirigible."[10] He may have been a "doyen of aviators" but he was probably not the inventor of the first parachute; Leonardo da Vinci had sketched one somewhat earlier. Baldwin was accompanied by two other aviators, James C. ("Bud") Mars and Todd Schriver, in addition to a Captain Brice (or Price, who may have been a mechanic.) Bud Mars was born James C. McBride, son of a Great Lakes steamboat captain, on 8 March 1876, and left home to become a stunt man and balloonist. He became associated with Baldwin and in 1910 turned from balloons to airplanes and became the eleventh pilot to be licensed by the Aero Club of America. His father was ailing after an accident and the son did not want him to worry about him and changed his name to "Bud" Mars to avoid recognition in the newspapers.

The Nerviest Chinaman In The World

So Said Thousands at Pasay Beach Yesterday

LAST TIME TODAY

Capt. Tom Gunn

The Pioneer Chinese Aviator

FLIES TODAY AT POLO CLUB

9 a. m. To 11 a. m.

Admission -P-1.00 & Up

←

Upper left, Tom Gunn, Chinese-American aviation pioneer from San Francisco, holds control column while his instructor, Glenn Curtiss, holds strut, dressed in white collar and tie. Photo was taken at Curtiss flight school, North Island, San Diego, 1913. Lower left: Tom Gunn in Manila after being presented silver trophy by members of Chinese community for his spectacular exhibition flights. Philippine flag flies from mast but someone holds American flag at rear. At right, advertisement which appeared in *Manila Times*, 12 April 1914. Seats were one peso "and up."

Todd ("Slim") Schriver was also one of Baldwin's proteges in lighter-than-air work. He began as a mechanic and was on loan to Glenn Curtiss in 1909 when he took his airplane to Rheims, France, for the first international air show. Schriver later learned to fly but was killed in a crash in Puerto Rico, 1913.

A photographic reproduction in the SCMP[11] of Baldwin's "aeroplane," is actually the second version of a plane he called the "Red Devil," in which Baldwin did his first exhibition work. Merrill Stickler, curator of the Glenn H. Curtiss Museum at Hammondsport, New York, believes the picture was actually taken at Hammondsport. It is also one of the first photographs reproduced in the columns of the SCMP since it was founded in 1909. The Red Devil, much like a Curtiss, was powered by a four cylinder forty hp Curtiss water-cooled engine. After tests, Baldwin changed the design and made it a pusher, adding the interplane ailerons and a head plane elevator similar to the Curtiss machines. The wings of the plane, according to the story in the SCMP, were thirty-eight feet long. The machine weighed 650 pounds, developed 60 hp and carried ten gallons of gas which, in the words of the SCMP reporter, "sufficed for a hundred miles."

After reading this advance publicity, residents of Hongkong were no doubt disappointed to read in the newspaper on 7 February that the Americans had been forced to cancel their engagements. Already told by the colonial authorities they could not use the race course in Happy Valley, on the Hongkong side, it was agreed that they would move their machines and equipment to the Kowloon side to Taipo, third station north on the Canton-Kowloon railroad.

The Americans usually charged the equivalent of $8,000 [probably Hongkong dollars] for an exhibition, but a misunderstanding arose over the gate receipts and equally important, over military regulations. A SCMP reporter who covered the story failed to clarify the situation but he blamed the entire incident on "red tape" which he thought was a pity, and concluded his remarks with this verbal ground loop;

> (It) deprived Hongkong of this its first, and perhaps only for years to come, opportunity of witnessing the conquest of the air by proved experts.[12]

The American team went on to Manila to continue their tour of the Pacific.

Three months before they came to Hongkong, a Belgian pilot, Charles Van den Born, was exhibiting a Farman at "meetings" in Saigon, French Indo-China (VietNam). He received his training under Henry Farman, an Englishman who had moved to France to build airplanes and to start a flight school. Van den Born was issued license No. 37 by L'Aero Club de France on 8 March 1910 and, by the fall of the same year, he considered himself qualified to start out on an exhibition tour. He was accompanied by his wife and a mechanic.

Over Saigon on 10 December 1910, Van den Born became the first man to enter the Southeast Asian Air Space. He was liberally rewarded by French colonials, Chinese business men and others who raised a purse of $7,000 [probably Straits Dollars]. He also exhibited at meetings in Bangkok a few days later and was royally received and liberally rewarded by the King of Siam.

Perhaps learning of the plight of the disappointed Americans, Van den Born crated three Farman biplanes and sailed to Hongkong where he arrived 28 February 1911. His first exhibition flights were to begin 19 March and he was to use a biplane of his own construction "of the Chanuet-Farman type with quite a number of improvements," said the SCMP.[13] In the event of inclement weather, the SCMP had been asked to repeat that people on the Hongkong side would see "by the blue flag, hoisted on the C.P.R.'s [Canadian Pacific Railway] flagstaff on Hotel Mansions . . . that flying is postponed, whereas the red flag on the same flagstaff will indicate that flying is to take place."

Since the race course at Happy Valley was also off limits to Belgian airmen, the exhibition had to be held in Shatin Valley on the Kowloon side. Nevertheless, the Governor of Hongkong, His Excellency and Lady Lugard, and Major-General Anderson had "kindly consented to extend their patronage to the aviation exhibition," said the SCMP.

The stage was set, the red flag had been hoisted on Hotel Mansions, and the first man to penetrate the Hongkong Air Space in an airplane was about to take his place in history and in a Farman pusher. The Governor and his party were waiting in an improvised enclosure, not too close to the Chinese spectators. And then the unexpected happened. Shatin Valley is often hit by high winds rising off Kowloon Bay, and circumstantial evidence suggests that on this particular day, the winds were so high that Van den Born decided to cancel his flight, even if it meant disappointing the Governor of the Crown Colony of Hongkong. Despite the fact that airplanes today weigh many tons, wind velocity is still a factor on take-off, but Van den Born's airplane weighed about 700 pounds, a veritable paper kite, and he knew that a strong cross wind could easily flip him over.

The subject of Van den Born's failure to take off was a painful one, and he is not mentioned the next day in the SCMP or for weeks to come. Finally, it is learned that he had left with one of his Farmans for Canton. On 18 April 1911, the Canton correspondent of the SCMP sent in this report:

> The aviation meeting terminated in a most sensational manner yesterday. An enormous crowd assembled to witness the flights of Mr. Van den Born, among them being many high officials. His excellency Fu-chi, the acting Tartar [Manchu] General, left for his yamen about six o'clock and a few minutes afterwards, while making his way through a dense crowd near the Provincial Assembly building, he was suddenly attacked and killed on the spot.

The assassin, Wang Sang-choi, who was later apprehended, said that "he had no personal feeling against the general, that he was simply against all the Manchus." Thus, Van den Born had become an involuntary agent in the beginning of the Chinese Revolution. The Manchu authorities considered him guilty by association and refused to allow him to remove his airplane. Confronted by this intransigeance, he managed to set the plane on fire, and returned to Hongkong.

Although 10 October 1911 ("Double Tenth") is celebrated as the beginning of the Chinese Revolution, the first blow against the Manchus was struck in Canton on the last two days of April, 1911, a fortnight or so after the assassination of Fu-chi. On this occasion at least seventy-two patriots, mostly young men, threw themselves at the local garrison troops and were killed or captured and then killed. They have since been known as the "Seventy-two Martyrs" and are buried in hallowed ground at Wong Fa Hill near Canton.

The second Chinese-American to come to China to fly was Tom Gym who early on changed his name to Gunn. He

Art Lym, San Francisco-born Chinese who learned to fly on North Island, San Diego, c. 1914. Beehive type helmet was used as late as 1917 by French pilots in World War I. Many early pilots wore leather puttees to protect legs in case of crash, and leather button shoes were the latest in fashion wear for men. Inset, upper right: close-up of Lym at control column in another "pusher." Straps across back of neck and around shoulders moved ailerons left or right by "body English." Wearing cap on backwards was sign of a real racing sport, ground or air.

After learning to fly the Curtiss "pushers," Art Lym went to the Curtiss plant in Hammondsport, NY to learn how to fly a "hydroaeroplane." The photo above shows that it used a single center float, but not shown here are auxiliary floats on the wing tips. Photo was taken on ice of nearby Lake Keuka. In photo at right, Lym braces himself against the cold with "ski mask."

learned to fly under Glenn Curtiss at North Island in San Diego. In the spring of 1914, he crated at least two airplanes, including an "amphibian," into a ship and sailed for Manila, stopping off a few weeks in Honolulu where exhibition flights were held at Schofield Barracks. From there he went on to the Philippines and began a series of exhibition flights, one of which, at least, was advertised in the *Manila Times*.[14]

Tom Gunn, now "Captain" Gunn, became such a popular idol in Manila that he was persuaded to take his act to other cities of the islands. Honors were heaped upon him and he accumulated loving cups, medals and decorations, the latter, presumably, from the Philippine government. With his decoratons pinned to a white uniform, he went to a studio to have a publicity picture taken which was handed out to reporters.[15]

On 10 July 1915 the SCMP reported that Captain Gunn was coming to Hongkong. This was more than a year after he went to Manila and nearly a year since the Great War began in Europe. Funds were being raised by Chinese merchants in Hongkong to purchase fighter aircraft for the Royal Air Force, and the Chinese leaders in the community wanted Gunn to help in the fund-raising campaign. Not only did he bring at least two airplanes, but also a business agent, Mike Newman. Exhibition flying had become big business, with agents and advance publicity build-ups.

In his first days in Hongkong, Gunn was kept busy giving speeches and attending banquets arranged by his Chinese hosts. On 18 July, a Sunday, Gunn's seaplane and biplane arrived from Manila and plans were laid for a two-day exhibition at Shatin.

But this time, the Governor-General, Sir Francis May, K.C.M.G., a new appointee, did not plan to attend the meet. He probably heard about his predecessor who had sailed across the bay in the royal tender to see Charles Van den Born fly his Farman, and that, as everyone remembered quite well, ended in a cancelled engagement.

On this occasion Tom Gunn took no chances with the winds off the bay. He used his amphibian, took off from Kowloon Bay and circled over the crowds in the stands, cutting off his motor, gliding in the usual manner, and then returning to the point of takeoff on the bay. He had no problems and at the end of the two-day event, Chinese admirers presented him with a huge floral horseshoe.

On 16 October Mike Newman announced that Gunn would be flying again soon, but first he wanted to talk to the Chinese Governor-General of Kwangtung Province in Canton about starting a flying school. On this visit to the headquarters, he was accompanied by Newman and the United States Consul in Canton.

About this same time a Wong Un-ki, a native of Kiangsu Province (north of Shanghai) had recently returned to China and was being referred to in the SCMP as "a famous flier" who had manufactured an airplane and named it *Fai Lung* ("Flying Dragon").[16] The "fai" here is a Cantonese corruption for the mandarin *fei*. After the American Volunteer Group began scoring against the Japanese in Burma in early 1942, someone probably remembered the "Flying Dragon" and called the Americans *Fei-hu*, "Flying Tigers!"

Lym Fu-yuan, better known in the Bay Area of San Francisco as "Art Lym," the last name a Cantonese corruption for the mandarin *lin* (meaning forest), was the third Chinese to learn to fly in California. He got his pilot's license from Glenn Curtiss in 1913, and also traveled to Hammondsport, New York, to learn how to fly a Curtiss "hydroaeroplane." He had his picture taken on the winter ice, high behind the control column, with a knitted muffler to protect his face against the wind. He carried copies of this picture with him to China. On 19 January 1914 he signed one which he presumably mailed to a friend in the United States.[17] On 11 February 1914, he was back in San Francisco and mailed a card to John McNamara, Curtiss plant manager at Hammondsport. These messages indicate that Art Lym went to China in late 1913 and returned to California, at least by 11 February 1914. He probably went to China to see what the opportunities were for a trained pilot and for starting a flight school—like Glenn Curtiss had in San Diego.

Some time in the summer of 1916, Lym returned to Canton and began a series of exhibition flights there, and later in Swatow, the big coastal city lying to the north of Canton. After Tom Gunn arrived in Canton in 1915, and after Lym returned in 1916, there are repeated references in the English press to these two fliers, both immensely popular wtih the Chinese people wherever they went or flew. Both were trying to start a flying school, but there never seemed to be any funds left in the provincial treasury for aviation after the army generals were through paying off the troops.

Tom Gunn, apparently at the request of Dr. Sun, returned to the United States in 1918 to solicit funds among wealthy Chinese business men for aviation in China. Presently, he was under arrest in Detroit; the police thought he was a dangerous alien because he said he represented a "revolutionary party" which, with America at war and Bolsheviks marching in Moscow, was probably considered a threat to the national security.

It took a little while before Gunn could prove his American citizenship, and friends convinced the Detroit police that he was no threat and he was released. How much money he collected was not revealed but he apparently returned to Canton with several training planes.

Meanwhile, after the death (of natural causes) of President Yuan Shih-kai in June 1916, the government in Peking foundered as badly as the first republic led by Dr. Sun Yat-sen in Canton in 1911-12. While he ruled, Yuan Shih-kai had made it clear he did not want Dr. Sun back in China. After Yuan died in 1916, no one threatened Dr. Sun and he returned in 1917 to Shanghai and later made his way to Canton. He was greeted with ovations from the people in the coastal cities.

In Canton, Dr. Sun made a new effort to organize a national government with the help of local military commanders. It was openly called a "military government" and Dr. Sun was named "generalissimo," a term probably used for the first time in the English press in relation to any Chinese leader. [Chiang Kai-shek, after several years in office, also came to be called Generalissimo, or "Gimo" (pronounced gee-mo).

On 24 January 1919, the SCMP reported that "since their return to Canton, Capt. Tom Gunn, Capt. Arthur Lym and Jung Wai-lung, the three well-known Chinese aviators, have been considering the establishment of an aviation school in this province (and) are looking for a suitable site for the school building and the Whampoa Military Academy building has been selected . . . it is expected that work will be commenced shortly."

Actually, it needed more than work; it needed to be rebuilt. It was destroyed in a fire in 1916 as the result of carelessness on the part of a local supervisor. The building was rebuilt in 1919-20, but just how much was accomplished by Gunn, Lym and Jung has never been documented, but they laid the ground work for the Cantonese Air Force. In 1924 the first cadet class was enrolled at the Academy and was graduated in 1925. Among the graduates was Wang Hsu-ming, better known to Americans in World War II as "Tiger Wang."

The flying school in Canton was handicapped not long after it got airborne by the untimely death of Captain Gunn, either as a result of a fall from a ricksha, or as the result of an "assasination plot," according to a feature story written for the *San Francisco Examiner.*[18] There was considerable street violence in Canton, as already indicated, during this period. For example, a Chinese engineer in charge of engine maintenance at Whampoa Academy, was robbed and murdered, a loss obviously felt keenly by both Gunn and Lym, for trained mechanics were in short supply.

The family of Tom Gunn arranged to have their hero buried in Chinese soil. He was a Cantonese by ancestry and birth, and like so many pioneers in the Cantonese community in America, they wanted to be buried in Canton when they died.

Art Lym stayed on in China, rising to major-general in the Cantonese Air Force, still active in World War II, the Grand Old Man of Chinese Aviation. He moved his family to Hong-kong after the Communist takeover of the Mainland in 1949, and there he died in 1970. His daughter, Mrs. Renee Robertson, brought his ashes back to the Golden Gate for burial, a reversal of an old tradition.[19]

Aviation in China is referred to for the first time in the 1913 edition of Jane's *All the World's Aircraft,* when it is reported that the government in Peking had ordered twelve Caudrons, already mentioned. In 1914, Jane's repeats the same information about the Caudrons with the additional information that a homemade aeroplane had been built at Nanyuan in 1913, but no clue to the builder.

The 1914 edition of Jane's also states that Art Lym had been appointed flying instructor to the "Chinese Army" and gives the names of several people allegedly associated with aviation in this order: T. Gunn, A. Lym, Prince Tsai Tso, Colonel Tsing, Lt. Pao, Wee Gee and Lt. Yao. The name "Wee Gee" is no doubt mispronounced or misspelled, but the other names, aside from Gunn and Lym, could be from north China and associated with the flight school at Nanyuan, but not Gunn or Lym, neither of whom was in China in 1914 except for a brief visit by Lym.

The name "Prince Tsai Tso" is probably an error for General Tsao Kun who was garrison commander (*ssu-ling pu chang*) at Nanyuan. Referring to him as "prince" was a polite gesture by the new rulers of the Republic, since the monarchy had been abolished. But there can be little doubt that he was related to the former Manchu rulers.

The 1914 Jane's concludes its report on China with the information that "about 40 officers were under instruction in June 1914."

Evidence that a flight school was in operation at Nanyuan at this early date comes from a story in the SCMP which reported that President Yuan Shih-kai had given $1000 [yuan] to the family of Ng Yung-chung "the aviator who was recently accidently killed."[20]

The 1914 Jane's included this information:

> In the 'White Wolf' operations, four of the Caudrons were employed and found very effective in locating the enemy. No casualties to the machines occurred.

It seems fairly certain that the Caudrons, if any saw action, were based at Nanyuan. But Jane's fails to explain what the "White Wolf" operations were. White Wolf was the pseudo-

Top photo: Katharine Stinson takes off from Tokyo's Aoyama Parade Ground in her Partridge-Keller, December 1916.

Lower photo: Katharine Stinson, surrounded by Chinese admirers and two turbaned Sikhs (left) at Nanyuan airfield after exhibition flight over south side of Peking 17 March 1918 in "Laird Looper."

nym for a bandit leader who began his depredations in the spring of 1914 and continued for nearly a year, burning and looting villages south of the Great Wall, west of Sian. His followers set a pattern for savagery which made Shensi Province famous for years to come as the home of the most cruel bandits in China.

The government campaign to capture White Wolf continued without success into the spring of 1915 when he seems to have been swallowed up into the loess dust and into history without a proper name. Contrary to Jane's, no military aircraft were used in the anti-bandit campaign. The only mention of aircraft appeared in the SCMP when "two aeroplanes suddenly appeared [in Sian] on Saturday, astonishing the people."[21] They were no doubt two Caudrons, but how they got there and how they got back, is not explained, nor is anything further heard about them. There was no airfield at Sian, and the pilots, of whatever nationality, probably put down on a country road outside the walled city. There was no race course.

The 1918 edition of Jane's carries information on the arrival in China of Katharine Stinson, the American aviatrix, and first woman in the China Air Space. Miss Stinson was a younger sister of Eddie Stinson, one of the engineers who developed the "Detroiter" in the mid-1920s and helped found the Stinson Aircraft Corporation. Miss Stinson learned to fly in 1912 when she was sixteen years old, and was along with her family in starting a flying school at "Stinson Field," San Antonio, which is still in use next to Brooks Field.

But it was the Japanese, not the Chinese who brought her to the Orient. Already in the forefront of aviation among Asian nations, the Japanese wanted to see anything that could fly. They had already seen Bud Mars, et al, and they were especially interested in seeing a woman fly an airplane. Miss Stinson took off in her "Partridge-Keller" from the Aoyama Parade Ground on 15 December 1916, and her loops and glides thrilled the crowds. From Tokyo she went on to Yokohama and cities in western Japan, everywhere being given a tremendous ovation. On her last stop at Nagasaki she busted the crank case on her 60 hp Anzani motor, and it took several weeks before repairs could be made.

It seems that while waiting for the repair job, she decided to have her mechanic, Frank Champion, mount a 50 hp Gnome spare engine in the second Looper she had along. After test flighting it in Nagasaki, she and her retinue sailed to Shanghai.

In Shanghai, Miss Stinson began an exhibition tour that would take her to Peking, Tientsin, Hankow and Nanking. Whether or not on her first flight over Shanghai, she made a bad landing on the Kiangwan race course, striking the white railing fence in a mild ground loop, knocking down several railings but doing only minor damage to her aircraft.

From Shanghai, she went to Peking and here she performed at a special exhibition seen only by a few people on the south side of the city, but plainly in view of her chief patron, General Li Yuan-hung, temporarily the president of China. The profile on Miss Stinson's tour in Jane's quotes an anonymous correspondent in China who said that "her exhibition included looping the loop, volplanes [glides] with engine stopped to within a few feet of the ground and then switching the ignition on again and rising over the heads of the assembled crowd." The correspondent considered all this an "extremely dangerous exhibition" for the spectators and "one that should have been discouraged." But he finally admits that her flights over China were "our most modern

Pilots and mechanics line up fo picture at Santa Monica befor take-off on world cruise in 1924 Left to right, Major Frederick C Martin, flight leader, and Alva L Harvey, mechanic on the *Seattle* Lt. Lowell Smith, and Arthu Turner, mechanic on the *Chicago* Lt. Leigh Wade, and Henry H Ogden, mechanic on the *Boston* and Lt. Erick Nelson, and Johr Harding, Mechanic on the *New Orleans*. Turner, who became ill ir Seattle, was replaced by Lt. Leslie Arnold. Douglas "cruiser" ir background. Below: World cruise at Kanatic, Alaska, 20 April 1924 waiting for new engine before tak ing off for China.

example of the science of aviation that we have had in this country." He failed to mention that she was the first woman in this country to fly an airplane.

Katherine Stinson returned to the United States in April, 1917, after America went to war with Germany. She later married Miguel A. Otero, also a pilot, in Mexico, and died at her home in Sante Fe, New Mexico, in 1976.

After she left Japan for the United States in 1917, her mechanic, Frank Champion, who was also a pilot, apparently was invited by the Japanese to stay and serve as a flight instructor. He was in the air on 31 October, flying a monoplane over Kochi, lost a wing at 4000 feet, and corkscrewed in.

Ruthy Tu was the second woman in the China Air Space, but she did not make her debut before 1932. She was probably associated with a Cantonese army through her father. In a contemporary photograph she stands by a marker with several Chinese ideographs which read: "Do you have the determination to resist?", i.e. to fight the Japanese invasion of China? She wears the uniform of a Chinese army officer but without insignia of rank, or a cap.

In the spring of 1924, Captain Georges Pelletier Doisy, a leading airman of France, touched down on the Shanghai Race Course en route to Tokyo. He was flying a Breguet XIX, accompanied by a crew chief who was also his co-pilot. But when Doisy landed on the race course, he damaged his landing gear, not seriously, but enough to delay his flight.

By a stroke of good fortune, a Colonel Chou, probably base commander at Lunghua, had met the captain on a visit to France. Colonel Chou was probably second in command for aviation in Chekiang Province where the garrison commander was planning to open a flying school at Shien Chiao. Jane's for 1927 says he has six Breguet XIX biplanes, two Morane Saulnier school monoplanes and two Morane "Penguines" for constructional aircraft.

Colonel Chou, after conferring with his superiors, offered to trade airplanes with Doisy, one Breguet XIX for another, a generous offer which was accepted. Colonel Chou is shown in a contemporary photograph with the undamaged airplane, flanked by Doisy and his copilot, and a third foreigner in black tie holding a cane who was probably the French consul. All three foreigners are wearing pith helmets, the unofficial badge of the white man in Asia, while Colonel Chou wears a straw "boater."

But Doisy did not attempt a direct flight to Tokyo across the East China Sea. Instead, he took the long way around by way of Mukden and Pusan. Two years later he set a record for a flight between Paris and Peking, a distance computed

at 10,000 kilometers, in sixty-three hours, thirty minutes.

On 4 June 1924 three United States Army planes landed on the Whangpoo River at Shanghai on a flight around the world. Up to this time no airplane had ever circled the globe. When word spread through aviation circles that the Americans were going for the honor, other airmen from England and France hurried to get ahead of them. All failed, and the American team, which started with four airplanes, came home with three, but the feat was accomplished.

The flight leader, Major Frederick C. Martin, was accompanied by three lieutenant pilots, Lowell Smith, Leigh Wade and Erick Nelson. The four airplanes and a backup took off from Santa Monica, California, on 17 March 1924. This was not the actual point of departure, but Douglas Aircraft Company was located here and wanted to have pictures taken to give the company a share in the national publicity. And to give more personality to the aircraft, each was named after a city, namely *New Orleans, Seattle, Boston* and *Chicago.* The aircraft were Douglas DWC "cruisers" with Liberty motors, open cockpit biplanes. Major Martin was accompanied by Staff Sergeant Alva L. Harvey, crew chief, and the other ships also had a crew chief.

After reaching Seattle, the wheels of the aircraft were replaced with floats, and the quartet took off on the morning of 6 April 1924. Major Martin and Harvey got lost in a fog bank off Alaska and mushed into a mountain, unhurt, but their aircraft washed up. They wandered over the tundra for several days before reaching a haven on the Bering Sea. Lt. Smith took over as a flight leader and the mission continued to Japan and China. In Shanghai, the crew stayed three nights at the old Astor Hotel located just below the Garden Gate Bridge, while their aircraft floated on the river nearby. From Shanghai they flew to Amoy on 8 June, where they stopped to refuel, and on 10 June to Hongkong and Saigon on the same day. From Saigon they flew to Calcutta where floats were replaced with wheels which had been sent ahead weeks earlier. The wheels were again removed for the flight across the north Atlantic and on this leg of the flight, Leigh Wade had engine trouble and turned back. He managed to catch up with the others for the final leg of the flight in another plane. Today, General Wade is the only surviving pilot and lives in Washington D.C. Alva Harvey did not reach China in 1924, but he got there over the Burma Hump in World War II as commander of the 40th Bomb Group (VH). He now lives in retirement, also in Washington, D.C.

General Wade attributed the success of the mission largely to Erick Nelson whom he characterized not only as a fine pilot but "one of the best mechanics that ever touched an airplane."[22] Nelson immigrated from Sweden to the United States in his youth.

Lufthansa, the burgeoning German airline of the 1920s sent two aircarft to China via the Soviet Union in 1926, arriving at Nanyuan on 31 August. This flight paved the way, four years later, to the founding of a Chinese-German airline called Eurasia Aviation Corporation.

The U.S. Marine Corps, which rushed a battalion of troops to Tientsin in 1927, was reinforced in 1928 by a squadron of pursuit planes which gave exhibitions of aerial acrobatics over Tientsin, and also demonstrated, for the first time, formation flying in the China Air Space.

On 1 August 1929, a front page story in the *Peking Leader,* reported that two French airmen, on a heading from Hankow to Peking, had been forced down at Chengchow. Apparently there was no damage but a large crowd of well-wishers in the French community of Peking had gone down to Nanyuan to greet the visitors and were disappointed. The two fliers had flown from Paris in a Breguet, probably via Indo-China.

Above: Lockheed Sirius lands on Lotus Lake, Nanking, China, 21 September 1931 with Charles and Anne Morrow Lindbergh in cockpit. Old wall around city of Nanking in background. Below: Lindberghs received by Chiang Kai-shek and Mme. Chiang. Picture was taken after Lindbergh had been decorated by the leader of the new government of China.

Francis C. Chichester became the first to fly an airplane, a Gypsy Moth seaplane, from Australia to China. Island hopping across the southwest Pacific, he arrived in Manila and went on to Formosa before flying across the Straits to China. He landed on the Whangpoo River at Woosung, near Shanghai, on 7 August 1931. After a few days of rest, he flew to Japan via a stop-over at Kagoshima and then on to Katsuura near Yokohoma. On 15 August, he failed to see a high-line on take-off and crashed. He was nearly killed. After his convalescence, he returned to his home in England. This is the same Chichester who sailed solo in yachts across the Atlantic and around the world in the 1950s and '60s.

On 19 September 1931, Charles and Anne Morrow Lindbergh coasted in for a landing on Lotus Lake in Nanking, and when this happened, the China Air Space was ready to explode with dreams of the future. Now, nothing was impossible. Ever since the first airplanes began to fly, there had come aircraft of all types and wingspans to demonstrate to the Chinese how to fly and how to travel by air from one city to the other. But the Chinese were not entirely

Weston Burt Hall, American aviation pioneer, speed racer and airplane salesman, in Nanking, 1931. Airplane is Fleet 2-seat trainer with Kinner engine, fuel tanks in upper wings. Fleet logo appears on rudder and under it U.S. commercial registration number.

convinced. They had seen quite a few of these new-fangled machines wrecked at the ends of fields and race courses and others which cracked up flying over mountains. It was obvious to everyone, surely, that traveling by river boat, train or sedan chair was much safer than by air.

But the arrival of the Lindberghs helped to dispel these fears. They demonstrated that the airplane, if handled by a crew of competent people, was just as safe as riding on a train or sedan chair. Furthermore, distance was no barrier. The Lindberghs had flown "North to the Orient." The Lockheed Sirius which brought them this far was a low wing monoplane, powered by a 680 hp Wright Cyclone motor designed in 1929 by John K. Northrop and Gerald Vultee. The one that flew to China was fitted with Edo floats.

The Lindberghs landed at a sad moment in Chinese history. It was the day after the Japanese army invaded Manchuria, and it was the year of one of the most disastrous floods in living memory. The Yangtze River was lapping over its wide banks all the way from Nanking to Hankow. Millions of Chinese were either homeless or starving. Some came to Shanghai to sell their children. No one could even guess at the extent of the tragedy or how to deal with it.

The Chinese were anxious to entertain their distinguished visitors, but Lindbergh asked that all entertainment be cancelled. He placed himself at the disposal of the government for air surveys of the flooded areas and on 21 September and later he flew several missions over the flood areas. observing the havoc caused by the great break in the Grand Canal to the north of Nanking.

From Nanking the Lindberghs flew up river to Hankow to continue their survey flights. They landed on the Yangtze River near the British carrier, *Hermes,* and since there was apparently no safe anchorage for the Lockheed, the British commander of *Hermes* offered to lift the Lockheed on board for the night. The next day, the Lindberghs got into their plane for take-off on the river and, as it was being lowered over the side by derrick, an unnoticed cable near the derrick tripped one wing and the plane flipped upside down as it fell into the river. Sailors on board *Hermes* jumped over the side to rescue the couple and later retrieved the plane, one wing badly damaged. A delay of days or weeks was inevitable. But a day or so after the mishap, the Lindberghs learned of the death of Dwight W. Morrow, father of Mrs. Lindbergh. They decided to return home by surface vessel.

In the decades between the wars—1918-1937—competition for the sale of civilian and military aircraft in China was keen. Most foreign salesmen looked on China as a vast market where great profits could be made, unaware of the low standard of living in a nation where the wheelbarrow was still the most common carrier.

The government in Nanking as well as provincial governments were the chief buyers of foreign aircraft, but up to the early 1930s, aside from the Lockheed which the Lindberghs flew, the Chinese buyers were not entirely convinced of the superiority of any particular aircraft, European or American.

The matter of superiority, however, was settled temporarily by the arrival in China of J.H. ("Jimmie") Doolittle, in April 1933. Now a major in the Army Air Corps reserve, he had been hired by Curtiss-Wright to go to China and demonstrate their new pursuit ship, the Hawk II (Navy F-11). In the days he spent at Shien Chiao, his spectacular exhibitions of aerial tactics, take-offs and acrobatics brought an order for fifty airplanes, the largest single order ever placed by the new government.

From Shien Chiao, Doolittle flew south to Canton and repeated his exhibition for Cantonese leaders, but sold no airplanes, probably because of the uncertain future of the provincial government. He returned to the United States and nine years later came back to China, this time in a parachute from a B-25 medium bomber after the "Doolittle raid" on Tokyo.

In August 1914, Japan joined the Allies, France and England and declared war on the Central Powers. A few days later she attacked the port city of Tsingtao, a German enclave on the east coast of China since the 1890s, a thorn in the side of the Japanese Navy.

China did not declare war on Germany until 1917 and when the Japanese launched their attack on Tsingtao in 1914, they were also attacking China. But China was a house divided, unable to drive the Germans out or to stop the Japanese from coming in. Moreover, the Japanese were aided by the British who were so anxious to see the Germans out of China that they sent 910 officers and men of the 2nd Battalion South Wales Borderers, and 450 NCOs and men of the 36th Sikhs to the aid of the Japanese. Despite the combined force, in which the German defenders were probably outnumbered ten to one, it was not until the morn-

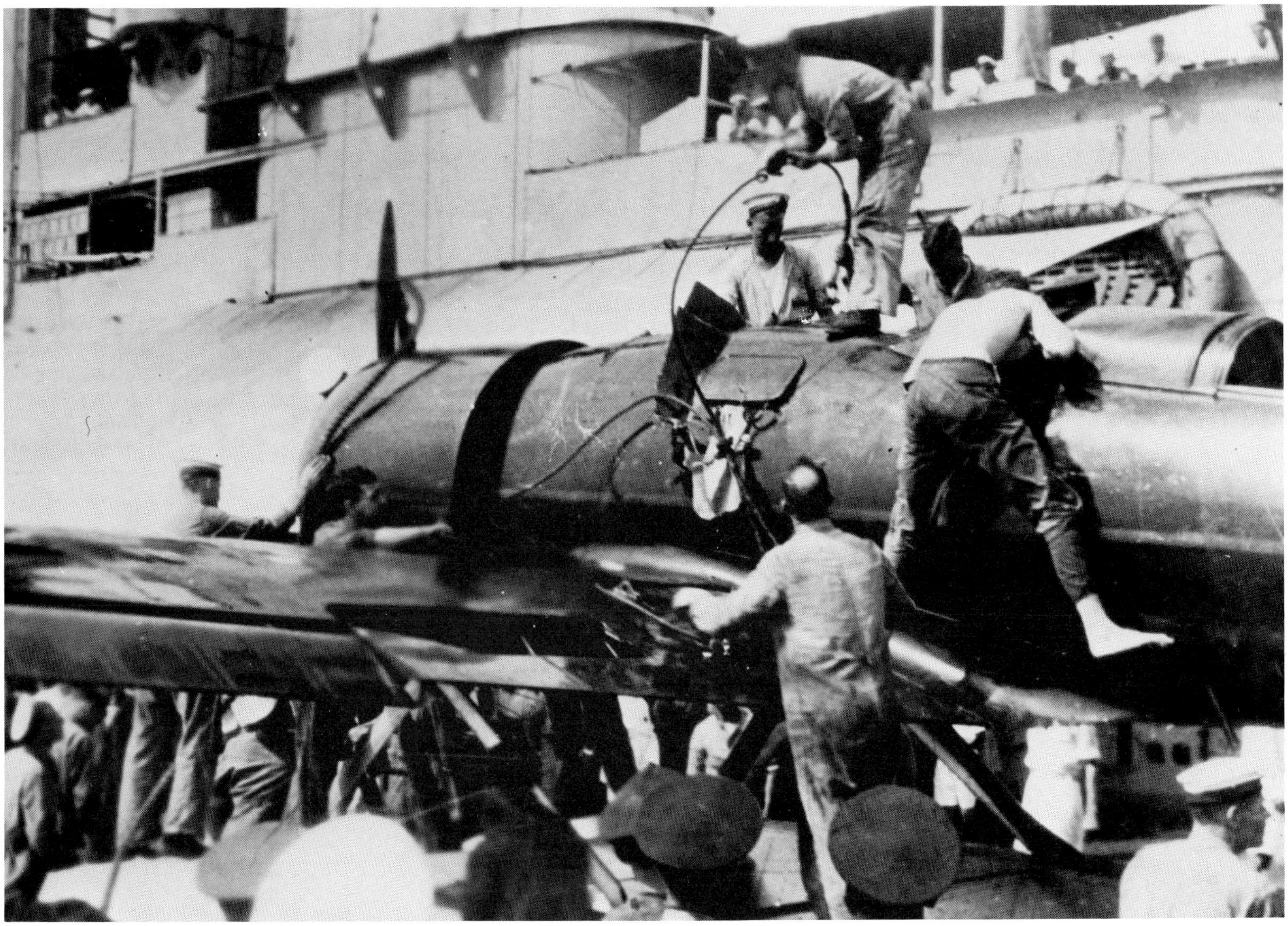

ing of 7 November that the Germans raised the flag of surrender.

During this seige at least one Etreck acted as a spotter for the German coastal artillery, and when the surrender came, the pilot flew his airplane across the bay and surrendered to the Chinese authorities. But he refused to give up his airplane and when the Chinese insisted, he tossed a match to it and that ended the first phase of German aviation in China.

In the attack on Tsingtao in 1914 several Japanese seaplanes off the tender *Wakamiya* were used for observation, the first Japanese aircraft in the China Air Space. A contemporary photograph of *Wakamiya* shows a Farman amphi-

Colonel Lindbergh, shortly after his Lockheed was retrieved from Yangtze River, inspects cockpit, still shoeless and wet from dunking in river. Cable around fuselage which lifted Lockheed back on deck is being uncoiled by crew of H.M.S. *Hermes,* an early British carrier which, in September 1931, was visiting Hankow when accident happened. *Hermes* went down in WWII.

A Kawasaki C-5 parked in C.N.A.C. hangar in Nanking, March 1935. This liaison plane was ordered from Kawasaki by *Asahi* of Osaka, one of leading dailies in Japan. In June 1933 it replaced an earlier Kawasaki A-6. *Asahi* accepted new aircraft in February 1934. It had maximum speed of 335 Km/h, the fastest airplane in Japan. On 20 March 1935 Captains Shinno and Tsukagoshi took off from Osaka on a heading for Nanking, via a fuel stop in Seoul, Korea, arriving Nanking in ten hours ten minutes. On the return flight, plane took off from Shanghai on 25 March and flew to Osaka non-stop in six hours twenty minutes. Foreign military attaches considered it a freak, not the precursor of the future Mitsubishi heavy bombers. Note rising sun markings on wing tips. *Asahi* in Japanese means "the Sun."

bian suspended from a derrick. It seems that none of the Japanese aircraft ever engaged the Etrick in combat, or that any bombs were dropped on the German fortifications.

The first bomb attack in the China Air Space was made against the Forbidden City of Peking. The details are sketchy. Bombs of undetermined size or make were dropped on a railway junction, Fengtai, on the southeast side of Peking, and narrowly missed a train leaving for Tientsin on or about 7 July 1917. Continuing, the SCMP, in a delayed story, explained further:

> The great sensation of the day, however, was the bombing of the Forbidden City by one of the Nanyuan aeroplanes shortly before eleven o'clock this morning [7 July]. The machine came slowly up from the south and circled several times above the Forbidden City with a wing tipped in the air. Suddenly the plane, upon whom the eyes of all Peking were trained, righted itself, cut down speed, and dropping something that was a mere flash against the blue sky. Then there was a little more circling and a second flash, and then came the third and last which was followed by a report heard throughout the Tartar City. Satisfied with its work, the machine turned its nose south and flew home. It has since been learned that the first bomb failed to explode, the second dropped in the Nan Ho [South River] near the Presidential Palace, and the third reached its mark in the Forbidden City. One soldier of Chang Hsun and one chair bearer were injured, and perhaps a few killed.[23]

The trouble began with Chang Hsun, a warlord who had managed to recruit a small army and boldly entered Peking before troops of the National government realized what had happened, Chang Hsun wanted President Li Yuan-hung to dissolve the national assembly in Peking and restore the monarchy. He also wanted twenty million dollars (yuan) to spare the city or he would destory it.

Circumstantial evidence suggests that Tuan Chi-jui, minister of war and commander of troops, hired a Swiss aviator, probably an instructor at Nanyuan, to drop the bombs on the city to demonstrate to the troops of Chang Hsun what the consequences might be if they continued to resist the government. Resistance in the city crumbled shortly, but silver bullets, not lead bullets, probably brought Chang Hsun down. He was never heard of again. His troops took off their uniforms and were absorbed into the population.

No suggestion is made in the press report of the type of demolitions used by this first bombardier. They were probably potato-masher type hand grenades thrown from the cockpit of the airplane, not large enough to do any harm to the Forbidden City, even in a direct hit. The raid was the first and last ever made on the Forbidden City. The airman who made it was either not asked to run another mission or he declined to volunteer for it. As a bombardier, his aiming, demonstrably, was not very good.

The first American to die in the China Air Space was Robert Short of Tacoma, Washington. A trained pilot, he found little call for his talents at home during the Depression, and he sailed to Shanghai, ready to fly or sell airplanes. He was engaged by the L.E. Gale & Company, representatives for Boeing aircraft in China.

On 28 January 1932, Japanese Marines attacked the Chinese 19th Route Army in Chapei, a northern suburb of Shanghai. The fighting continued for six weeks. During this time Japanese airmen had no opposition and took complete control of the China Air Space.

On February 24, Robert Short was assigned to fly a Boeing airplane, single seat fighter, for delivery to Nanking. Shortly after take-off he saw three Mitsubishi BIM type 13-two seaters, from the Japanese carrier *Kaga,* over

Memorial raised to Robert Short by Chinese Government after his death in battle with Japanese airmen over Soochow on 22 February 1932. Memorial was built over his grave near entrance to Hungjao aerodrome on west side of Shanghai. Reading down, Chinese ideographs are *Shiao Te* (Short) *i-ssu chih* (his sacrifice), and at bottom, *mu*, for tomb. Picture was taken on day Chinese artist was carving English name on base of memorial.

Soochow. Short perhaps wanted to show what a Boeing fighter could do because his machine-guns were loaded with live ammunition. He attacked one of the planes but three other Mitsubishis came in to help out and Short was shot down. Yoshiro Sakemago received credit for the aerial victory, the first for a Japanese in the China Air Space. One Japanese gunner was killed in the duel with Short.

The Chinese held funeral services for Robert Short in Shanghai. His mother, Elizabeth Short, was invited to the ceremonies. Her son was buried near the Hungjao airdome and a handsome stone was erected. Every year from 1933-1937, Chinese airmen held memorial services at his grave site on the anniversary of his death.

In a nation which had long expected little but exploitation by foreigners, the dramatic death of Robert Short, in its defense, stunned the Chinese people. As his funeral cortege moved up Bubbling Well Road toward Hungjao, half a million people lined the street and the name of "Short-ta" passed from one lip to another, followed many times by the Chinese expression for, "What a pity!"

Carl Nahmacker, airplane and car salesman in Shanghai, shortly before test hop in 1933 which ended in his death. At right, a commercial plane with Chinese ideograph *yu* on fuselage, meaning mail plane.

Jimmy Doolitle prepares demonstration flight of Curtiss-Hawk II at Shien Chiao (near Hangchow), spring of 1933. In distance at right stands Ford trimotor, probably the plane of Chang Hsueh-liang, at Shien Chiao for engine overhaul.

II

The Chinese Air Force

After he resigned as president of the Republic of China in early April 1912, Dr. Sun Yat-sen spent the next decade or more struggling to diminish the power of the warlords who had taken over the country in the absence of a new federal authority. Dr. Sun had helped to topple a dynasty with high-sounding slogans which sounded reasonable to Chinese livng in San Francisco and to students at Peking University, but which held little appeal to the Old One Hundred Names *(po-hsing)* working in the rice paddies of south China. As a result, there was no unifying philosophy for political action, and the nation became fragmented into provincial and regional power centers, some led by former provincial governors, some by former bandits, later known as "warlords."

In what might be called the "Second Revolution," Dr. Sun worked steadily to build up a new political consciousness to defeat the warlords, and by 1923 he had increased his personal influence to the point where the Communist Party in the Soviet Union thought he might be a useful tool in the expansion of world revolution. In 1923, Soviet agents went to see Dr. Sun in Shanghai and out of these early contacts, arrangements were made by the Soviet Union to help Dr. Sun's party, the Kuomintang, to develop a new base for the assumption of power in China. One of the means to this end was to develop not only a well-trained army, but a new air force under Soviet instructors and with Soviet airplanes.

The training of cadets for the new air forces went on simultaneously with the training of army officers at the Whampoa Military Academy located on the outskirts of Canton, an old institution which had carried over from the Manchu Dynasty.

The Soviet advisors and aviation instructors began to arrive in Canton in the fall of 1924 and more followed in 1925. The flight instructors brought a number of airplanes, all biplanes, probably R.IMS general purpose types, but all Russian versions of De Haviland's 9A designed in the last year of WWI.

The air force training section in Canton laid the foundation for the Chinese Air Force. A dispatch from Canton in SCMP reported that "thirty airplanes are shortly due in Canton to strengthen the air force. Mr. Remy, the Russian chief of the Aviation Bureau, who left Canton for Russia and Germany some time ago to purchase the planes, is expected back shortly."[1]

On 20 April 1926, the same newspaper reported that "Mr. Remy, Russian chief of the Bureau of Aviation, who resumed his office but a few weeks ago, has resigned. He left Canton yesterday for the north."

A few months later, the *Hongkong Daily Press* reported that the Kuomingtang Party in Canton "was busy reconstructing the nine airplanes recently imported from Russia. A number of Chinese mechanics from Hongkong have been engaged to strengthen the mechanical force of the aviation corps . . . Russians are prepared to pilot these machines if the KMT [Kuomintang] are unable to man them."[2]

During this frenetic period of Chinese history from 1924 to 1927, the *Hongkong Daily Press* had a correspondent in Canton who wrote under the pseudonym of "China Boy." On 17 February 1927, China Boy had this to say about aviation:

> Soviet agents in Canton are not so helpful to KMT comrades as they usually profess, and even the KMT press agencies have complained. Some $800,000 [Hongkong] alleged to have been put in the hands of the Soviet high commission [in Canton] has yet to be accounted for as the aeroplanes (sic) for which this was the payment have not all been delivered.

The only evidence that Russian pilots actually flew Chinese marked aircraft in this early period appears in correspondence appearing in the *Central China Post* (Hankow):

> I was very interested to read in your paper a few days ago of the airplanes belonging to the Southern army, the aviators of which mistook the city Kian for Changsha and alighted there to be immediately captured by the anti-Reds [i.e. by a force loyal to Chiang Kai-shek] . . . When I read the news I little thought that we here in Paoching [also as Paoking] would see a similar experience repeated. But the thing has happened and I am still wondering how such a thing could happen."
>
> At about 10:30 a.m. on August 19th I heard engine noise and later three airplanes circled the city, the first ever to be seen in this part of China. People flocked to the roof tops as well may be imagined. At one point the planes flew away only to return and fly lower than on the first approach. I then realized that they were looking for a place to land although there was no airfield around Paoching.
>
> Two of the planes came down on a river bank, both damaged, while plane No. 3 landed farther downstream, not quite as badly damaged. One of the pilots was slightly injured, and all were Russians with the exception of a Chinese aviation officer. Each machine had the name of Sun Yat-sen written on the wing and was numbered No. 1, 2 and 3 respectively. Also, on the underside of the wing was a circle with the KMT star in the center. Shortly after the landing the Russians were escorted into the city where they were feted and made much of. They stayed there two days and this morning they set off for Changsha

This Boeing 247 may be same plane flown by Royal Leonard in rescue of Chiang Kai-shek at Sian in December 1936.

Administration buildings at Hungjao air base, built by Chinese government on west side of Shanghai, as it looked after Japanese bombing attack in first days of "Shanghai Incident," 1932. Base has since been leveled for housing development.

not by airplane but by the good old Chinese way, i.e. sedan chair."[3]

The presence of a Chinese "aviation officer" with this flight suggests one of the planes, at least, was a two-seater. The story also makes clear that some aircraft were airborne in support of the Northern Punitive Expedition in 1926-27, but the only ones that were newsworthy were the ones that cracked up along the march.

The story in the *Central China Post* also reveals, for the first time, that the new government in Canton had adopted an insignia for military aircraft, the white pointed sun inside a blue field which is still in use with the Chinese Air Force, Republic of China. Although the correspondent from Paoching says the name of Sun Yat-sen appeared on the airplanes, he probably means the Chinese ideographs for Sun's intimate name, *Ming Te* [Brilliant Virtue].

In the 1930s various types of aircraft in China were marked by Chinese ideographs on the side of the fuselage, but military aircraft, as noted in photographs dated to 1931, were using the circle sun above and below the wing tips. Apparently none of the aircraft that landed at Paoching had any markings on the tail section, but photographs taken in 1933 show the adoption of thirteen alternate blue and white horizontal stripes on the fin, a marking adopted from U.S. military aircraft, that is, the stripes, but not the color scheme.

The correspondent who sent the above report to the Hankow paper was probably a missionary living in Paoching, a city which lies in the hill country about seventy-five miles northwest of Hengyang in central Hunan Province. There was an air strip here for emergency landings in WWII, but it has disappeared and the name Paoching has been changed to Shao-tung.

The fact that the three Russian pilots had to land on a river bank, almost surely the Yuan River, suggests that they had flown off course and run out of gas which was then, and still is a good reason for a forced landing. Since the correspondent fails to mention any armament on the planes, they probably were observation planes for one of the Kuomintang armies advancing north on Changsha through the Siang River valley.

After the Nationalists or Kuomintang established a new government in Nanking in 1927, a school for cadet training was begun. General Hsieh Mang, a former air commander, said he was graduated from the Nanking flying school in Class C, in 1929. His instructors were using mostly Curtiss JNs ("Jenneys") for primary training. The school also had a few British Moths and Junkers single engine trainers.[4]

The first class of cadets from the Nanking school, which was graduated in 1929 as Class A, may have participated in air strikes made in 1930 when Nanking armies of Chiang Kai-shek put down two rebellious warloads, Feng Yu-hsiang and Yen Hsi-shan in north China. The Rebel headquarters were bombed with light aircraft believed to have been Junker transports, and also American light bombers. Headquarters compounds of the rebels in Kiaifeng and Tsinan were bombed and the Chinese defenders, unaccustomed to seeing a machine that could not only fly but drop explosive charges, were terrified out of all proportion to the actual danger. Several pursuit aircraft also flew over Tientsin on 27 August 1930 and machine-gunned East Station, endangering if not killing a number of civilians. The following day the same planes dropped bombs of undetermined size and number on Peking as a warning of worse things to come unless the rebels capitulated, which they did.

Whang Ping-han commanded the Nanking squadron in the campaigns of 1930. Like the Army Air Corps of the United States, the Chinese Air Force was not a separate organization from the army. Actually, the Chinese Air Force,

Silorsky S-38, twin-engine amphibian, Wasp engines, believed to be first airplane purchased for transportation by Nationalist government. Photo taken in Nanking, summer of 1931.

Right: Foot soldier *(pu-ping)* attached to Nanking division fighting on Shanghai front in fall of 1937. Soldier wears ammunition bandolier around shoulder and two potato-masher type hand grandes copied from German make of WWI. German officers helped train these divisions but were recalled to Germany when war began. Author took photograph behind Chinese front in mid-August 1937.

as an independent organization, did not materialize before 1941 when it was commanded by General Chou Chih-jou. Hence, the command center of the air force continued to be the *hang k'ung yuan ming hui,* variously translated as the Aero Commission or Bureau of Aeronautics. The rescue patch first worn inside the jackets of foreign pilots flying for China in late 1937, was issued by the Aero Commission.

The success of the small air force in the Honan campaign against the northern warlords convinced the leaders of the Ministry of War of the need for expanding the army air arm. With this in mind, a national aviation conference was convened in Nanking in April 1931, although the word "national" was still somewhat of a euphemism, for it was mainly a meeting between two cliques of politicians, one from Canton and the other from Nanking. A big air show was also staged by agents of foreign aircraft firms including British, German and American.

However, the conference to merge the Nanking and Cantonese air forces came to a halt on 29 April when a sharp difference of opinion arose between the two factions. Unofficially it was probably a question of how aircraft funds were going to be allocated, but something else was afoot here which no contemporary Chinese historian has ever explained. Almost on cue, delegates from Canton, and Cantonese pilots who had been in training at the Nanking aviation school, fled the city and returned to Canton.

The rupture in relations between Nanking and Canton raised the fear of civil war between the two factions. General Chang Hseih-chang, director of the Aero Commission, was a Cantonese and he left with the others. His place was taken by Whang Ping-han, the same who had commanded the air group in the operations against the northern rebels. Whang was on a tour of the United States to observe progress in military aviation when he was recalled. On his return to Nanking he ordered twenty light bombing-observation planes from American firms, and nine Junkers K-47s, a pursuit ship, from Germany. In addition, he declared an embargo on the importation of all military aircraft into Canton. This was ignored by the Cantonese but when the Nanking government managed to seize six German-made aircraft consigned to Canton, civil war loomed large on the horizon and there is little doubt that it would have materialized except for the Japanese invasion of Manchuria on 18 September 1931, at which time both sides in the dispute agreed to concentrate on the defeat of Japan.

But the rift between the two factions was never healed entirely. It was a clash of personalities as well as of politics and money. But no one broke the commandment of Courtesy (*li*). For example, Colonel Ting Chi-hou, a parachute jumper at the Nanking aviation school, and a popular hero, joined the Cantonese faction in the split, but when he left Nanking, he found it hard to leave behind two new parachutes which had been issued to him. He wrote to Whang Ping-han, the new director, to explain that the circumstances of his departure had been so sudden he had found it difficult to return the 'chutes and he asked the honorable commander to accept his apologies. General Whang replied, praising Ting for his foresight in equipping himself with the type of gear necessary for any successful mission which, in the event of war, would no doubt prove of great practical value.[5]

The construction of air bases and air strips in China progressed rapidly in the early 1930s and considering the conservative attitude of a people who had gotten along for 3000 years without anything in the air more alarming than a

In 1935 Sun Tung-kang, in white gown, was first Chinese pilot to fly solo from Berlin to Nanking. In background stands the single engine Junkers which carried him across the deserts and mountains of Mideast, India and Burma.

Douglas O2MC-4-1s waiting for orders to attack rebels in Fukien Province, 1933. Planes were flown by pilots from Shien Chiao Aviation School.

Junkers K-47 fighter, used briefly in training at Shien Chiao, later moved to Nanking for special training purposes.

Chinese nose-gunner on Soviet Katiuska SB-2, c. 1938.

Chinese pilots in spare time playing gramaphone, c. 1937.

kite, the development of airfields was a remarkable phenomenon. For example, at least as early as 1929, two French fliers, en route to Peking from Hankow, could find a strip to land on in Chengchow, an important railway junction but never, up to this time, considered a place to land a plane.

These early fields, even though the runway was scarcely long enough for a basic trainer, were built by local warlords and garrison commanders. Thus, by 1935, there was scarcely a city of any size which did not have a field large enough for a single or multi-engined aircraft. But, until the Americans came to China in 1942 to activate a new air force, most of the fields were sod or earth, and most of them were bad in wet weather.

In 1933, the Chinese people took pride in watching the progress of a Chinese pilot, Sun Tung-kang, who flew a single engine Junkers from Germany to Canton via the Middle East and India, the first Chinese to cover that distance. The following year, a big Chinese air show was to be staged in March. But the air show did not live up to its advance billing, for when it took place only nine aircraft flew up from Shien Chiao and buzzed the west side of Shanghai en route to Hungjao airdrome. Four of the aircraft were gifts to the government from private citizens and trading companies.

In November 1933, a rebellion against the Nanking government broke out in Fukien Province on the southeast coast of China. This was a separatist movement fomented by a clique headed by Eugene Chen, an overseas Chinese who had tried but failed to find a place for himself in his ancestral land, and by General Tsai Ting-kai, a popular hero who commanded the 19th Route Army in Shanghai in the winter of 1932 and embarrassed the Japanese Navy by his spirited defense of Chapei.

But the Fukien rebels were also becoming involved with local Communist guerrillas and when Chiang Kai-shek saw the direction the rebellion was taking, he gave the rebel leaders a week to cease and desist or he would "destroy the jade and the stone alike." Temporary air strips were rushed to completion close to Wenchow and Chuchow and light bombers from Shien Chiao went into action. This, as one Chinese writer explains, "had a disheartening effect on the morale of the harrassed army, and hastened their final defeat . . . ."[6]

In 1936, there was an outpouring of sentiment, and of funds, to celebrate the 50th birthday of Chiang Kai-shek, on 31 October. Popular subscription raised millions of dollars to buy aircraft, mostly American. At a big celebration held in Nanking at the Ming Palace Airdrome, 200,000 people came to watch about a hundred aircraft fly formation, and dipping to salute a huge picture of the General.

Contributions for the airplane fund had been coming in all year. More than $800,000 (yuan) was raised in the province of Fukien alone to purchase aircraft in honor of the 50th birthday anniversary of the General.[7] Although a bit late, Chinese-Americans in Seattle, Washington, sent $27,575 in 1937 to purchase aircraft for the new air force.

The *China Weekly Review* reported that in order to "encourage popular interest in aviation," the Chinese Flying Club would offer free airplane rides for the next two months. The Flying Club was in the news a few months later when it managed to get Wu Te-chen, mayor of the Municipality of Greater Shanghai, to speak at a ceremony held at Lunghua airdrome on 18 June.[8]

There were no experienced pilots outside of government service and if any gypsy pilots were thinking of giving free rides, they had not consulted with the Aero Commission in Nanking. The Nanking government did not have jurisdiction over any Flying Club within the International Settlement of Shanghai, but it did have jurisdiction over the Lunghua Air Space. Since nothing further is heard about the club, it can only be assumed that the leaders had been grounded, indefinitely.

Pilots of Chinese Air Force, Mao Ying-chu (left) and Huang Kwang-han who claimed destruction of Japanese destroyer near Shanghai in 1937.

III

Aviation In North China

The cultural split that divided north China from south China was one of long standing and in aviation the two halves were poles apart, each developing its own air force and, equally important, each paying for it with no help from the other. The Cantonese faction could defy the Peking government because it was based on a rich province and because it could rely on donations from overseas Chinese in America and Southeast Asia, most of whom had immigrated from Canton and Fukien provinces and remained attached to their ancestral roots. The fact that foreign nations maintained diplomatic relations with north China, not Canton, should have given the northerners an advantage in aviation but it did not.

After World War I, which had spurred the development of aviation in the West, there was a surplus of both commerical and military aircraft, and manufacturers sent salesmen abroad in search of customers. The British were the most successful in China. They made a deal not only to sell aircraft but to take over the training program at Nanyuan (outside of Peking) which, since its inception in 1913, had barely managed to stay airborne.

The SCMP broke the story with a one-column headline, "Aeroplanes for China," with this report:

> G.W. Fradsham, acting agent general of the Peking Syndicate Peking, has signed on behalf of his firm, with General Ting [Shih-yuan], director-general of communications, a contract for the supply of several Handley-Page aeroplanes and capable of carrying 20 passengers and fly at the rate of 100 miles per hour. The Peking Syndicate is the Handley-Page agent.[1]

A report sent to the State Department by Rodney Gilbert, dated July 1926, states that the contract with Handley-Page was for six passenger planes, and with Vickers Ltd. for forty Vimy commerical aircraft, forty Vimy trainers and sixty-five Avro trainers.[2] However, there appears to be no reliable information available on the actual number of airplanes on order or how many reached China.

Meanwhile, the new director of training at Nanyuan, Group Captain F.V. Holt, on loan from the Air Ministry, was on his way to Peking in the spring of 1919 accompanied by a staff of flight instructors and mechanics.

The several commerical aircraft purchased by the Chinese were to be used on the first commerical air and postal service between cities in north China. But the pilots were British, and after a few trial flights, it was found that people still preferred riding a train, or a sedan chair, to flying, and the commercial angle was phased out.

The base commander at Nanyuan was probably K.Y. Chen, and Ting Chin, also known as Mouhan K. Tinn, was Holt's counterpart in the training program. Wang Ken, a West Pointer (class of 1918) was also associated with the Aeronautics department in Peking from 1920-21 although he was not a pilot.[3] But these three men were probably responsible for developing a satellite field on the northwest side of the city.

One of the problems Group Captain Holt faced in the training program was the lack of student cadets with any practical knowledge of mathematics. Also, some were tempermentally unsuited to flying, and some had physical defects. Actually, most of the cadets were there because they had family connections and money, but little talent for flying. Holt tried to solve the problem diplomatically but he never did. It was a clash of cultures, and American and Italian flight instructors who came later in the 1930s, ran into the same problem of "face" which made it difficult to wash out incompetent cadets.

The British instructors at Nanyuan remained less than two years. Holt returned to England in September 1921 and most of the instructors went with him.

The flight school at Nanyuan was carried on by Chinese instructors, but little was accomplished owing to a shortage of funds and the continued unsettled political environment. The 1928 edition of Jane's reports that "the original Peking Flying school at Nanyuan was closed down (and) not a single Chinese pilot is qualified to take passengers on a Vikers 'Vimy' commercial aircraft."[4]

But the main reason the school was closed was by order of Chiang Kai-shek who had established the national capital at Nanking. Here a new aviation school was started in 1927 and completed in 1928. Here the training program was carried on until 1932 when it was moved to the field at Shien Chiao near Hangchow.

An intelligence report from Peking sent to the U.S. State Department in 1921, describes briefly the history of the flying school at Nanyuan and then adds this information: "100 pilots were trained "up to the standard of pre-war flying certificate,"[5] presumably meaning pre-1917.

The Manchus, who originated in Manchuria, gave their English name to the area. After the Manchus were deposed in 1911, there was no strong voice to command the people in this abundant land "to hear and obey," and government directives after 1912 were coming first from Canton, later from Nanking, and by 1913, from Peking. Though shakey, the national authority continued to be exercised from Peking until 1928 when the national capital was moved to Nanking under the Kuomintang or "Nationalist" government.

The power vacuum in Manchuria was an open invitation to former Manchu army commanders to conquer part or all of the country. The one who finally succeeded was Chang Tso-lin. In the several years it had taken him to overcome his rivals, he had never taken much interest in aviation, since guerrilla campaigns are not entirely suited to holding airfields and protecting airplanes.

By 1920 Chang Tso-lin was master of Manchuria and one of the first moves he made to assure himself of more power was to start a flying school. This was the machine of magic, the shortcut to prestige. According to a report submitted to the U.S. War Department by Captain Parker C. Tenney, assistant military attache in Peking, Chang Tso-lin organized the first units of an air force in 1920 and turned the command over to Chao Keng-yun.[6]

In 1923 Chang Tso-lin, now called "Marshall," placed his No. 1 son, Chang Hsueh-liang, in charge of training although he too had no pilot's certificate. But he had enough airplanes to divide his air force into three squadrons although the squadron strength is not indicated.

Early in June, 1928, Chang Tso-lin was returning from his

A Schreck SBA 310 (French) two place sport amphibian with 120 hp engine, used in training program at Chinese naval base, Tsingtao, 1930s.

latest forray into north China, bringing his troops with him on a troop train while he rode in the last car. As the train passed under an overpass of the Japanese-owned South Manchuria Railway, his car was blown up and he was killed. Fifty years later, no one has come forward to prove or point any fingers at the killers although most sources blame the Japanese.[7]

On the Old Marshal's death, his No. 1 son, Chang Hsueh-liang, took over the reins of power, and the air force. He was soon being referred to as "the Young Marshal." A contemporary photograph in Jane's shows a French pilot, Georges Pelletier Doisy, visiting with Chang Hsueh-liang on the air field.[8] Doisy was on a flight from Paris to Tokyo and had already stopped over in Shanghai.

In July 1931, two months before the Japanese invasion of Manchuria, E.P. Howard, U.S. commercial attache in Shanghai, made a trip to Manchuria and sent in the latest information on the status of the Manchurian Air Force, usually referred to by the Chinese as the "Northeast Air Force."[9] Howard found that there were twenty-five Potez 25 A.2 (450 hp Lorraine—Dietrich engine) purchased from France in 1929 of which thirty were in good condition with seven waiting parts. In addition, he found one Breguet 19 in good condition, and one Dewoitine D-27 C, a single seat fighter, allegedly purchased for $28,000 after being used in Mukden as a demonstrator for one year.

German aircraft found at different fields included nine Junkers A-35-L (listed in Jane's under Sweden as K-53s), six of which were in flying condition and the others waiting for parts.

Aircraft purchased from Japan included three Army Type D-S, a copy of a Breguet with metal fuselage and German-built 600 hp B.M.W. engine. Also four old Japanese copies of French Newport-Delage, single seater, 300 hp German B.M.W., all second-hand equipment with several years on them before being sold to the Young Marshal.

Howard also found one Vought-Corsair (P.W. Wasp) waiting for repairs, and five Gypsy-Moth (wooden fuselage) in commission, with two more to be put in condition when parts arrived from England.

There were twenty-five Caudron trainers (129 Hispana-Suiza engine) in condition out of an original lot of sixty. Six or seven of these were still used for training but cadets avoided them if possible.

Finally, the French had sold the Young Marshal twelve Schreck S.B.A. amphibians, a two seater, used by their Navy, some equipped with 180 hp and some with 320 hp Hispano-Suiza engines. This is all that remained of an original order of sixty-two, and only six of these were in flying condition.

On one field Howard found six Potez 36s (40 hp Salmson engine) cabin monoplanes, which were purchased in 1930 but only one ever became airborne. Someone had sold the Young Marshal an underpowered airplane for cargo hauling. Finally, around *Tung Ta* airfield, Howard found two Fokker D-7s (B.M.W. engine) which were taken up on occasion but only around the field.

Howard estimated that the field at Mukden had about forty-five aircraft in flying condition, including training types, to which fifteen could be added with a few days of work on the engines. There were about 140 obsolete machines, useless except for cannibalizing, and included in the batch were fourteen old Vimy transports and six Avros with LeRone engines, some of which were probably liberated from Nanyuan by the Old Marshal on one of his campaigns into north China.

The Chinese officer who escorted Howard around the field at Mukden was "very scornful of the Japanese and particularly of their Breguet copies which he said, had an excessive landing roll and were much too slow and cumbersome."[10]

At the time Captain Tenney wrote his report in 1931, using Howard's notes, Tenney thought there were five Chinese pilots who could fly just about any machine, and about twenty who could handle trainers up to Corsairs, but none skillful enough for pursuit equipment.

Thus, after more than ten years of training there were no all-around pilots who could be trusted to engage in combat. And that is the way it was when the Japanese, without warning, launched their attack on Manchuria and almost overnight destroyed or captured most the "Northeast Air Force." A few planes managed to fly into north China.

Meanwhile, there were Chinese pilots learning to fly in other parts of China, namely in Canton. In 1928, three Chinese airmen flew an airplane of unidentified American make from Canton to Mukden, with stops for gas. Landing in Mukden was now possible because the Young Marshal had raised the flag of the Kuomintang over his headquarters, signifying his allegiance to the newly-founded government of Chiang Kai-shek in Nanking.

One of the three airmen who made the historic south-to-north flight was Chang Hsuei-chang, later referred to in the Chinese press as the "Lindbergh of China."

Not long after Chang's flight, Chang Hing-wan (no relation) and Chang Yuk-hang, both Cantonese, flying an amphibian of American make, logged 3,340 miles on a flight from Canton to Hankow, Nanking and Shanghai.

In addition to aviation schools in Canton, north China and Manchuria, schools were begun in southwest China, in Kwangsi Province and still farther west at Kunming, capital of Yunnan Province. The school in Kwangsi was funded by provincial authorities and located in Liuchow (WWII, a fighter base of the U.S. 14th Air Force.) Lam Wei-hsing, who probably learned to fly in France, was director of the school when it opened in 1924. Two of the instructors were former pilots of the Royal Air Force, and they used British trainers for the most part. These probably included three Armstrong-Whitworth light bombers, two Atlas observation light bombers, three PV6 pursuit trainers, seven Avros, and twelve Avians, in addition to primary trainers. It was one of the most ambitious training centers in south China for a time and allegedly pinned wings on 150 pilots. The program was still being pursued into the 1930s, but slowed up in 1933 when the provincial government ran out of funds and the British instructors withdrew. Chinese instructors, however, kept the program going until funds were again available to rehire the British, who returned in 1934.

Farther to the west lies Kunming, once a walled city lying north of Kunming Lake, a check point for incoming aircraft. An airfield was built between the lake and the city as early as 1923, when provincial authorities began buying trainers for a flight school. Since the most direct route from Yunnan to a sea port lay over the French-built narrow gauge railroad to Hanoi, it could be expected that the school at Kunming would favor French instructors and French aircraft. Jane's in 1926 says the French mission was under a Captain Arbitre, and the training school had half a dozen Caudron G-3s and Breguet biplanes.

By 1931, the Kunming school was facing problems of an unspecified nature which a new director, Chang Chih-kao, was appointed to correct. He began by ordering six American made primary trainers and engaging an American instructor to assist the French, but the American soon withdrew. One hundred or more pilots were allegedly awarded wings from 1931 to 1934, but of this number, no one was sure how many could solo.

In west China, heavily populated Szechwan Province was ruled by Governor Liu Hsiang, and one Liu family followed the other in power in the 1930s and into the 1940s before the collapse of the National government in 1949. The first Liu opened a flight school at Peishiyi, a field lying west of Chungking and revamped into a fighter base in WWII. Like the school in Kunming and Liuchow, training standards were low and political influence was high in the choice of cadets.

Another provincial aviation school was opened in Changsha in south central China in the 1930s. This was run entirely under Chinese supervision. The main runway at Changsha, lying north to south along the tracks of the Hankow-Canton railway, was destroyed in 1938 to prevent landings by Japanese aircraft.

Small aviation schools were operated at Tsingtao and Tsinan in Shantung, and still another at Nanchang in Kiangsi Province. Until Italian instructors began to train cadets at Nanchang in the mid-1930s, none of the other schools mentioned above ever played a significant role in the development of aviation. Nanchang, however, became a major training center and site of an assembly plant for Italian aircraft.

The school at Tsingtao had seven French-made Schreck amphibians, one with 180 hp Hispano-Suiza engine, the others with 320 hp engines of the same make. All were said to be hand-me-downs from the training school initiated by Chang Tso-lin in Mukden, and had been shipped or flown out of Manchuria to escape capture by the Japanese Army in September 1931.

The school at Tsingtao had two hangars which were connected with the harbor by concrete ramps, in addition to an administrative building and residences for the staff. Only a handful of pilots got their wings here, probably under Admiral Hsieh Kung-chien.

Fragmentary sources describe a naval college at Mamoi near Foochow where, in 1916, an aeronautical branch was added to the school curriculum. The main effort here was to study aircraft design and related aeronautical problems. China's first airplane was probably built here, but, with an imported engine.

A Junkers W-33 with logo of Chinese Air Force under wing but no stripes on tail section.

IV

Central Aviation School

In distance, new administration building for Central Aviation School at Shien Chiao. In foreground, cadet barracks, c. 1933.

At midnight on 28 January 1932, Japanese Naval Landing Party (marines or "Bluejackets") attacked the Chinese suburb of Chapei lying adjacent to the International Settlement of Shanghai, on the north. The aim of the Japanese was to "teach" China a lesson and to halt a national boycott of Japanese goods which was directed from Shanghai. The Chinese were boycotting Japan because the Japanese Kwantung Army had invaded Manchuria on 18 September 1931 and occupied the main cities and railroads.

The Chinese were expecting trouble in Shanghai from the Japanese, uncertain just when it might come or how it would start. To be on the safe side, the Nanking government had stationed the 19th Route Army, a Cantonese unit, in Chapei, opposite Hongkew, an adjoining suburb but part of the International Settlement and the part which the Japanese were supposed to defend against outside attack, that is, from the Chinese.

The Naval Landing Party units moved out of their barracks in northern Hongkew at midnight of 28 January and struck west against the 19th Route Army with the Shanghai-Nanking Railway station as their first objective. But in one of the great surprises of contemporary history, the 19th Route Army not only held its lines but struck back and forced the Japanese to retire from their first penetration of Chapei. In the next six weeks, before being outflanked to the north of the city, the 19th Route Army held its ground in the face of superior enemy artillery and air power. It came to be called the "Shanghai Incident."

In the brief battle for air superiority around Shanghai, the Chinese found themselves ill prepared to challenge the superiority of Japanese pilots and aircraft, and most Chinese aircraft were destroyed on the ground or parked on the line for lack of parts. The Japanese airmen scoured the fields at Shien Chaio, Soochow, Nanking, Hungjao and Lunghua, looking to catch any exposed aircraft.

But when the war ended in May, the Chinese were not

中央航空學

CENTRAL AVIATION SCHOOL

SHIEN CHIAO, HANGCHOW, CHINA.

CABLE ADDRESS:
"CENTRAVIA"

CODE USED:
BENTLEY'S SECOND

TELEPHONE: 3073
3074
3075

REF. NO.

June 20th, 1934.

General C. J. Chow,
Commandant, Central Aviation School,
Shien Chiao.

Dear General Chow:

I desire to have Mrs.E. Holmes visit me for a few days as my house guest in the Awakening Village.

Will you please advise me if you have no objections to her visiting us.

Very truly yours,

M. R. Knight.

Letter written by M.R. ("Reed") Knight to Chou Chih-jou asking for permission to allow Mrs. E. Holmes to visit the Knights in "Awakening Village," where quarters had been built for foreign staff at Shien Chiao. Although the Japanese were well aware of the school program, no visitors or newspaper reporters could approach the base without authorization. The abbreviated Chinese ideographs in the lower right are written by General Chou approving the request.

entirely without aircraft. They had flown many planes to the far west to avoid total destruction, and the Japanese airmen, flying from a carrier off Chungming Island in the Yangtze estuary, did not have the range to follow them. As a result, the Chinese ended the brief war with 270 miscellaneous aircraft on report, although scarcely a third were combat ready and most of these were two-seater biplanes.

But Japanese aviation, especially naval aviation, had shown that it was superior in every branch to the Chinese. Japanese pilots had been trained in carrier takeoffs for several years; the Chinese had no carriers, not even a tender.

Not long after a truce was signed in Shanghai between China and Japan on 5 May 1932, the Nanking government made plans to create an up-to-date school of aviation under American supervision. T.V. Soong, Harvard-educated and now minister of finance, turned to Edward P. Howard, United States trade commissioner in Shanghai, and to Julian Arnold, commercial attache. Howard had an avid interest in aviation and among his acquaintances was been chosen as the site for the new training center and the official stationary calls it "Central Aviation School." The name of the village near the base was called Shien Chiao, loosely translated as "Awakening Village."

Since there was no housing available immediately on base, the Americans were given quarters in hotels overlooking West Lake in Hangchow and they drove back and forth to work for several months.

Before the party left to board the *Hoover* in San Francisco, Major Jouett wrote a letter from New Orleans to Knight, then in Kansas City, in which he explains that he had sent him an advance of $600 and that he, Jouett, had been appointed by the Chinese government to direct a mission for the "purpose of establishing a military aviation training center, to reorganize the Chinese air force, and to be military aviation advisor to the President of China."

Further, although the training mission would be composed of "civilians," they would retain their reserve commissions in the U.S. Army Air Corps. He also advised Knight that he expected the same loyalty from him that any officer

First student nine-ship formation ever attempted with Fleet trainers at Central Aviation.

Major John H. Jouett, U.S. Army Air Corps, retired. "Jack' Jouett had been retired not for any disability but because the army had been ordered to cut back. These were Depression times and among the first to feel the axe were field grade officers. This was China's gain, and when Jouett began looking around for instructors and mechanics, he found qualified people, like himself, who had been prematurely retired.

After inspecting official records, Jouett picked a cadre of first-class instructors and mechanics. They boarded the *President Hoover* 17 June 1932, each man with a contract in his pocket for three years and first class passage to and from his station. The party landed in Shanghai 8 July.[1]

One of the instructors was Morris B. Knight, usually referred to as "Reed" Knight, and, according to the contract he signed, he was to be paid $491.90 on the 15th day of every month for total of $15,000 less $700 already advanced. In addition he was to get $85.33 per month house allowance.[2]

An old air base a few kilometers east of Hangchow had in the army would get. In his penultimate paragraph, he warns him that the nature of the mission was to be kept top secret "until we arrive in China."[3]

Major Jouett correctly guessed that the Japanese would learn about the mission, all too soon.

In addition to Knight, others who accompanied Jouett on the *Hoover* were L.R. (for Roy) Holbrook, Ellis D. Shannon, Christopher Mathewson, son of famed baseball pitcher Christy Mathewson, Harry T. Rowland, W.C. ("Foxy") Kent, and Roland L. Sansbury. Several brought their wives, and Jouett, in addition to bringing his wife, also employed a private secretary, Alma Wade. Several months later in 1933 the cadre of instructors was augmented by the arrival of John Schweitzer and Harvey Greenlaw. The latter was a West Point man, class of 1920, and after completing his contract with Jouett in 1935, he remained in the Far East and in 1941 became chief of staff to Chennault and the First American Volunteer Group. His wife, Olga Sowers of California, accompanied him to Shien Chiao and later to Kunming and kept a diary of the First American Volunteer Group.

A year after the A.V.G. was deactivated in 1942 she wrote a book on her experiences called *The Lady and the Tigers.*[4]

In addition to pilot instructors, Jouett brought along or later hired mechanics and engineers, including G.B. Clark, Paul G. Zimmerman, H. ("Nemo") Poncetti, Frank Gay, T.L. Taylor, a flight instructor, James A. Coughlin, mechanic, W.A. Spencer, mechanic and parachute rigger, W.H. Pennington, engineer, Edward Wingerter, who was put in charge of engine overhaul, and Donald R. Stevenson, who supervised aircraft overhaul. John May, and Cecil Folmar, both flight instructors, joined the school in 1936.

Greenlaw actually replaced Christopher Mathewson, who was nearly killed in a crash of a Sikorsky-38 at Lunghua.

W.H. Pennington died of typhoid and his place was taken by Spencer who became chief engineer. Dr. Eldrige Adams, flight surgeon, was another replacement following the resignation of Dr. Harold Cooper.

Chinese officers attached to the training school staff included Dr. Nelson Feng, flight surgeon, Col. C.F. Fong, in charge of Chinese engineering staff. Major Li Pei-li, assistant engineering officer, Major Lee, line chief, Captain Yi Chou-ma, assistant engineering, W.P. Tsao, operations officer, Liu Chao, cadet commander, Kao Chia-hsing, squadron commander, and Wang Tien-hsing, squadron commander.

Major Jouett, who was quickly promoted to "colonel" by the Chinese, brought with him U.S. Army flight manuals and teaching aides, all in English. The task of translating the most pertinent materials was turned over to a civilian aviation enthusiast, P.Y. Shu, born in Changsha, who finished his undergraduate work at the University of Michigan.

When Jouett got to Shien Chiao, according to Knight, he found what was left of the Chinese Air Force after the massacre of 1932. Chinese pilots who had been trained in air academies around the country were nearly all field grade officers, many of whom should have been washed out. Jouett reduced most of them to company grade but their pay scale remained the same as field grade officers, which was a matter of both "face" and life style.

Jouett sought to make the flying school a Chinese Randolph Field. The field itself had been laid out in the 1928-29 period shortly after the government of Chiang Kai-shek picked Nanking as the nation's capital. The runway had to be lengthened but it was not an all-weather field.

There were two small steel Junkers hangars which had been built in the late spring of 1932, and several buildings destroyed or damaged by the Japanese in 1932 were restored.

But there was no drinking water, electricity, machinery or repair shop, and most of the airplanes in the hangar were dilapidated Moths which had been flown in from fields in the west of China after the armistice was signed in 1932 between the Chinese and Japanese at the end of the "First Shanghai Incident."

Work got under way on a large hangar as well as a new school building for class room instruction. A deep well was dug for water, a radio station was erected for homing aids, and an operations office, woodworking shop, engine overhaul shop, airplane repair shop and dope room were eventually completed, and finally an auditorium for lectures and programs.

In addition, a clinic for the base surgeon and three Chinese student surgeons was erected before the end of the year. This was supervised by Dr. Eldrige Adams who had the responsibility of explaining to the Chinese surgeons how to choose cadets with a good sense of equilibrium, strong muscles, balance, and perfect eyesight. Jouett considered his service invaluable.

While construction work held first priority in the early months at Shien Chiao, flight instruction got under way by 17 September, with refresher courses for officers, especially for the purpose of selecting people who could be used as instructors to assist the American staff. Fifty of these officers were picked and forty-four passed the physical given by the flight surgeon (equivalent to the U.S. Army 64). Of the forty-four, a total of seventeen washed out in flying and one was continued for further instruction. The twenty-seven survivors required an average of four hours and twenty-minutes to solo. Their qualifications were fair, and the nine who were retrained as assistant instructors, were rated as equivalent to fifty-hour students at Brooks Field. Of the nine, seven improved with experience, one was relieved, and one was about to be relieved in November.

About one hundred cadets reported for training and an equal number were to be replaced every six months. John

Blackburn Lincock III light bomber, one of two supplied to China. British manufacturer, according to Jane's (1932), built only four. This photo taken at Shien Chiao. Cubicle at far right was Chinese guard post.

空軍特別黨部本校第二區黨部全體委員

執行委員會幹事 執行委員 羅良　執行委員 石英　執行委員會助理幹事 萬嘉鵬　監察委員 崔滄石　執行委員 張柳雲　執行委員 李瑞彬　執行委員 徐寅　執行委員 胡偉克　執行委員 李克元　監察委員 張韶舞　執行委員會錄事 漆海波　監察委員 沈開寰

A section of air cadets who attended 2nd class of trainees at Central Aviation School. This photo was taken from class annual 1933.

Schweitzer, on a visit to Taiwan in 1970, was entertained by his former cadet-students and was given a souvenir platter as a memento of his visit.[5] On the platter are inscribed the dates of the several classes which were graduated from Shien Chiao, namely, the first, in 1933, the second in the spring of 1934, the third in December 1934, the fourth in June 1935, the fifth in January 1936, and the sixth in October of 1936. But Schweitzer was no longer at Shien Chiao in late 1935. These classes were given their wings under Chou Chih-jou and a few American instructors who continued with the school after the others left, their contracts expired.

The first class of eighty-eight was divided into fifteen sections of five or six cadets each and assigned to individual American and Chinese instructors. Less than half completed the course.

Instructors at the school found the Chinese cadets to be intelligent, highly motivated and well-disposed.

Reed Knight recalls that the 1st Bombardment Squadron moved to Nanking for training and used Italian Fiat BR3s. He said there were ten of these on the field at Nanking and "it was a struggle to keep seven or eight operating at once." He also recalls seeing about thirty Hawk IIs on the line when he left the school in 1935. These Hawks had supercharged Curtiss Wright engines and were made at the company plant in Buffalo, New York.

Schweitzer recalls that the Curtiss Hawk biplanes used at Shien Chiao, known to the Navy as F-11s, were powered with 715 hp Cyclone radial cooled engines, 12 cylinder. Each carried two 30 caliber guns, one in each lower wing. These F-11s arrived at Shien Chiao in the following order: eighteen between 15 March and 8 May 1933; seventeen between 19 June and 19 July 1933; and fourteen between 14 August and 1 September 1933. The cost of each was $12,000 less the engine.

Above: Chinese bomber pilots, all dressed in flight suits, pose with their American instructor, Ellis Shannon (center), at one of air bases in Nanking where 1st Bombardment Squadron was being trained, c. 1933. Below: Chinese cadets at Shien Chiao line up for inspection with Fleet trainers on line in background.

For primary trainers, according to Knight, the school used the Consolidated Fleet with 125 hp Kinner engines, and there were 120 of these at the school in 1935. Twenty Douglas 02Mc-4s were either on order or already assembled, and there were twelve Northrop light bombers for a 1st Bombardment Squadron. Ellis Shannon was the supervisor for bombardment training.

In addition to the flight school, another school was opened to train engineers and mechanics. The mechanics in this section were supervised by William A. ("Al") Spencer, who became chief engineer after Pennington died of typhus. Most of the instruction was carried on in Chinese by trained mechanics. Spencer also supervised parachute rigging.

It was estimated that operation of the Central Aviation School cost the Nanking government a million dollars a year. This did not include aircraft, but it included salaries, quarters for foreign staff, barracks pay and mess for the cadets.

Colonel Jouett's Chinese counterparts in the training program were Mao P'an-tsu (Peter Mow) who was succeeded by Chou Chih-jou.

After Colonel Jouett returned to the United States, he wrote an article for *Asia Magazine*.[6] He reveals no secrets but extolls the Chinese airmen who had, since the Japanese invasion of China began, "given a good account of themselves." He made special reference to three pilots, Lian Hong-yuan, Chang, Shang-kan, and Chou Tung-feng (his spellings).

Although the Jouett mission had fulfilled its contract in June 1935, the school was carried on by Chinese instructors and three American holdovers, Holbrook, Kent and Spencer. A new superintendent was found, George E. A. Reinburg, a West Pointer (class of 1930), and he continued for several months until stricken by a heart attack and died on the base 5 December 1936.[7]

After the Jouett mission left China in the summer of 1935, Captain Claire L. Chennault, in command of tactical training at Maxwell Field, Montgomery, Alabama, received a letter from Holbrook, an old friend, who asked Chennault to recommend some new pilot instructors and engine specialists for the school at Shien Chiao. Among the several that Chennault recommended were Sterling Tatum, John ("Luke") Williamson, William C. ("Billy") MacDonald, Rolfe Watson, an armament specialist, Sebie Biggs Smith, a mechanic and John Holland, also a mechanic.

Chiang Kai-shek on viewing stand at graduation ceremony for one of the flight school classes at Shien Chiao. Photo probably taken 1934. At far left, Colonel John Jouett, director of training, and next to him, Chang Hsueh-liang "the Young Marshal" of Manchuria. Behind him is probably a member of the Italian training mission. The Generalissimo took deep interest in developing aviation for China and never missed a graduation ceremony at Shien Chiao. Mme. Chiang stands directly behind her husband.

Sebie Smith remembers it this way: "I think it was some time in March of 1936 that we all piled into an old Keystone bomber, that is, Holland and I, Watson, MacDonald, and Williamson, and flew to Washington. Tatum flew up in an Air National Guard plane from Birmingham. We were taken to the Chinese Embassy, explained what it was all about, shown the contract, and if we agreed to go, to sign up. We did. They gave us one month's salary in advance and travel

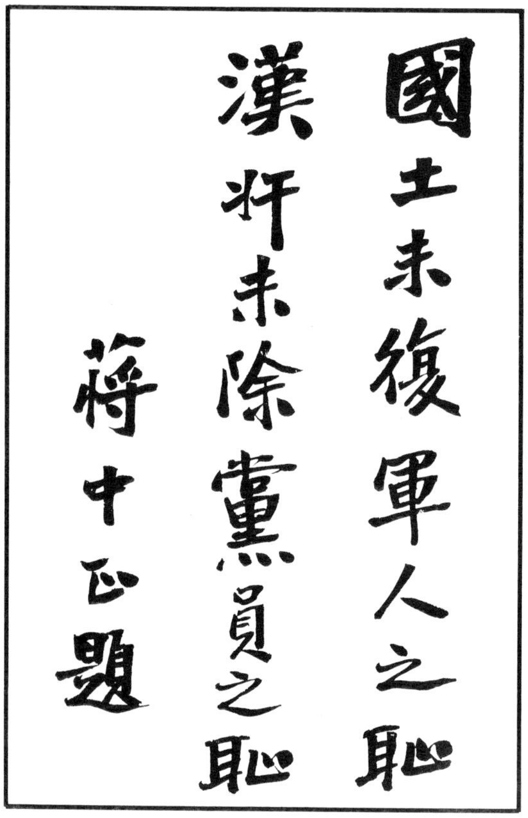

Chinese modes of calligraphy are as varied as the calligraphist. Here are two samples, written by famous men of China which appear in a school album published by a cadet class at Shien Chiao in 1933. One at left is by Chiang Kai-shek, one at right by Lin Sen, president of Republic of China. The latter also includes his personal "chop" at lower left. The message from Chiang exhorts his future fliers: "Shame to soldiers if nation [north China and Manchuria] not recovered. Shame on party members if traitors are not eliminated." The message from Lin Sen reads: "Flying on high cloud" (i.e. your horizons are unlimited).

expenses—the most money any of us had ever seen."[8]

They were all permitted to resign from the Army Air Corps with an honorable discharge. Holland and Williamson got married. Tatum was already married, and, together with Watson, MacDonald and Smith, the group sailed out of Victoria, B.C. on 11 July 1936 on board the Canadian Pacific liner, *Empress of Russia.* In Kobe they were greeted by Japanese immigration agents who already knew they were coming.

In Shanghai, they were met by Roy Holbrook who had arranged a number of parties for them, including one hosted by the Aero Commission and another by William D. Pawley. After a week in Shanghai, Holbrook took the "advisors" to meet Generalissimo and Mme. Chiang Kai-shek. Instead of having them go to work right away, Mme. Chiang wanted the group to learn something about China and Chinese history. She turned them over to W.H. Donald, the former Australian newspaper correspondent turned advisor.

After a week of "instruction" from Donald, the advisors returned to Shien Chiao, and Sebie Smith recalls the first assignment he had was to assist Holland and Watson in assembling a shipment of about thirty Hawk IIIs, and an equal number of Curtiss Shrikes, an attack ship.

Jouett fails to mention in his article in *Asia* that the flight school at Shien Chiao was moved temporarily to Loyang in North China in the spring of 1933. Most of the American instructors, together with the training planes, flew to the new base in stages. The move was made to avoid surveillance by Japanese observation planes. The Japanese had no right in the China Air Space but they trespassed at will.

Loyang was a trifle out of range for the Japanese, or at least they did not run any reconnaissance flights that far west, but, according to "Al" Spencer, Loyang in spring was "cold, windy and sandy, and living conditions were terrible."[9] The combination of wind and sand made training difficult, while maintenance on engines was way out of proportion to normal conditions. After a few months, the school was moved back to Shien Chiao, bringing along the Fleet trainers, Douglas O-2MCs, four Junkers pursuits and one Armstrong-Whitworth. Everyone was glad to be back in

The 1st Bombardment Squadron of Chinese Air Force trained at Nanking under Reed Knight (left) and Franklyn G. Gay who stand in center of picture, flanked by cadets. Behind them stand Douglas O2MC-4s and Italian Fiat BR-3 bombers, and one Fiat fighter. Chinese government had twenty BR-3 bombers on order, and eighty-five O2MC-4s on order when training program at Shien Chiao started in late 1932.

Shien Chaio, said Spencer, and the training program went ahead without further interruption, despite Japanese surveillance.

Continuing, Spencer had this to say:[9]

"When there was a graduation, the Generalissimo, with Mme. Chiang, and Dr. H.H. Kung and T.V. Soong and ranking generals were there. We would have a ground inspection of all aircraft followed by a march in review, ending with a fly by of all graduated pilots. This was always impressive. There were also gunnery and bombing demonstrations. By the end of 1934 the school had graduated 150 students, and 200 would enter every six months for a twelve month course. Of course half of them washed out.

"But the school began to have some troubles too, due to the complaints of students who were washed out and caused their families to lose face.

"We had several visits from Count Caleazzo Ciano and his wife Edda, daughter of Benito Mussolini, and several ranking Italian officers began to show up. It seems the Italian government had made some rash promises to the Generalissimo if he would get rid of the Americans.

"The Italians later took over the fields at Loyang and at Nanchang, but their training was a farce. Their planes were coming apart at the seams and they had to borrow the American planes to carry on. It almost seemed like they were trying to sabotage the Chinese Air Force.

"Most of the students we graduated were good pilots and good leaders but there just weren't enough of them. They took the brunt of the early fighting over Shanghai and Nanking and many of them died heroes. This mess was turned over to Chennault, but by that time he didn't have much to work with."

In the above interview, Spencer refers to Claire L. Chennault, formerly in charge of tactical training at Maxwell Field and one-time commander of the 19th Pursuit Squadron at Hickam. While at Maxwell in the mid-1930s Chennault also organized the "Three Men on a Flying Trapeze," an aerial precision act which performed at national air shows and state fairs to glamorize the role of the Air Corps. It was at an air show in Miami, Florida, in December 1936, that a visiting Chinese delegation saw the performance and also met Chennault. The leader, Peter Mow, indicated that China could use a man with Chennault's talents but this was 1936 and he was not ready to give up a career in the Army Air Corps. However, a few months later, he was informed that he was being retired on a physical disability: bad hearing. Since he had been in touch with former team mates of the old Flying Trapeze, "Luke" Williamson and "Billy" MacDonald, who were already in China, it took him little time to make up his mind to accept the offer of the Chinese to come to China and serve as an advisor to the Chinese Air Force.

Chennault was discharged from the U.S. Army Air Corps on 30 April 1937, and he left shortly for San Francisco to catch the Dollar liner, *President Garfield.* In Kobe, Japan, he was met by Billy MacDonald. After a few days of sightseeing ashore, and noting the flimsily-built Japanese houses and factories, the two went on to Shanghai where they were greeted by W.H. Donald, 31 May 1937, and taken to the Park Hotel for dinner with former friends, most of them from Maxwell Field.

Chennault came to China on a three-month contract at U.S. $1,000 per month. He stayed eight years, first as an advisor, 1937-1941, second as organizer of the First American Volunteer Group (Flying Tigers) 1941-42, third as commander of the United States China Air Task Force, 1942-43, and finally as commander of the 14th Air Force, 1943-45.

Meanwhile, after parties and introductions in Shanghai, Chennault was taken to Shien Chiao and later to meet Generalissimo and Mme. Chiang Kai-shek. What they wanted was a frank report on the state of the Chinese Air Force, air bases and training facilities, American and Italian. P.Y. Shu, interpreter and translator of text books at Shien Chiao, was assigned to Chennault and became a long time associate and friend of the American advisor. Billy MacDonald and Sebie Smith were also assigned to accompany the inspection team, and Mme. Chiang arranged for two Douglas BT-2s to be put at their service. Chennault flew one and MacDonald the other, with Smith and Shu riding in the back seats.

The inspection tour, begun in late June 1937, covered Canton and Nanchang in south China, Hankow in central China, and Sian in north China and the team of inspectors was ready to leave for Peking when word came of the Japanese attack on Marco Polo bridge, 7 July 1937—the beginning of the seven-year war between China and Japan.

Chennault and party flew back to Hankow to wait for instructions, and he was asked to fly to Nanchang and continue his inspection work. After a couple of weeks at the dusty Nanchang field, he and the director of training at Nanchang, Peter Mow, were summoned to Kuling, where the Chiangs were expecting them. Peter Mow, a Soviet-trained pilot, was asked by the Gimo to report on the strength of the Chinese Air Force, and he was forced to

admit that while there were 500 airplanes on report, only ninety-one were ready for combat. The Gimo nearly exploded.

Mme. Chiang turned to Chennault to ask if he agreed and he said he did. His forthrightness established a rapport with Mme. Chiang which continued the rest of his life. When he died in 1958, Mme. Chiang was among the mourners at Arlington National Cemetery. Obviously, the Generalissimo had begun to suspect that the training program under the Italians at Nanchang and Loyang was not going well, and when Chennault confirmed Mow's figures, the truth was out. It is not certain whether the Gimo told the Italians to leave, or whether Mussolini, knee-deep in occupation problems in Abyssinia, decided it was time to call them home. Whatever happened, they were seen no more in China.

Not long after the Italians withdrew, aircraft factories in Italy got an order from Japan for seventy-seven Fiat Br-20s, but they saw only limited service in China.

After the war began between China and Japan in 1937, Japanese planes destroyed the field and buildings at Shien Chiao. Most of the student cadets at Shien Chiao and at Nanchang took trains to the west as far as they could ride and then went by bus to Kunming. There had been a provincial flight school here since the 1920s, as noted earlier. After the United States entered World War II, some of the cadets from Kunming, went for advanced training to Thunderbird and Luke fields in Arizona. On 27 March 1942, the first class at Thunderbird was given wings by Shen Tieh-hsieh, a former staff officer in Nanking. This class and later classes were largely absorbed into the Chinese American Composite Wing (CACW) which was activated in Kueilin late in 1943 under the command of Clinton D. ("Casey") Vincent.

During the three years when the Americans were in charge at Shien Chiao, there were several air shows and demonstrations of new aircraft by American sales people. Generalissimo and Madame Chiang Kai-shek were usually invited. One pilot who came to demonstrate the Hawk II in 1933, as mentioned earlier, was Jimmy Doolittle.

Reed Knight remembers Frank Hawks at Shien Chiao. Hawks, like Doolittle, had broken speed records and recently had established the fastest time between the east and west coasts of America. He came to China to sell Curtiss Condors, a big, for that time, two engine prototype with extra fuel tanks on the side of the fuselage. The Chinese were interested in purchasing a plane for evacuation of wounded troops from the fighting front, which, at that time, was in Kiangsi where the Red Army was locked in battle with Nationalist troops. Hawks took the Condor up and flew around for a while and then prepared for let-down, the Gimo and staff watching.

But Hawks was having a problem. The bell on his landing gear was rattling and he inserted his handkerchief behind it. When he came down the warning bell did not sound and he forgot to lower the gear. With no wheels, he skidded the length of the field, bending the props and damaging the underside of the fuselage. That ended the future of the Curtiss Condor as a mercy ship in China. The Gimo mumbled to his aides that if an expert like Hawks could crack up that easily, his Chinese pilots would do no less.[10]

Dr. H.H. Kung, minister of finance, bought the Condor from Curtiss for about $10,000 and had it repaired for civilian flying. Reed Knight supervised the repair job at Shien Chiao. Foxy Kent and Al Spencer did some part-time flying for Kung, hauling Chinese bank notes from banks in Shanghai to banks in west China.

When the Young Marshal began visiting Nanking in the early 1930s, he often had his Ford Trimotor parked on the line at Shien Chiao for maintenance. He bought a second Ford which he later gave to the Gimo. Reed Knight flew it to Chungking twice with the Gimo as passenger.

Above: After fulfilling their contracts at Central Aviation school in the early summer of 1935, most of the American staff, led by Colonel Jouett, returned to the United States. Farewell parties were given by the Chinese staff and after one of them, this souvenir picture was taken of both Chinese and American staff. Members of the American staff are seen, seventh from the left and reading right, they are: Edward Wingerter, mechanic instructor and Mrs. Wingerter; Reed Knight, flight instructor and Mrs. Knight; Harvey Greenlaw, flight instructor; John Schweitzer, flight instructor; Harry Rowland, deputy to Jouett; Alma Wade, secretary to Jouett; General Chou Chih-jou, school commandant; Colonel and Mrs. Jouett; G.B. Clark, test pilot and engineering officer; Dr. Eldridge Adams, flight surgeon and Mrs. Adams; John Jouett Jr., a visitor; Thomas Taylor, flight instructor; Al Spencer, assistant in engineering in charge of Chinese mechanics; H. Poncetti, mechanic; and Frank Gay, mechanic. Not present from the original group were Holbrook, Shannon, Kent and Sansbury. Below, a Hawk II over Shien Chiao with G.B. Clark in cockpit.

Above: These Chinese Air Force cadets had come to Luke and Thunderbird fields in Arizona in 1942 for advanced training. Here they are given a day off to visit a movie set in a mining town east of Phoenix, and have their picture taken with the star, Irene Dunn (center). Visiting brass included, in back at left, Major Barry Goldwater, U.S. Army Air Corps, and in center, T.V. Soong, head of China Defense Supplies, in Washington D.C. Below: Graduation ceremony and picture taking for class of Chinese pilots at Luke or Thunderbird in Arizona. Each student pilot has inscribed his name in Chinese, probably for the American director (center).

Jack Schweitzer (up front) with Chinese cadet in Fleet trainer. About 120 Fleets were purchased by Chinese army in 1930s.

German Junkers fighter, K-47, at Loyang, May 1933. Roy Holbrook under wing at left.

Export Curtiss Falcon NR310E, sold to China in 1930s.

Italian Savoia-Marchetti trimotor, probably at Nanchang air base, 1936.

A DeHavilland Moth 60 trainer, sold to China in mid-1930s.

An Italian Fiat bomber in hangar probably at Nanchang. The ship had a wood longeron covered with plywood. The Fiat engine developed 950 hp on 12 cylinders, and had a range of six hours. It had a wing span of fifty-three feet, cruising speed of 120 mph and top speed of 149. Benito Mussolini, the new premier of Italy, probably gave it to China. It was already obsolete.

A French Farman transport in front of hangar at Shien Chiao, c. 1935.

These Douglas O2MC4 two-place reconnaissance and observation planes were used at Central Aviation School for basic training. Logo on fuselage appears to be early version of Nationalist white pointed sun in blue field. This photo, taken in 1933, may be earliest evidence that Chinese Air Force had adopted this logo for wing tips and fuselage.

Some modifications were made on this airplane and it was referred to as "the Chinese Northrop." Horizontal stripes on tail are Chinese Air Force.

Junkers K-47 fighters in front of hangar at Loyang, spring of 1933.

Bellanca 29-90B fuselage being moved by Chinese workers with bamboo poles. Picture was probably taken in Canton in spring of 1938.

A new Vought Corsair V-65C still with manufacturer's numbers on empennage with blue and white stripes on fin, but no white pointed sun in blue field visible under wing. Logo was probably painted on top of wing, Chinese government had seventeen Corsairs on order in 1932.

Head-on view of Vultee V-II attack bomber. Location in China uncertain.

Italian Breda used at Shien Chiao, early 1930s

A French Farman F. 303 with three Gnome engines, probably only one to reach China in 1930s. Location of this photo uncertain.

Above: Purchase of twenty American-made Bellancas by Chinese was made in complicated deal with Air France and this 29-90B still has name and logo of Air France on fuselage when it was being reconstructed in China, c. 1938.

Hawk75H, the "China demonstrator," which was presented to Claire L. Chennault in 1937 by Mme. Chiang Kai-shek. This photo taken at Curtiss plant on 16 April 1937. Chennault seldom flew this plane. After Chinese government moved to Hankow in December 1937, the Hawk was flown mostly by Elwyn Gibbon, a member of the 14th Squadron.

Bellanca 29-90B ready for test flight. Chinese Air Force markings are on fin, c. 1938.

Harold Welch, a test pilot, in front of Curtiss 19R trainer. He was later killed in crash at Hankow, c. 1938.

Lt. E.D. Dorsey, U.S. Army reserve, was demonstrating a Boeing pursuit P-26 when he crashed to his death at Hungjao air base west of Shanghai, 27 November 1934. At left, Chinese inspect wreckage of airplane.

Rolfe Watson, third from left, front row, was on detached service as armanent instructor at Nanchang flight school when it was operated by Italian mission in 1936-37. One of Italian instructors is seen third from right in front row.

The last group of foreign aviation instructors to reach China in 1936 were introducted to Generalissimo and Mme. Chiang Kai-shek at Kuling, a mountain resort and summer home of the Chiang's. Instead of going to work immediately, Mme. Chiang asked that they spend a week in Kuling, learning about Chinese culture and history under the tutelage of Old China Hand W.H. Donald, former Australian newspaperman turned political advisor. The six aviation specialists seen here are, l. to r. Sebie Smith, Rolfe Watson, Roy Holland, Lake Williamson, Donald, Sterling Tatum and Billie MacDonald. All have been to a Chinese tailor for double-breasted white "duck" suits, but Donald wears British "shorts."

Al Spencer, engineer and parachute rigger at Shien Chiao, here puts on his own parachute preparatory to a flight in Fleet trainer, c. 1933.

Harry Sutter, engineer and radio man trained in Europe, who assisted aviation program in Nanking and later in Kunming.

H. ("Nemo") Poncetti, one of mechanic instructors at Central Aviation School.

John May, engineer and test pilot at Central Aviation School.

Right: Colonel G.E.A. Rhineberg, a West Pointer, became the American commandant in charge of training at Shien Chiao in summer of 1936 but died 5 December of a heart attack.

Left: John Schweitzer, flight instructor at Shien Chiao, in cockpit of Curtiss Hawk II.

Nelson Feng, Chinese flight surgeon at Shien Chiao, c. 1934.

Reed Knight, flight instructor at Shien Chiao in mid-1930s, seen here in front of Northrop light bomber, a Curtiss Wright 710 hp Cyclone engine.

Frank M. Hawks, noted air racer, came to China in 1934 to demonstrate Curtiss Condor, an airplane represented in China by William D. Pawley.

Royal Leonard, personal pilot for Chiang Kai-shek, and author of *I Flew for China* (1938).

Cecil Folmar who taught bombardment tactics at Central Aviation School.

V

Central Aircraft Manufacturing Company

When the revamped aviation school was laid out at Shien Chiao in 1932, the government built a machine shop to make minor repairs to aircraft. This was handled by a new corporation called Central Aircraft Manufacturing Company, founded by William D. Pawley, New York, with the Chinese government holding a majority stock interest.

Pawley originally came to China to sell airplanes for the Curtiss-Wright export corporation, but advanced beyond the status of a salesman to become an important link in the progress of aviation in China. The new corporation built a plant of several brick-steel buildings within the Shien Chiao complex, and photographs taken in 1934 reveal an extensive facility for foreign and Chinese employees.

The Italian government was anxious to find a market for its aircraft in China, too, and not long after the Americans concluded arrangements to build at Shien Chiao, a group of Italian aviation people, represented in China by Colonel Mario de Bernardi, began negotiations leading to the opening of an assembly plant at Nanchang (referred to earlier.)

The factory or assembly plant, was supposed to turn out not less than one hundred aircraft a year. The Italians agreed to bear all costs of operation, as well as of construction. In theory the assembly plants at Shien Chiao, Nanchang, and a third one built on the North River in south China, were supposed to turn out 200 aircraft annually.

The CAMCO factory at Shien Chiao was in business a little more than three years before it was destroyed by Japanese bombers. Pawley collected as much of his tools and equipment as he could and shipped everything by train to Hankow where a temporary factory was established. Here he remained nearly a year, mostly assembling Vultees and Bellancas and installing oversized tires on them in order to cope with the muddy field at Hankow.

The operation at Hankow was in turn interrupted by the approach of Japanese ground forces in mid-October 1938, and Pawley and his employees picked up their machine tools and lathes and moved by truck to Kunming, several hundred miles to the southwest, over mountains, valleys, muddy roads and narrow bridges.

But Kunming was only a stopping place. With the Nationalist government moving to Chungking after Hankow fell to the Japanese in October, 1938, the Chinese Aero Commission decided that the place for a new factory should be close to Burma where it could be supplied by railroad from the port city of Rangoon to Lashio, and from Lashio over a new road called the "Burma Road" running northeast through the mountains.

The new CAMCO factory was built on a river dividing west China from east Burma at a remote village called "Loiwing" (pronounced Lo-ee-wing).[1] Building materials were rushed north from Rangoon and within a few months, the new facility was ready early in 1939. Initially it was meant to assemble and maintain Curtiss aircraft sold to the Chinese government, but in 1941, the plant helped to maintain the P-40s being flown by the First American Volunteer Group.

William D. Pawley, president of Central Aircraft Manufacturing Company and one of leading salesmen of American aircraft to China in 1930s.

The business of recruiting technicians, mechanics, armorers, instrument and propeller specialists was a formidable one, but Pawley, with years of experience and worldwide contacts in aviation, was able to recruit a staff, most of them at $350 per month and expenses. There was no housing nearby and Ivaloo Watson, widow of Rolfe Watson, recalls that her husband paid local Burmese-Chinese workmen ten U.S. dollars to build a *basha* which the couple lived in for several weeks until frame houses could be completed for the foreign staff.[2] Diesel engines provided electric power for the factory, and lines were run from the Diesel to staff quarters and soon there were not only lights but small refrigerators for Carlsbad beer.

Chinese and Americans who came to work here found it an interesting place. Mrs. Watson, one of the first American women to travel over the Burma Road from Kunming to Loiwing, remembers this period of her life and looks back on it with nostalgia and pride. Asked to write a brief account of her experiences she replied with this narrative which she calls "My Trip Across the Burma Road." Her story begins:

"When we arrived at Kunming, we learned that the airplane factory owned by Mr. Pawley and the Chinese government was in the process of moving to Loiwing. Sebie

Buildings of Central Aircraft Manufacturing Company at Shien Chiao. Railway tracks probably off main line between Shanghai and Hangchow. Photo taken in October 1934.

Main plant of Central Aircraft Manufacturing Company at Shien Chiao (near Hangchow). Photo taken 29 December 1934. Speaker under cornice mainly for air raid warning.

A Curtiss Hawk III in front of hangar shortly after completion of repair job at Central Aircraft Manufacturing Company, Shien Chiao.

Interior view of Central Aircraft Manufacturing Company's modern plant built at Shien Chiao in 1934.

Lacking fork lifts or tractors, Chinese workers use bamboo poles to move fuselage of this Bellanca 29-90B to repair shop, while a Vultee V-11, an attack bomber (foreground), waits its turn.

Northrup 2EC attack bomber after assembly in Pawley plant at Shien Chiao, c. 1935.

This is probably most widely used photograph ever taken of the Burma Road which connected northern Burma with west China in World War II. Some turns could only be made by backing up and working around the bend. Accidents were commonplace and in many instances, wrecked vehicles were pushed over the side to keep road open. But not all of the road was as tortuous as this particular stretch, which lay west of Paoshan.

and Lucy Smith, Rolfe and I had arrived in Kunming by way of Hongkong, Haiphong and Hanoi.

"After two or three weeks in Kunming, Rolfe took a convoy of trucks over the Burma Road. The distance of approximately 600 miles from Kunming to the border was made in five days, and four and a half days coming back to Kunming. Rolfe took Chinese workmen and supplies for the factory. He had to carry enough gasoline to reach Lungling in the mountains about 425 miles west of Kunming.

"Monday, Rolfe, Lucy and I were at Colonel Chennault's house in Kunming having lunch when an air raid alarm sounded and the three of us hurried out to the cemetery, while Chennault remained behind to finish a steak he was eating. This was old hat for him.

"That day the CAMCO management asked Rolfe to take another convoy over the Burma Road including two trucks and a passenger car with the women as passengers.

"Rolfe, Lucy (Smith) and I went shopping for groceries and cooking utensils to prepare meals along the way. The groceries consisted of crackers and cookies, canned fruit, particularly Del Monte peaches, and other foods which we could get at the stores in Kunming, supplies which came up the French-built railroad from Indo-China. And within a day or two we were off, up and down the famous Burma Road. There were only a few places where a truck could meet another truck, seems to me, one in the morning and one in the afternoon. The road was still under construction and repair. There was no machinery whatsoever. The workers used pick axes, shovels, and baskets balanced across their shoulders to move earth and materials.

"We left Kunming on April 12, 1939, Rolfe driving the first truck which, besides myself, carried our cook and his wife whom he married the night before we left. He paid forty Chinese dollars for her, and she had a girl about two years old with her. Then there was also the wife and son of the cook at the club house. Our food supplies also included hard boiled eggs and a gallon bottle of boiled water, which was for everyone.

"In the second truck were Chinese workmen and supplies for the factory. Lucy Smith, E.M. Smith and "Murph" Garrold rode in the car, Smith driving. Seems I started out in the car but shifted to the truck Rolfe was driving.

"The first night we slept in the courtyard of a hotel. We stayed at Hsia Kwan the next night, not far from Lake Tali.

"There were no rest rooms along the road, and the road was becoming more mountainous all the time. Some times we drove around one mountain half a day before there was a place narrow enough to put down timbers for a bridge to get to the other side of a river. The trucks were in low gear most of the time, either going up or coming down the winding roads.

"We met Sebie Smith who was returning from Loiwing, and he took the other truck, and his wife, Lucy, who had been with us, then changed trucks and rode with him. At night I slept in the front seat of one truck and my husband in the front seat of the other.

"The fourth day we reached Paoshan. There were cobble stone streets here with pigs and chickens running about everywhere. The people were rather dirty and ragged. Many had large goiters owing to lack of iodine in the local wells. In Paoshan I saw the first Chinese woman I had ever seen with bound feet.

"We passed over the Mekong River suspension bridge, a one lane bridge guarded by soldiers. This part of China was the center of the opium industry, too, and we saw acres and acres of poppies growing.

"The next night we were at Lungling, 425 miles from Kunming. Here we managed to buy gasoline for our trucks.

"The next day we reached the Burma border where there was a pole across the road and guards. There was a customs officer here too, and a large godown next to a tea house. We saw some Shan people, one of the tribes of Burma.

"After we got through Namkham in Burma, we saw the place where Dr. Gordon Seagrave had his famous hospital.[3] It was still seventeen miles by road to Loiwing but only five miles across the paddy fields.

"The first time Rolfe was in Loiwing he contracted with a local crew of workmen to build us a house for ten U.S. dollars. It had bamboo siding woven with bamboo floor and grass roof. There were no nails. Everything was tied together with rope or fibre. The kitchen was built the same way with one extra room for the Chinese cook and his family. There was a mud stove of three sections with grates across the sections and a galvanized canopy over the stove to keep the grass roof from catching fire.

"The house had one large room, one bedroom and an ante room. There was electricity furnished by engines at the factory. The house was raised about three feet off the ground. We had to have a platform built through the floor for the refrigerator because the wind often picked up the grass roof for a few inches and that pulled up the floor boards, upsetting the refrigerator. The house was pretty at night. Until our bedroom furniture arrived, we set up our cots in the living room. We had a water bearer to carry our drinking water which we always had to boil.

"This bamboo house looked good to us after six days on the Burma Road. We lived in it several weeks until frame and stucco houses could be built for the foreign staff."

In the above account, Mrs. Watson mentions the names of several people who were employed at Loiwing. In addition, she also recalls the names of Andrew Sargeant who served as line chief, assisted by a Chinese mechanic, Bacon Shaw; Charles Hunter, factory foreman; Albert Anderson, technical representative for Curtiss Wright; Mervyn ("Murph") Garrold, in charge of fuselage work; Walter Sobel, welding; Sebie Smith, instruments; and Watson, armaments. M.D. ("Doc") Walsh was comptroller.

Other Americans who worked at the assembly plant in Loiwing were Andrew Sobol, Frank Crane, William Stocker, Frank Lackell, William Schultz, Andrew Sargeant, and (?) "Buzz" Felio. Several of the men had their families with them and were living in the cottages built on the hillside overlooking the valley where the factory and air strip were located.

The plant remained at Loiwing from 1939 until the early spring of 1942 at which time Japanese bombers were able to reach it from bases in Thailand. Despite the defensive net of the American Volunteer Group, enemy bombers managed to slip through now and then and it was obvious to Pawley that he had to abandon still another factory.

Although most of the foreign staff returned to the United States after Loiwing was destroyed, several, including Crane, Sobol and Lakell, accompanied Pawley to Mysore, India, where he was engaged by the Indian Government to build a factory, which came to be known as Hindustani Aircraft, Ltd. After the U. S. Army Air Corps began to spread out in India in 1943, the plant was purchased by the Army and Pawley returned to the United States. He later served as U.S. ambassador to Peru and finally to Brazil.[4]

DC-2 for China National Aviation Corporation arrives in Loiwing in western Yunnan where Central Aircraft Manufacturing Company had assembly plant (building at right). This photo, probably taken in 1940, shows Chinese employees greeting plane, waiting for mail from home. Assembly plant came under attack by Japanese bombers in spring of 1942 and had to be abandoned.

Above: Dr. Gordon Seagrave, missionary doctor and author of "Burma Surgeon," built this hospital at Nankham, on "Burma Road" between Lashio and Paoshan. Facade is of stone.

Below: M.D. Walsh, plant manager for Central Aircraft Manufacturing Company. This is probably a hotel in Hangchow. Note profusion of potted plants set out by Chinese gardners.

Above: Robert W. Fausel, test pilot for Curtiss Aviation Division, brought Hawk 75Q and CW-21 interceptor to Burma in winter of 1939 and had both ships assembled at Pawley plant in Loiwing. In photo above, he stands in front of bamboo hoist using block and tackle to lift wings to fuselage. All work was done under bamboo mat shed until factory building was completed. Below, right: Fausel prepares to take off from Loiwing in the CW-21, for Kunming, early February 1939.

Ivaloo Watson, wife of Rolfe Watson, an armament specialist who worked for the Pawley organization at Loiwing in western Yunnan Province is seen in above picture in front of truck she drove with over "Burma Road" from Kunming to Loiwing, a distance of several hundred miles of mountains and valleys. In top photo, she sits on porch of bamboo "basha" which Chinese carpenters built for US$10. Fifty gallon drum on top of ladder at far right is probably for shower. After a few months, foreign-style housing was completed on hillside as seen in upper right. Houses were built on poles to allow air to circulate under floor.

VI

Aircraft Sales And Salesmen

One of the more successful airplane salesmen in China in the 1930s was A.L. ("Pat") Patterson, a representative for several companies. Patterson says that he sold 250 Fleet trainers, six Sikorsky S-38s, sixty North American NA-16s, eighty Vought Corsairs, sixty Ryan STC-4s, two Boeing 247s, four Curtiss Condors, and millions of dollars of miscellaneous aviation equipment and supplies to China. He also says he would have sold fifty Seversky P-35s to the Chinese had not William Pawley, agent for Curtiss, convinced the Chinese that the Hawk III was a better buy for the money.[1]

Below follows a list of aviation dealerships, most of them with head offices in Shanghai, early 1930s.[2]

Arnhold and Company, Ltd., (British)

Sassoon House, Shanghai.
(a) Blackburn Aeroplane and Motor Co., Ltd.
(b) British Russell Parachute Co., Ltd.
(c) Callon Limited, (airplane dope);
(d) Chance Brothers and Co., Ltd. (glass and field lights).
(e) De Havilland Aircraft Co., Ltd., (Moth planes and Gypsy motors).
(f) Reid and Sigrist, Ltd. (aeronautical instruments).

Bay, R.V., (Italian)

Palace Hotel, Nenking Road, Shanghai.
(a) Aeroplani Caproni.
(b) Isotta Frachini (airplane motors).

China Airmotive Company, (U.S.)

343 The Hamilton House, Shanghai.
Miscellaneous American accessories, materials, parts et cetera.
A.L. Patterson was manager.

Compagnia Italiano D'Estremo Oriente—Marelli, (Italian)

190 Kiukiang Road, Shanghai
(a) Magneti Marelli (ignition parts and field lights and accessories).
(b) Salvator, (parachutes).
(c) Soc. An. Fiat (planes of all kinds, engines and airport accessories).
(d) Soc. Italiana Ernesto Breda (planes of all kinds).

Consolidated Aircraft Corporation, (U.S.)

302 The Dollar Building, Shanghai.
(a) "Douglas" airplanes.
(b) "Northrop" airplanes.

Far-East Aviation Company, Ltd., (British)

435 Szechuen Road, Shanghai.
(a) Armstrong Siddeley Motors, Ltd. (aircooled motors).
(b) Armstrong Whitworth Aircraft, Ltd. Sir W.G. (planes).
(c) Brown Brothers, (aircraft accessories).
(d) Bruntons (Musselburgh), Ltd. (aircraft materials).
(e) B.S.A. Guns, Ltd., (guns and ammunition).
(f) Cierva Autogiro Co., Ltd. (autogiros).
(g) Cirrus Hermes Engineering Co., Ltd., (planes and aircooled motors).
(h) Comper Aircraft Company, Ltd. (sports and training planes).
(i) Handley Page, Ltd. (heavy aircraft).
(j) Irvin Airchute Co. Ltd. (Canada), (parachutes).
(k) S. Lewis, (flying clothing).
(l) Monarch Engineering Co. Ltd. (bombs).
(m) Reynolds Tube Co., Ltd. (tubing).
(n) A.V. Roe and Co., Ltd. (all kinds airplanes).
(o) Saunders-Roe Ltd., (amphibians and flying boats).
(p) Short Bros. (Rochester and Bedford), Ltd., (flying boats).
(q) Thos. Sinton and Co., Ltd. (aircraft materials).
(r) Smith's Aircraft Instruments, (aircraft instruments).
(s) Sparten Aircraft, (commerical and training planes).
(t) Titanine-Emaillite, Ltd. (airplane dope and paint).
(u) Westland Aircraft Works (airplanes all kinds).
(v) Williamson Manufacturing Co., Ltd. (aerial cameras).

The L.E. Gale Co., Fed. Inc., U.S.A., (U.S.)

49 Nanking Road, Shanghai (Head-office at Hankow).
(a) Waco Aircraft Company, (airplanes of all small types).
(b) United Aircraft Exports, Inc. only for Hunan, Hupeh, Honen, Szechuan, Kansu, Shensi, and Shansi, (all types of planes; Pratt and Whitney motors; Hamilton standard Propellers).

Hosken Trading Co., Inc., (U.S.)

133 Yuen Ming Yuen Road, Shanghai.
(a) Fairchild Aviation Corporation, (Aerial cameras, photographic, training and commercial planes, aerial surveys).
(b) Granville Brothers Aircraft Corporation, (high speed planes).
(c) Switlik Parachutes Company, (parachutes).
(d) Pioneer Instrument Company, (aerial instruments).

Intercontinent Corporation, (U.S.)

353 The Hamilton House, Shanghai.
(a) Curtiss-Wright Corporation, (military and commercial planes and motors).
(b) Curtiss-Aeroplane and Motor Co. (airplane motors).
(c) Curtiss-Wright Airplane Co., (military and commercial planes).
(d) Sperry Gyroscope Company, (sound locators, direction indicators, maritime instruments, search lights).

The Jardine Engineering Corporation, Ltd. (British)

50 Yuen Ming Yuen Road, Shanghai.
(a) Vickers (Aviation) Ltd., (military aircraft and supplies).
(b) Supermarine Aviation Works, (military and commercial flying boats).
(c) Bristol Aeroplane Co., Ltd. (military airplanes and aircooled motors).
(d) Vickers Armstrongs, Ltd. (guns, bombs, etc.).

L.A. Lewis, (British)

205 The Hamilton House, Shanghai.
(a) Kinner Aircraft and Motor Corporation, (small airplanes and air cooled motors).

May-Kee (China), Ltd., (Swiss)

20 Canton Road, Shanghai.
(a) Soc. Idrovolanti Alte Italia ("Savoia-Marchetti" seaplanes and landplanes).

Reiss, Massey and Co., Ltd., (British)

88 Museum Road, Shanghai.
(a) Hawker Engineering Co., Ltd. (military and training planes).

Siemssen and Company, (German)

451 Kiangsi Road, Shanghai.
(a) Junkers-Motorenbau G.m.b.H., (airplanes and airplane motors).

The Skoda Works, (Czech)

24 The Bund, Shanghai.
(a) The "Avis" Airplane Factory, (commercial and military planes).

United Aircraft Exports Corporation, (U.S.)

205 The Hamilton House, Shanghai.
(a) Chance-Vought Corporation, (military planes).
(b) Hamilton Standard Propeller Co., (all-metal-propellers—fixed and adjustable).
(c) Irving Airchute Co. (parachutes).
(d) Sikorsky Aircraft Co., (Amphibian planes).
(e) Pratt and Whitney Aircraft Co. (air-cooled motors).

Vaclav Havlicek, (Czech)

649 The Hamilton House, Shanghai.
(a) Czechoslovak Military Airplanes Factory ("Letov" military airplanes).

Boeing Aircraft Company, (U.S.)

Park Hotel, Shanghai.
(a) Boeing Aircraft Co., (pursuit and transport planes).
(b) Stearman Aircraft Co., (training and sport planes).

VII

Commercial Aviation in China

(China National Aviation Corporation

Chinese leaders in government and business grasped the potential for commercial aviation just about as quickly as their counterparts in Europe and America, but they lacked the technology to develop this potential. As early as 1920, Hollington K. Tong, a young newspaperman in Tientsin, wrote a comprehensive article outlining the need for commercial aviation and air mail service, but the report, although not unheeded, was not implemented for many years.

One of the first attempts to organize a civilian airline was made in 1921 by a Sino-British company in Peking, using Handley-Pages flown by British instructors at Nanyuan. The airline operated between Nanyuan and Peitaho, a summer resort town on the Gulf of Chili frequented by foreign diplomatic staff and missionaries. But when the tourist season ended, so did the demand for tickets, and the airline was discontinued.

In the next decade, little was done to encourage civilian aviation. For one thing, the country was not ready for it. There were too many political and economic questions to be resolved first. Finally, after Chiang Kai-shek had consolidated power to form a new government, he chose Nanking, not Peking, as the capital for China, and interest in civil aviation was revived.

Clement M. Keys, a naturalized American citizen born in Canada is often called the "father of commercial aviation in China." Keys began his career in aviation by getting control of the Glenn H. Curtiss Company which had prospered in World War I, but was floundering in 1920 when orders fell off and there was little cash to meet the payroll. Facing bankruptcy, the company sold out to Keys and Clarence Dillon, a New York banker, for $659,000. A new company was organized, the Glenn H. Curtiss-Wright Company, which later became the leading supplier of aircraft to China, particularly trainers, and it was Keys who first saw that there was money to be made in commercial aviation in China. However, it wouldn't come easy. The new government, only recently established in Nanking, was jealous of its prerogatives and it was determined to move cautiously.

An associate of Keys, R.O. Hayward, arrived in China in December 1928 and met with Sun Fo, minister of railways. Since railways were included under the general heading of transportation, it was to this section that Hayward was directed. Later, the railways were reorganized under a new department called the Ministry of Communications *(Chiao t'ung-pu)* and for the next decade, most airline contracts were supervised and enforced by this department.

In the beginning, the Chinese were mainly interested in establishing an air mail service, and all early airplanes carried the ideograph *yu* on the fuselage, an abbreviation of the Chinese word for "post office" *(Yu-cheng-yu).* An American company was set up by Keys called "Aviation Exploration" and through it, he guided the negotiations, via Hayward, which led to the creation on 12 April 1929 of China National Aviation Corporation *(Chung Kuo Hang-k'ung Kung-ssu).* Under the terms of the contract, Aviation Exploration received exclusive air mail rights for ten years on three main routes, and the Americans promised to operate "up to date first class efficient equipment." In short, China did not want to be used as a dumping ground for used airplanes and parts.

But Sun Fo had not consulted with the aviation bureau of the Chinese Army, and when the army learned of the contract, Chang Ching-yu, chief of the bureau, wanted to have it cancelled. Although voicing lofty sentiments about China's sovereignty, the real issue was power. Who was in charge here, the Ministry of Communications, or the aviation bureau? A rather painful period followed for the American representatives. The Chinese were arguing about controls when they did not have airports or even passenger planes to control. But the Americans finally agreed to raise $1,000,000 for airport construction in the main cities along the Yangtze River and at Lunghua on the Whangpoo River in Shanghai.

By May 1929, the Ministry of Communications had replaced the authority of the Ministry of Railways in aviation matters, although no one could be sure yet who was in charge. Nevertheless, one of the first official acts of the new ministry was to sign an agreement with the Stinson Aircraft Corporation of Detroit to begin a mail service between Shanghai and Nanking. The planes used were land planes which did not require a long runway. A new field was quickly completed at Hungjao on the west side of Shanghai, and the Ming Palace airdrome in Nanking was made ready to take the Stinson "Detroiters."

The deal made by the Chinese with Eddie Stinson violated the contract made earlier with Aviation Exploration directed by Keyes, but the latter, to preserve peace, made believe he saw no evil.

Three SM-If Stinsons, a four-passenger, high wing monoplane, were shipped to China accompanied by two pilots and a mechanic. The planes were uncrated and assembled, probably at Hungjao, and the first went into service on the Shanghai-Nanking run on 8 July 1929. All things considered, this was a remarkable achievement in the time since the contract was signed.

Meanwhile, C.N.A.C., according to its original contract, had until 17 October 1929, to become airborne on its Shanghai-Hankow route. The pilot Keys sent to China to make the run was Harry G. Smith, and the most important decision Smith made after he arrived was to fly the Loening which could land on water with its wheels up, or on land with its wheels down. It was a six-passenger airplane using a 525 Pratt & Whitney Hornet engine. Smith also picked up a top-notch pilot to help him, E. (for Ernest) M. Allison, who later became senior pilot for C.N.A.C.

Finally, on 26 August 1929, China Airways, a wholly-owned subsidiary of Intercontinent Aviation Inc., the parent company of Aviation Exploration, was incorporated under the provisions of the China Trade Act of 1922, with Frank B. Price as its first president. The company made a dry run on 12-13 October 1929.

Photo taken at Stinson plant in Detroit, c. 1929. Writing at far right on crate reads "Better buy a Stinson than wish you had."

Below: CNAC Pilot Hewitt Mitchell in uniform, Shanghai, April 1936.

Hewitt F. Mitchell, flew this Stinson on Chungking-Chengtu route for C.N.A.C., in first six months of operations. In this picture, taken 1 February 1933 at Suining, Szechwan Province, he stands fifth from right, while E.M. Allison, operations chief for airline, stands beneath propeller. In his diary, Mitchell has this to say: "I was forced down [at Suining] by bad weather and damaged part of the landing gear in some rough places on the military airport recently constructed there. I had no co-pilot along that day and my passengers were five Chinese women and a baby. There never had been an airplane land there before and of course thousands of people gathered around . . . . After four days Mr. Allison arrived from Shanghai with the necessary new parts. We jacked the ship up gradually with an auto jack, made the necessary repairs, and on the seventh day flew it back to Chengtu . . . ."

Stinson Detroiter at Nanking with *yu* ideograph to signify "mail" plane, c. 1934.

The Martin M-130, better known as the "China Clipper" for Pan American World Airways over Hongkong, c. 1936.

But China National Aviation Corporation was almost lost in this corporate reshuffle. Here was a contract made with the Stinson company to carry mail between Shanghai and Nanking, and another contract with China Airways to carry mail farther up the Yangtze River. Actually C.N.A.C. was under a cloud of disapproval both because of opposition from Chang Ching-yu and because it had failed to "furnish, equip, police, maintain and have ready for operation adequate airports."[1]

China Airways began regular service between Shanghai and Hankow, 525 miles, on 21 October 1929, with Birger Johnson flying in the open cockpit of the Loening, goggles and helmet, carrying one pound of mail and two passengers, Mr. and Mrs. Sun Fo. Allison flew in the opposite direction over the same route the same day, and daily service continued the next ten days of October when operations were halted while the company waited to get paid.

Payment was due 15 November but it was not made until mid-December at which time flight schedules were resumed. But political infighting both among the Chinese and among the Americans, held up further progress. The stalemate was finally broken on 8 July 1930 when a new contract was written and under the terms of this contract China Airways was joined to China National Aviation Corporation, and the Stinson-operated airline was absorbed by CNAC.

The new corporate structure, under Chinese law, was authorized to raise $3,000,000 in capital, the Ministry of Communications to hold 55 percent of the stock, and the American partners 45 percent. The contract authorized three routes: Shanghai to Chengtu via Nanking, Kiukiang, Hankow, Ichang, Wanhsien and Chungking; 2) Shanghai to Canton via Ningpo, Wenchow, Foochow, Amoy and Swatow, and 3) Shanghai to Peking via Hsuchow, Tsinan and Tientsin. In the next several years CNAC expanded into a fairly modern airline, adding to the Loenings with DC-2s and Douglas Dolphins.

In 1934, Dai Enki, managing director of CNAC, issued a statement outlining new airplane purchases for the line. Delivery was expected early in 1935 of a tri-motor Ford for the Chungking-Kunming flight, followed by a 14-passenger Douglas, the DC-2, for the Shanghai-Peking service. And next to arrive would be one or two Fairchilds for the Yangtze River service, probably Shanghai to Hankow. The Douglas was going to carry a steward who would serve a snack lunch with tea or coffee. Tourists on the Peking flight were also promised a glimpse of the Great Wall and the Ming Tombs.

Meanwhile, Keys had lost heavily in the stock market crash of 1929, and by 1933 he was looking for a way out of his obligations. A new airline, called Pan American Airways, heard about his plight and early in 1933 Pan Am acquired the American-held shares in CNAC. Harold M. Bixby was sent to China from New York to run the airline.

It was the intention of Pan Am to use CNAC as a feeder line for its trans-Pacific Clipper service getting underway from San Francisco to Manila. Edwin C. Musick arrived in San Francisco on 8 January 1937 with a new Sikorsky which he flew up from Miami to augment the Clipper fleet already flying between Alameda, California, and the Philippines.Pan Am was projecting a total over-water flight in 1937 of a million miles. By the spring of 1937, arrangements had been completed for the first Shanghai-to-San Francisco flight with connecting flight from Shanghai to Hongkong where the Clipper would meet the Shanghai plane.

By the end of 1938, most of the major cities in east and central China had come under Japanese occupation. CNAC continued to fly between cities in west China and as far northwest as Lanchow. Supplies for the government in Chungking were brought into Kunming over the French railway to Hanoi, but this route was cut off in the summer of 1940. For several months supplies were trucked in over the Burma Road from Lashio, but this was totally inadequate, CNAC then began flights to Lashio. But this traffic was suspended when British authorities in Burma, acceding to Japanese pressure, withheld permission for trucks or airplanes to enter Lashio. This blockade was lifted after six months, but Lashio, lying far to the north of Rangoon, on a railway head, was not capable of keeping up with the demands for the transport of supplies into China. Finally, on 23 November 1941, a DC-3 for CNAC pioneered a new route across the Burma "hump" and landed on a Chinese field at Liang-ho about seventy-five miles north of Loiwing in western Yunnan Province.

Two months after this pioneer flight, China became an ally of the United States in the war against Japan. CNAC now became a partner of China in a larger sense than the shareholders of the corporation, Chinese or American, had ever anticipated. More planes were added to the company fleet, including several C-54s. A big hangar and warehouse were built at the end of the main runway at Kunming Air Base, and civilian pilots braved the same dangers as the U.S. Army pilots in the troop carrier commands and the Air Transport Command. From 1942 to 1945, ten CNAC pilots lost their lives. Several were lost in the years 1938 to 1942, either to enemy action or to other causes.

The Japanese blockade of east China made shipment of new aircraft to China more difficult and CNAC was scarcely able to keep up with the rate of attrition experienced between 1937 and 1939. An inventory of its aircraft on 1 January 1939 would show only seven airplanes in service, namely, two DC-2s, one DC-3, one Ford Trimotor, two Consolidated Commodore flying boats, and one six passenger Haviland Dragon Rapide. Some were lost in the next two years but additional aircraft were flown in across the Hump, including three used Curtiss Condors.

After the end of World War II, the company shook off the dust of west China and re-established service between cities of east, north and south China, and airplanes, mostly war surplus, were added to the company fleet including five Convairs. But, as the Red Armies of Chu Teh advanced into south China, there were fewer and fewer cities to fly to, and suddenly there were none. At this point, almost the entire

CNAC temporary shed hangar at Chengtu holds Stinson, 1933. One of men in picture is Italian, representing League of Nations in study of China's economic problems.

CNAC passenger transportation at Chengtu, 1934.

In early 1930s, CNAC planes flew through Yangtze Gorges above Ichang to reach Chungking.

Jimmy Doolittle on visit to Peking in May 1933 after demonstrating Hawk II at Shien Chiao. Airplane is Stinson Detroiter used on Shanghai-Peking run for CNAC in 1933 and flown by Hewitt Mitchell. On this flight, Doolittle flew in right seat.

CNAC fleet of airplanes flew into Hongkong pending the question of their ownership.

(EURASIA AVIATION CORPORATION)

In the late 1920s, German engineers and businessmen, long in the forefront of European aviation, were eyeing China as a future market for airplane sales and services. Anxious to impress the Chinese with the quality of their aircraft, two transports from the growing German airline, Lufthansa, were flown to China, and landed at Nanyuan outside Peking on 31 October 1926. But the political situation in 1926, especially around Peking, was not conducive to airplane sales.

Lufthansa eventually made application to the Nanking government to form an airline in China, and after the government ratified a new contract with China National Aviation Corporation, it later made an agreement with Lufthansa for a second airline to be known as Eurasia Aviation Corporation. The logo for the aircraft on this line featured the Chinese ideograph *ya*, the first in the word *Ya-hsi-ya*, a transliteration of the word "Asia" in Chinese.

The new corporation was capitalized at $1,000,000 with the Chinese holding two-thirds of the stock, and the Germans the other third, but the Germans were quite generous about repayment terms.

The Germans had plans for an airline connecting Berlin to Peking via the Soviet Union and Chinese central Asia. In 1933, one of the Lufthansa pilots flew a single engine Junkers on the proposed route over the Soviet Air Space, and instead of landing in Peking, he came directly out of the northwest into Shanghai, but no further flights were authorized by the Soviets.

Most of the Eurasia fleet was made up of tri-motored Junkers 52s and single engine W-34s. On 11 September 1935, a new Junkers arrived at Lunghua after a flight across the "southern route," via Athens, Cairo, Bagdad, Karachi, Allahabad, Calcutta, Rangoon, Hanoi and Canton. The air-

Charles L. Sharp, pioneer in aviation, top pilot and later an executive with China National Aviation Corporation, seen here in cockpit of "Kiukiang," one of early Loening amphibians used on commercial routes along cities of Yangtze River.

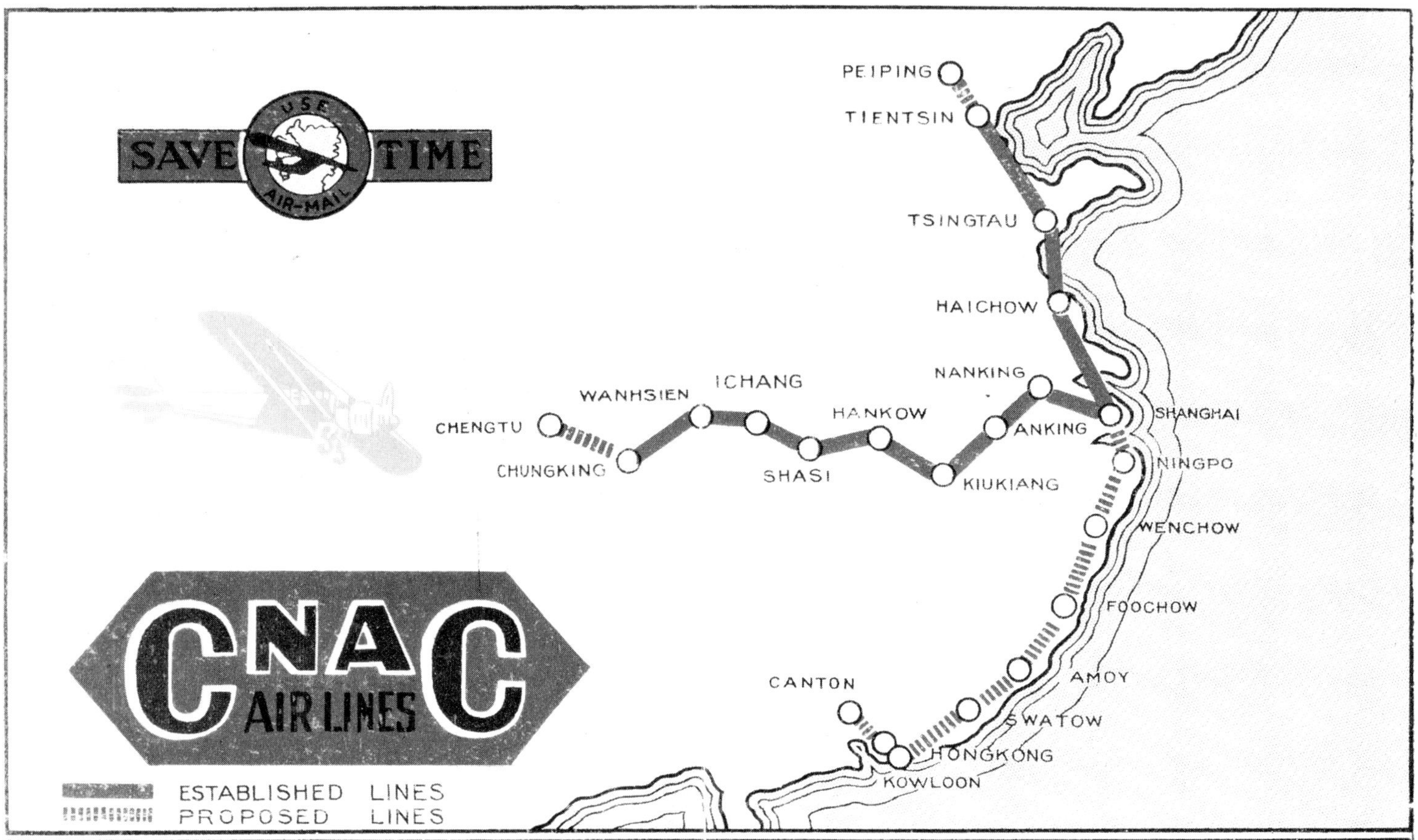

SHANGHAI HANKOW AIR LINE

DAILY EXCEPT MONDAYS

TIME TABLE

WEST BOUND READ DOWN			EAST BOUND READ UP	
8:00 A.M.	DEP.	SHANGHAI	ARR.	3:05 P.M.
10:15 A.M.	ARR.	NANKING	DEP.	12:50 P.M.
10:30 A.M.	DEP.		ARR.	12:35 P.M.
12:15 P.M.	ARR.	ANKING	DEP.	10:50 A.M.
12:20 P.M.	DEP.		ARR.	10:45 A.M.
1:20 P.M.	ARR.	KIUKIANG	DEP.	9:45 A.M.
1:35 P.M.	DEP.		ARR.	9:30 A.M.
3:05 P.M.	ARR.	HANKOW	DEP.	8:00 A.M.

FARE TABLE

ONE WAY				ROUND TRIP
$ 48.	SHANGHAI	TO	NANKING	$ 84.
$ 112.	"	"	ANKING	$ 196.
$ 152.	"	"	KIUKIANG	$ 266.
$ 200.	"	"	HANKOW	$ 350.
$ 64.	NANKING	"	ANKING	$ 112.
$ 104.	"	"	KIUKIANG	$ 182.
$ 152.	"	"	HANKOW	$ 266.
$ 40.	ANKING	"	KIUKIANG	$ 70.
$ 88.	"	"	HANKOW	$ 154.
$ 48.	KIUKIANG	"	HANKOW	$ 84.

FLY THE GORGEOUS GORGE TRIP

SEE THE GRANDEUR OF THE YANGTZE RIVER GORGES FROM THE AIR—ONE OF THE WORLD'S WONDERS. FOR SCHEDULES AND RATES APPLY TO ANY C.N.A.C. AGENT OR TRAVEL AGENCY

This advertisement is undated but it probably ran in a Shanghai newspaper or magazine in the summer of 1933.

plane was a Junkers D-ANYK, a low wing with three 650 hp engines and top speed of 180 miles per hour, carrying seventeen passengers. It was scheduled for Shanghai-Chengtu service.

Eurasia held its own against CNAC in north and west China, running a service from Shanghai to Lanchow, 1,059 miles, via Nanking, Loyang and Sian. The other route went from Peking south to Canton, 1,100 miles, via Taiyuan, Loyang, Hankow and Changsha, somewhat along the tracks of the Peking-Hankow and Canton-Hankow railroads.

In 1935 Eurasia also inaugurated a new service between the railway head at Paotou in west Suiyuan Province, and the city of Lanchow on the Yellow River in south Kansu Province, flying a single engine Junkers.

Hans Koester, the pilot on the W-34, was an avid camera fan and took many photographs of the Great Wall, the Yellow River and Ordos Desert. Later an article by Koester

E.M. Allison, senior pilot for CNAC, poses in baggage compartment of Douglas Dolpin, probably at Lunghua on Whangpoo River. Two Wright engines dominate wings. Plane was first used on flights between Long Beach and Catalina Island. CNAC bought plane as first replacement for Loening amphibians which launched air mail service to cities along Yangtze River in early 1930s. Chinese ideograph at back, left, reads *yu* for postal service, and two up front read *fu-chien* usually written as Fukien in 1930s, a coastal province south of Shanghai.

DC-2s for China National Aviation Corporation on line at Lunghua airdrome, 1937, Whangpoo River in background. Hangar with letters on roof was built in 1934 as first permanent facility, and larger hangar at left in 1945.

Loening amphibian for C.N.A.C. at Lunghua on Whangpoo River, south side of Shanghai. The Chinese ideograph on fuselage is *yu*, the first ideograph in the Chinese word for postal service.

Harry G. Smith, first operations manager for China airways which was merged with C.N.A.C. Smith later flew private airplane of Chiang Kai-shek.

Early nose hangar for C.N.A.C. amphibian at Lunghua, c. 1931.

One of first Loening amphibians flown in China for C.N.A.C.

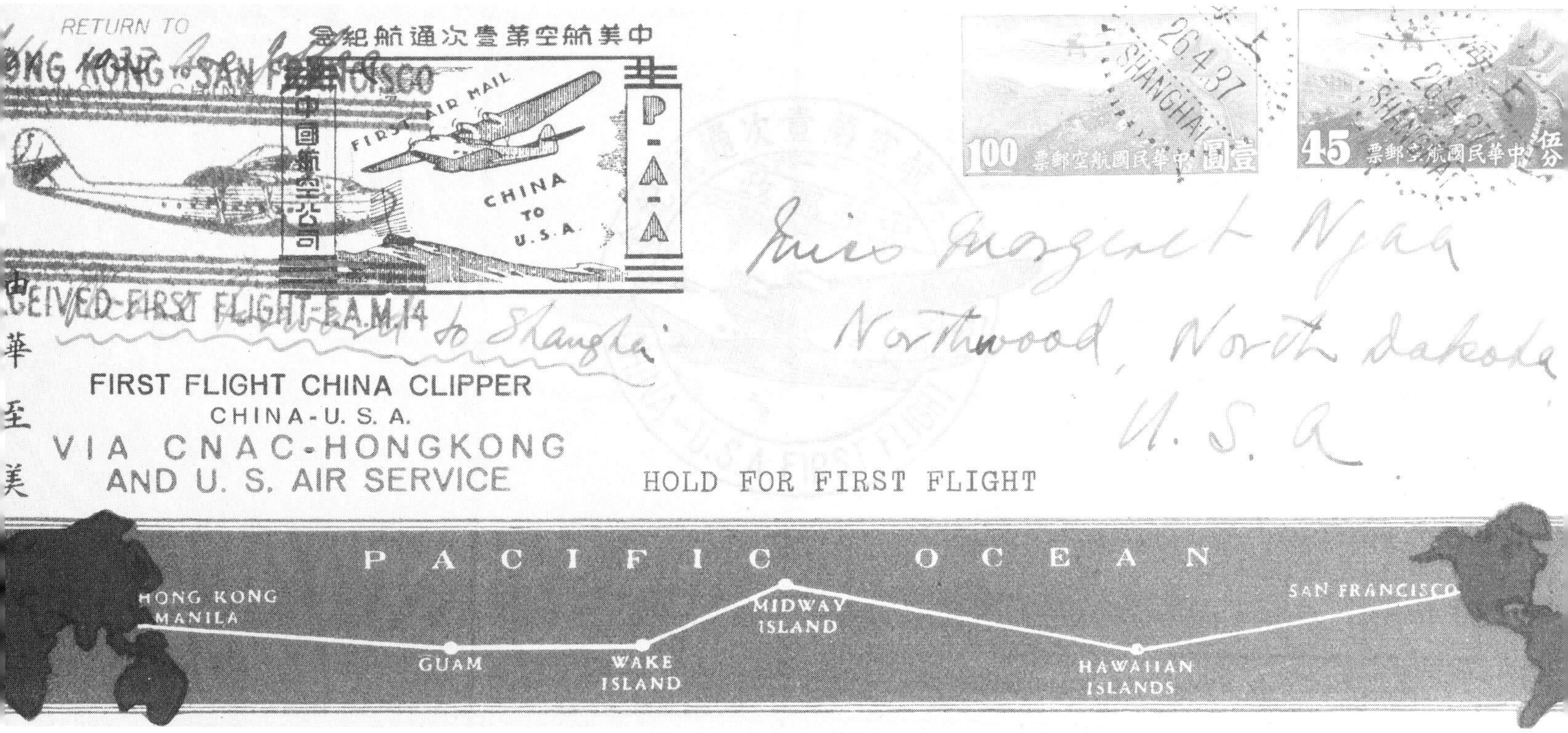

A Chinese philatelist and friend of the author asked to send this letter on the first scheduled flight of the China Clipper to an address in the United States. It was addressed to "Miss Margaret Njaa" (Rosholt) and the envelope returned to Shanghai by her parents who understood the arrangement.

appeared in the *National Geographic Magazine*.[2]

When war broke out in Peking and Shanghai in 1937, Eurasia moved its head office to Sian. The *China Weekly Review*[3] reported that a new tri-motor Junkers had arrived in Sian from Berlin via Syria and Sinkiang. It was probably the last new airplane to join the Eurasia fleet in China.

Eurasia was continuously expanding its air service, but it never caught up with CNAC either in volume of business or passenger sales, probably because CNAC had a head start and had been assigned the most lucrative routes between cities. By the summer of 1936, the last "normal" year before the war began in 1937, Eurasia was flying more miles in China but making less money than CNAC. But there was ample room for both airlines and the competition would have served China well but for the intervention of hostilities between China and Japan. After the war began, flight schedules had to be revised and new facilities built, but as the Japanese armies advanced into central China, the size of "free China" diminished and by September 1938 most of the Eurasia flights in east and north China had to be suspended. New routes were opened in west China, and twice-weekly service was operated between Kunming and Hanoi until the summer of 1940.

But the political situation in Germany was also changing and Eurasia pilots and crew chiefs were being urged to return to Germany for service in the Luftwaffe. The company became entirely Chinese-owned on 1 August 1941 after China broke off diplomatic relaitons with Germany. The airline was never revived after World War II.

(SOUTHWEST AVIATION CORPORATION)

In Canton, the principal city of south China, there had been frequent stories in the press since the 1920s of establishing an airline to connect Canton with the larger cities in south China, mainly in Kwangtung, Kwangsi and Yunnan Privinces. After several false starts an air service was launched in 1934 under the auspices of the governors of Kwangtung and Kwangsi called "Southwest Aviation Corporation." The airline used only Chinese pilots and crews although it was favored by a first class maintenance shop located north of Canton, operated by Charles May, an American.

The *China Weekly Review* had this to report:

> The three large Stinson airplanes recently purchased from America by the Southwestern Aviation Corporation have been assemblied at Canton. They are being placed on the Canton-Hanoi, and Canton-Nanning service."[4]

The Japanese occupation of Canton in 1938 forced the dissolution of the airline and it was not revived after WWII.

(CENTRAL AIR TRANSPORT CORPORATION)

A fourth commercial airline, known as Central Air Transport Corporation, was launched in Chungking by the Ministry of Communication on 1 March 1943. The company was a step-child of Eurasia Aviation Corporation. In the general reshuffle of flight schedules and diminishing profits, no new aircraft were added to the Eurasia fleet after 1937, and by 1941 there was only one in condition to fly and that was parked at Kaitak in Hongkong. After the first wave of Japanese bombers swept over Hongkong on 8 December 1941, the Eurasia plane, a six passenger Junkers W-39, took off for the mainland. The Ministry of Communications used it for charter flights and by 1942 it had logged more than 150,000 miles without a ground loop.

Central Air Transport, starting with the one Junker W-34, gradually added more airplanes and by 1945 the company had three Lockheed Hudsons Type 3-A 29s, a twin-engine monoplane, Shortly after the end of WWII, private investors were able to buy twenty percent of the stock in the airline for $5,000,000. With new capitalization, the company acquired twelve C-47s from U.S. Army surplus in India, and in July 1946, 150 additional C-47s and C-46s, the latter mostly for parts.

The author on 24 December 1935, preparing to board Eurasia Aviation Corporation's "Wilhelm Schmidt" at Sian for flight to Shanghai. Schmidt was sales agent for German aircraft in Shanghai. Sheepskin coat, made by Chinese tailors on Mongol border, was worn by author on 500-mile trek by camel caravan across Gobi desert.

Captain Johann Risztics, pioneer pilot for Eurasia Aviation Corporation, in photo taken at Nanking, 1931. Plane is two-place trainer.

Below: Junker JU-52 on field at Tientsin, September 1934. Ship was on tour of Chinese bases for Eurasia Aviation Corporation, a subsidiary of Luthansa. Emblem of Nazi party on tail section was first seen on any aircraft in Far East, but Chinese had it removed before ship went into service.

Right: Count Wulf-Diether Graf zu Castell-Rudenhausen, Eurasia pilot, steps back into cabin door to pose for picture by author, 24 December 1935.

Below: Eurasia's single engine Junkers W-33 used on feeder lines in west and northwest China. Chinese ideograph on fuselage reads *yu* for postal service. Photo probably taken at Sian.

Photo by author taken 23 December 1935 through window of single engine Junkers W-33 on airfield at Paotow. Onlookers were Chinese, not Mongols. Chinese ideographs on wing of aircraft are last two in word *Ya-hsi-ya,* meaning "Asia." This is Eurasia Aviation Corporation's airplane. Building in distance is not hangar, but at right, part of old wall that once surrounded city of Paotow. There was no runway here, only hard surface on wind-swept desert. Camels are two-humped Bactrians found throughout Asia.

By 1946 the airline had established service to twenty-six cities, mostly east and central China. Like the pilots of CNAC, who flew their aircraft into Hongkong after the last air fields were occupied by the Red Army, CATC pilots also flew most of their aircraft into Kaitak on 9 November 1949, pending a decision of their ownership.

Meanwhile, another airline was launched in north China in 1937 with Japanese backing and Japanese aircraft, first referred to in news dispatches as "Huitang Corporation," flying routes from the cities of Tientsin to Changte (in Jehol Province,) Tientsin to Changpei, also north of the Great Wall, and Tientsin to Peking. It was probably the Japanese equivalent of an air transport command, since there was little civilian passenger traffic to places like Changpei.

Commercial aviation had also taken wings in Manchuria. According to a brief story in *Far Eastern Review,* air lines in Manchuria in 1934 covered 2,300 kilometers with a total mileage flown of 2,700,000 kilometers, all under the auspices of the "Manchuria Aviation Company."[5]

(CIVIL AIR TRANSPORT Inc.)

In China, after the end of World War II, the nation faced a crises in transportation. Bridges all over the nation had been destroyed or damaged by bombers of the 14th Air Force, and hundreds of railway locomotives and cars had been destroyed by fighter planes and medium bombers. River boats and steamers which once carried millions of people over the great river systems of China were sunk, burned or beached. China had suffered not four years of war but seven, and scarcely a family had not been adversely affected.

Claire L. Chennault, now a civilian, who had commanded the 14th Air Force in China, recognized the desperate conditions faced by the Chinese people and realized that the quickest way to bring relief to the needy and orphans of war was to fly it in. During the war Chennault had become acquainted with Whiting Willauer, a young business man who had been working in China Defense Supplies in Washington D.C., and the two agreed to form a partnership to organize an airline.

The *CAT Bulletin,* a house organ published twice a month for employees of Civil Air Transport, carries a brief story by Willauer on the early history of the airline which reads, in part, as follows:

> In the fall of 1945 when General Chennault and I decided to form a partnership with the purpose of organizing a new airline in China, we were two men with ideas, know-how and ideals, but very little cash . . . . By October 1946 we had obtained an operating franchise . . . . Our resources were dwindling fast and we knew that at least three months would have to lapse before we could have any planes . . . . Money was needed to send pilots and crew chiefs to Honolulu and Manila to pick up the aircraft . . . . We had figured that it would take about $250,000 of working capital to tide us over until CAT could support itself, and that was why the General and I and our Stateside associates had long before made a deal for financial support from the Flying Tiger Line. Louis Prescott was sent to China to survey the situation and to act as comptroller of the funds when furnished. When he was accidentally murdered [in Manila] there was no one available in the Flying Tiger organization to take his place and therefore the line backed out.
>
> I don't think either the General or I will ever forget the shock when we received the cable that the much needed financial support had been withdrawn. We were left with an airline franchise but only enough money between us to carry on . . . . In thirty-five days of intense negotiations with Chinese bankers we raised the necessary funds. Our esteemed ex-Chairman, Wang Wen-san, and our friend L.K. Taylor worked with us day and night until this was done.
>
> Looking back on it all now, General Chennault and I are glad that it happened this way, except for Louis Prescott's sad and premature death. For through our association with the Chinese banking group CAT became a truly Sino-American venture, and it is basically because of this partnership with our Chinese associates that CAT grew and now holds a permanent place in commercial aviation as "The Orient's Own."[6]

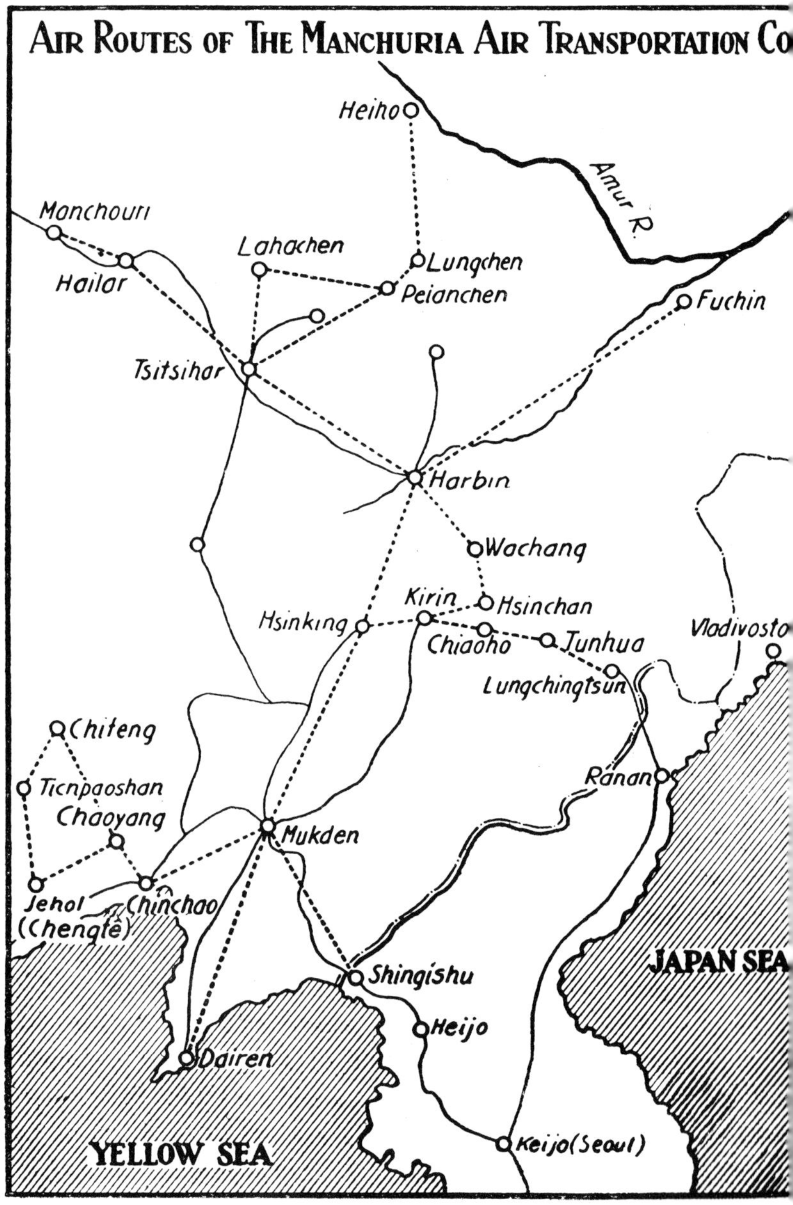

Map copied from *Far Eastern Review* dated March 1934. The air lines shown here were all operated by Japanese military or semi-official airlines.

Louis Prescott was killed by a stray bullet, meant for someone else, while reading a newspaper in the lobby of the Manila Hotel.

According to Chennault in *Way of a Fighter,*[7] the United Nations now agreed to provide a loan of $2,000,000 for the purchase of aircraft and other equipment, all of which would serve as collateral on the loan. On 25 October 1946, after concurrence of the Nationalist government in Nanking, an operating agreement was signed by the Chinese National Relief and Rehabilitation Administration (CNRRA) on one side, and Chennault and Willauer on the other. CNRRA represented the United Nations in this instance, and the new airline became known as CNRRA Air Transport.

In his account, noted above, Willauer makes clear that the new airline was founded earlier with the backing of Chinese bankers in the early months of 1946 when a frantic effort was made to get the airline airborne. There was a ground echelon of former mechanics, all Chinese, who had served in the 12th Air Service Group in Kueiulin during the war, and Dorothy Yuen (Leuba) recalls that Colonel Richard Wise, former C.O. of the 12th A.S.G., was trying desperately to keep these valuable workers together.[8] By a coincidence, Wise was on active duty in Shanghai at this time in connection with war crimes trials, and in his spare time, served as "operations officer" for the budding airline. Other pioneers who assisted in this critical period, aside from Dorothy Yuen, were Phyllis Altamire, John M. Williams, and Clyde D. Farnsworth, a newpaper man, and first editor of the *CAT Bulletin.*

Relief supplies and services were en route to the interior of China by the end of January 1947, and by October that year the United Nations airline was carrying a million ton-miles per month, and by April 1948 this had risen to two million a month.

The United Nations contract expired in October 1948 and the assets of the company and loan obligations were taken over by the partners, Chennault and Willauer, with the help of the Chinese bankers. A new airline, called Civil Air Transport (CAT), was born. It flew the Chinese flag but was incorporated under the laws of Delaware.

CAT flew in competition with CNAC and CATC. There was business for all three airlines, but CAT probably had the advantage in top-notch pilots, nearly all former fighter pilots who had served in the U.S. Navy, Marines or Army Air Corps.

But, by the fall of 1949, CAT's far-flung operations were being diminished by the advance of Red Armies into west and northwest China, and finally into east China, which forced the company to move its headquarters from Shanghai to Canton. By early November, nearly all was lost to the Communists except the island of Hainan and Taiwan and lesser islands along the east coast of Fukien Province.

The other airlines, CNAC and CATC, both partly or entirely owned by the Chinese (Nationalist) government, were uncertain about their future and seventy-one aircraft, mostly C-47s and C-46s, began to descend on Kaitak. In addition, there were probably eleven aircraft parked on the field either on scheduled flights or under routine engine checks and repairs, making the grand total of eighty-one CNAC and CATC aircraft. Since a new government had been proclaimed in Peking a year earlier, the question was: to whom did these eighty-one aircraft belong, the government in Peking or the government on Taiwan which had organized and developed the two airlines?

Part of the problem of ownership was solved by the "defection" of eleven airplanes flown by Chinese co-pilots back to the mainland.

The situation on the mainland left the partners, Chennault and Willauer, facing a double crises: what were they going to do with their own planes and pilots? and what could they do to prevent the seventy aircraft of CNAC and CATC from being removed to mainland China? As for Civil Air Transport's future, "Whitey" Willauer put it this way: ". . . it was like a railroad with no track . . . no routes and no business (and) to top it all the Hongkong government cancelled the CAT radio permit."[9]

Cancelling the radio permit meant no CAT airplanes could enter the Hongkong Air Space. They could leave but not come back. This was resolved shortly by an appeal made to London by the U.S. State Department for restoration of the radio permit, which was granted.

It was important now for the partners, Chennault and Willauer, to prevent the seventy remaining aircraft of CNAC and CATC from falling into Communist hands. At least that is the way they saw it, and the Nationalists on Taiwan were of the same opinion, since the aircraft could be used for an air assault on the new government rising from the ashes on Taiwan.

Maj.-Gen. Claire L. Chennault and Whiting Willauer, co-founders of Civil War Transport, pose for picture for cover of CAT *Bulletin* published 1 December 1949. "Cat" logo for airline was later changed.

While the Peking government was negotiating for the recovery of the aircraft, seven of the CNAC and CATC planes were sabotaged, although the demolitions used were probably potato-masher type hand grenades and did no serious damage. The saboteurs were never apprehended. Rumors on the Kowloon side held Nationalist sympathizers responsible.

Chennault and Willauer, meanwhile, had taken steps through the Hongkong courts to ask for an injunction against the removal of the aircraft as well as padlocking company warehouses filled with parts and engines. But before this could be formalized, Communist agents, with the concurrence of the Hongkong authorities, loaded two river steamers with supplies and shipped them to the mainland.

Civil Air Transport C-46s on flight line at Sunya air base, south tip of Hainan Island, after evacuation from mainland China early November 1949. Concrete runway was built by Japanese in WWII.

Sterling Bemis, CAT pilot of this Cessna 195 at Sining, capital of Chinghai Province, prepares to take off for Lanchow. Chinese mechanics have brought pail of water or other liquid. Man at left may be station manager, S.F. Kao. Cessna 195 carried five passengers or 1000 pounds. Six were on order. Four had arrived in April 1949 and two more were on the way when Mainland China fell to Communists. The Cessnas were flown to Taiwan and spent the next years flying between cities of the island. MOC is abbreviation for Ministry of Communications. Photo by author, August 1949.

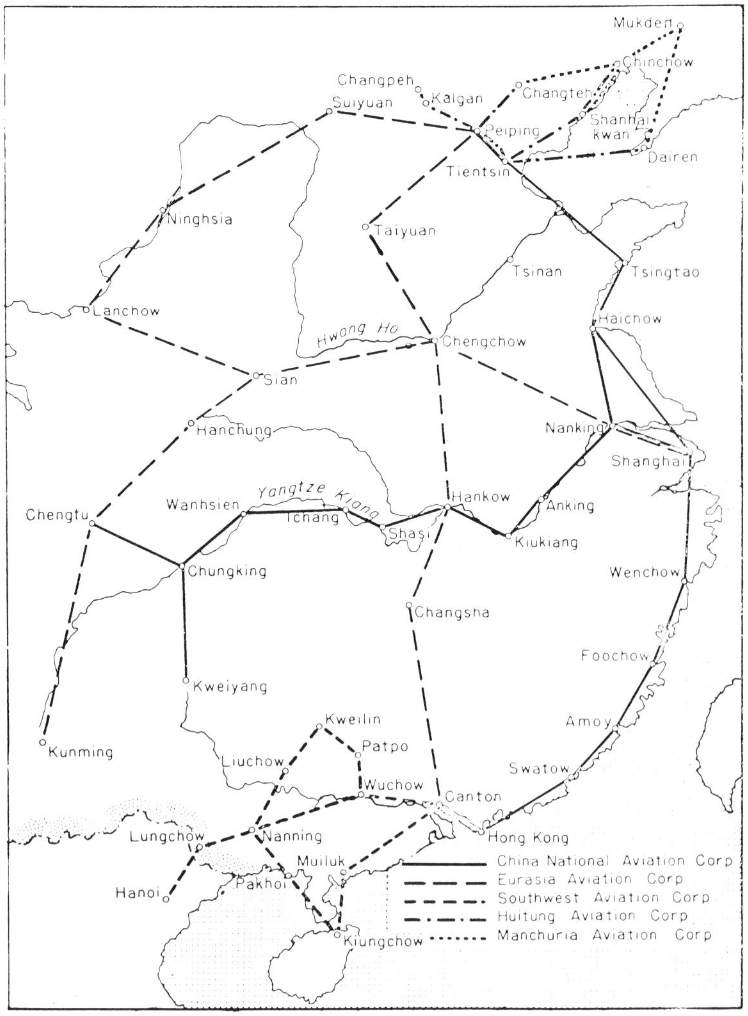

Commercial airline routes in China in summer of 1937, last year of normal operations before Sino-Japanese war began in July-August 1937. Map taken from "Air Transport in Pacific Area" by Sydney B. Smith in *Institute of Pacific Relations*, summer 1941.

The next step for CAT was to enter into negotiations for the purchase of the Chinese and American shares in the two airlines which would make both subject to United States law. Negotiations were being conducted at the same time with the Chinese Embassy in Washington D.C. and the State Department to sell the shares to Chennault and Willauer which was eventually accomplished.

On 15 February 1950 this notice appeared in the *CAT Bulletin:*

> Civil Air Transport Inc., a corporation organized under the laws of the state of Delaware, U.S.A., hereby announces that on December 11, 1949, all the property and assets of CNAC and CATC were sold to Chennault and Willauer, a partnership, and were subsequently resold to the Civil Air Transport, Inc. which is now owner of all aforesaid assets. All employees of CNAC and CATC are hereby notified that CAT Inc., considers their employment terminated as of December 12, 1949.

Meanwhile, after a month of litigation, the Hongkong Supreme Court awarded all CNAC and CATC aircraft to the plaintiffs, that is, to the Peoples Republic of China. The Department of State in Washington, D.C. now engaged the law firm of William Donovan to fight the case. After the decision went against CAT, Donovan filed an appeal with the Privy Council in London and more than a year later, the Council overturned the decision of the Hongkong court and awarded all CNAC and CATC aircraft standing on Kaitak to Chennault and Willauer. By this time the airplanes, which had been exposed to salt air for nearly two years, were unfit to fly without overhaul. The British authorities in Hongkong were most anxious to have them removed but insisted that none should attempt to fly out, fearing complications from the Peking government and/or anti-aircraft guns beyond New Territories. As a result, an aircraft carrier of the U.S. Fleet in Japan, sailed to Hongkong and picked up five Convairs and sailed for San Diego, where they were to be reconditioned at the Consolidated Vultee factory.

Thomas G. Corcoran, a Washington attorney, who had acquired a minority interest in CAT air lines in association with several other investors, including James Brennan, sent Brennan to Hongkong to become vice-president of the airline, and an office was opened in Burbank, California, to handle the sales of the C-46s and C-47s which were being shipped to the West Coast. The new salvage company, Aviation Parts & Equipment Inc., was managed for a time by Brennan and later by Robert Lee, a son-in-law of Chennault.[10]

A number of C-47s were also shipped to a salvage yard in Manila, but the majority of aircraft, including the five Convairs in San Diego, were reconditioned in California and sold to buyers around the world, especially in Central America. The war in Korea, which broke out in June 1950, raised the value of all aircraft, old and new, and most of the former CNAC and CATC aircraft sold for more than cost price, although the cost of reconditioning probably balanced the appreciation caused by the war.

As for CAT itself, it was "running out of track" before all the CNAC and CATC aircraft put down at Kaitak. The company was facing a financial crises which could be solved only by an infusion of more capital and new air schedules. Through old friends and former air corps associates, Chennault appealed to the State Department for help to keep the airline alive. Paul G. Helliwell, an attorney in Miami and former chief of intelligence in Kunming during World War II, urged the State Department, through the recently-created Central Intelligence Agency, to buy the airline and use it as a "cover" for covert operations in the Far East. After discussions at the highest levels, the CIA agreed to buy out the partnership of Chennault and Willauer. On 10 October 1949 CAT began flying for the CIA, and on 1 November documents were signed giving the CIA ownership of Civil Air Transport, Inc. Chennult remained on the board of directors and Willauer was president for a year or so before he was appointed United States ambassador to Honduras.

With the CIA behind it, CAT airlines, in the 1950s became a major auxiliary carrier in the movement of supplies to United Nations forces in Korea, and, in 1954, it assisted French colonial forces defending Dienbeinphu, dropping food and ammunition, even as many of the same CAT pilots

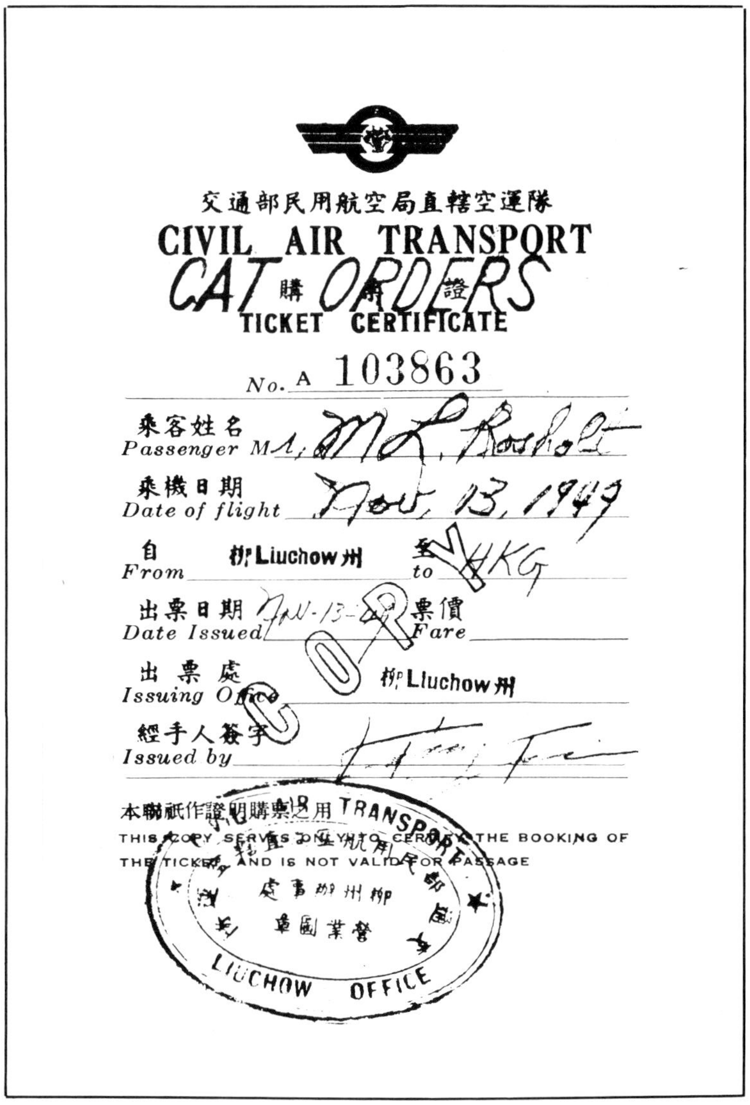

交通部民用航空局直轄空運隊
CIVIL AIR TRANSPORT
CAT ORDERS
購 票 證
TICKET CERTIFICATE
No. A 103863
乘客姓名 Passenger Mr. M. L. Rosholt
乘機日期 Date of flight Nov. 13, 1949
自 From 柳 Liuchow 州 至 to HKG
出票日期 Date Issued Nov-13-49 票價 Fare
出票處 Issuing Office 柳 Liuchow 州
經手人簽字 Issued by
本聯祇作證明購票之用 THIS COPY SERVES ONLY TO CERTIFY THE BOOKING OF THE TICKET AND IS NOT VALID FOR PASSAGE
LIUCHOW OFFICE

had done a a few years earlier in the siege of the city of Taiyuan in north China by the Red Army of Lin Piao.

The airline introduced a luxury flight, called the "Mandarin," in a modified DC-4 on 20 April 1950, and by 15 October 1958 a DC-6B joined the Mandarin flights.

On 16 February 1968, a Boeing 727 crashed on landing in Taipei. That marked the beginning of the end for CAT. The crash, in which eighteen passengers, three flight attendants and four crewmen died—but not the American pilot—gave the Chinese an excuse to cancel the airline charter and to begin liquidation proceedings to replace it with China Airlines, a wholly-owned Chinese corporation.

A CAT can live only so long, but this one had lived an exciting life.

VIII

Flight of the Tin Goose

NC 432H, U.S. Commercial license assigned to Ford Motor Company when this trimotor was built, was sold and shipped to the "Young Marshal" of Manchuria in early 1931. In the summer of 1931 it was flown from Mukden to Peking to avoid capture by Japanese forces threatening to overrun Manchuria. Above, the "Tin Goose," as everyone called her, is parked in front of hangar at Nanyuan air base, south side of Peking, with Carl Kjus, crew chief, hurrying back into hangar for something he forgot. At left, perhaps one of first air views ever taken of Temple of Heaven on south side of Peking, and at right, the "Goose" flies over city of Peking.

The Ford Motor Company, in 1926, began production on a monoplane called the "Trimotor" because it had three engines. The early models were fifty feet long with a wing span of seventy-four feet., and seated twelve passengers.

A second model had three Pratt & Whitney Wasps, each capable of 1275 hp., with absolute ceiling of 18,500 feet and maximum cruising range of 659 miles. Since it was covered with corrugated metal, it was often called the "Tin Goose."

Chang Hsueh-liang, the new ruler of Manchuria after the death of his father in 1928, ordered one of these new Fords in late 1930. Earlier he had been using a Junkers F-13 for his personal transportation. This was one of the most popular German airplanes built after WWI and a favorite of Lufthansa through the 1920s.

When the Ford Motor Company shipped the Trimotor off to China, it also sent a crew to fly and maintain it. Carl Kjus, an ex-airmail employee from Chicago and Maywood, Illinois, was hired as the mechanic, with Perry Hutton, pilot and Edward Hamilton, co-pilot.

During the nearly four years he was in China, Carl Kjus kept up a fairly regular correspondence with his parents in Maywood. In a letter written to them from Muken in 1931 he fails to explain where the Tin Goose was assembled but he seems to suggest that within a month after his group reached Mukden, the Goose was ready to be preflighted.[1]

On 18 April 1931, Kjus was writing from the Grand Hotel in Peking to tell his family that the Goose was parked at the field, presumably Nanyuan, waiting for its chief passenger, the Young Marshal. The flight down from Mukden to Peking had come off without a hitch.

On 2 May, the Goose was in Nanking. Kjus writes that the plane left Peking on a Wednesday, landed at Tientsin, and left the following day for Nanking, a trip made "non-stop," as he explains, in five and a half hours. The runway on the field at Nanking was soft and when the Goose began taxiing back to a hardstand, she went down. The next day about 200 workers were called out "and they just about carried her into a hangar."

On 17 May, while still in Nanking, Kjus was expecting to leave the next week "as the big show is all over"[2]

The crew was anxious to get out of Nanking "as there is not a damn thing to do . . . and the hotel is terrible."

On 27 May 1931 Kjus is back in Peking, with the Goose, nd on 9 July he is writing again from the Grand Hotel. Meanwhile, the crew of the Goose had logged only sixteen ours of flying time, presumably since leaving Mukden. here was no hangar on the field in Peking big enough to old the bird, but Kjus was happy to report that she was now eing moved into a new hangar. A few weeks earlier a wind torm had nearly blown the tail section off and Kjus spent a veek repairing the damages. He tells his family that he is oping to get back to Mukden soon, but right now the 'oung Marshal had come down with typhoid fever, and erry Hutton had chicken pox.

By 22 August, Kjus writes that the crew of the Goose, nxious to get in some flight time, was becoming a nervous vreck, sitting around doing nothing. He heard that Colonel indbergh was on his way to China and hoped "we get a hance to fly him around a bit . . . . "[3]

His next letter from Peking is dated 25 September 1931, ust a few days after the Japanese invaded Manchuria on 18 September, and as Kjus explains, "shot up the city [Mukden] illing several hundred people."

On 23 October he writes from Bridge House Hotel in Nanking where he has been for two weeks. He says the Goose has been doing quite a bit of flying lately, five trips etween Peking and Nanking this month, and the trip was eing made in one hop since the Goose carried 500 gallons f gas. He also writes that "the boss has ordered another Tin Goose which should be here Nov. 15," and he was anxious o see it because he hears it is a "pretty hot plane."

On 24 March 1932, Perry Hutton and his crew boarded a oastal ship for Hongkong where a new Goose was being picked up." Kjus said it was called the "Club model, and old for U.S. $70,000." The crew was expecting to fly it back o Peking in one day with a stop in Hankow for gas.

By 22 April 1932, the new Goose had arrived and the crew has assembled it, presumably at Kaitak. Kjus says "it sure is a peach." It had balloon tires, oleo shock absorbers, a two-way radio communications system, and a top speed of 155 mph.

A letter describing the flight of the new Goose to Peking is missing, but Kjus writes from Peking on 4 September to say that "things are getting a little shaky around here, as the Japs are making it hot for the Marshal." Kjus was not scared, he says, because "all we have to do is quit our jobs and walk over to the American Guard which was 600 Marines, enough to take care of the Japs and the Chinese as well . . . . "

The crew was flying the Goose to Shanghai in the morning and the next letter available from Kjus was written on 1 January 1933, again from Peking, when he says that "we made a three week trip with the marshal to Hankow, Nanking, Shanghai, and Hangchow." In this letter he also mentions Frank Cole for the first time although he fails to make clear what Cole's status was in the aviation community. He was called a "mechanician," a word meant to describe a flight engineer or crew chief, and he appears in a group picture with other airplane people taken at a party in Peking, circa 1932.

On 24 April 1933, in a letter from Shanghai, Kjus reveals that the Young Marshal has left for Europe and has turned over his Trimotor to the Nanking government, although it is not clear whether this was the first or second Goose. In

Chang Hsueh-liang, third from right, (civilian gown), better known as "the Young Marshal," stands in hangar at Shien Chiao. His American crew has probably brought the "Tin Goose" here for an engine change. C.B. Clark, test pilot and engineer for Central Aircraft Manufacturing Company, stands second from right, and Peter Mow, senior aviation officer of Chinese Air Force in early 1930s, stands at center left, in uniform and puttees.

Most of the men in this picture were pioneers in the field of aviation in China, and some flew the *Tin Goose*. They have come together, in black ties—mandatory at the time—to have a party, most likely in a suite at the Grand Hotel de Pekin, c. 1932. The lady in the center behind the table is probably a White Russian refugee from Manchuria and the two girls at her side may be her daughters. Seated at far left is Carl Kjus, mechanic on the *Tin Goose*. The lady in the chair and man next to Kjus are not identified. In back row, l. to r., A. (for Allan) M. Alexander, newsreel cameraman for Hearst; Paul Heise, a motion picture executive, and Letha and Perry Hutton, pilot on the *Goose*. In horn-rimmed glasses is Jimmy Elder, Jr., an Englishman and boyhood friend of the Young Marshal. He holds his hand on shoulder of Julius Barr who became pilot after Hutton left a couple of years later. Man on edge of chair at right is Frank Cole, a mechanic and later with Royal Leonard, pilot for the Generalissimo. Later, Julius Barr also flew the Generalissimo's private airplane.

Shanghai, which Kjus describes as a "wild city," he has heard there were "big butter and egg men from the states" who were talking about starting an airline between Shanghai and Manila.[4]

Meanwhile, Perry Hutton resigned as chief pilot on the Goose and left China. His place was taken by the co-pilot, Harry G. Smith, whom Kjus had known from the air mail service days in Chicago.

His next letter, dated 1 August 1933 is written from Kuling, a mountain resort town where Chiang Kai-shek had a summer house. Prior to arrival in Kuling, Kjus had been in Hangchow for six weeks overhauling the Tin Goose, although the repair shop was not in Hangchow but east of the city at Shien Chiao where there was an "aviation school there and lots of equipment." The plane returned to Nanchang on 24 July, and the crew was allowed to come to

Kuling by train and bus to the foot of the mountain where they transferred to sedan chairs for the ride up the mountain side!

There were several hundred foreigners in Kuling in summer, mostly missionaries, and a few business people from Nanking and Shanghai. There were tennis courts "but the main sport was hiking." Kjus hoped the General [Chiang Kai-shek] would stay here a couple more weeks because it was so pleasant, and at night cool enough to use a blanket.

By 15 October 1933, Kjus is back in Nanchang for a week and then on to Shanghai for three days. Harry Smith was now in the left seat of the Goose, flying Chiang Kai-shek. In the next three months, Kjus apparently did not write home. Finally, on 27 February 1934, he wrote from Nanchang to say that he had flown about eight hours on Christmas day and managed to get to Shanghai for New Years Eve, and "we have been doing quite a bit of traveling the last two months. A week in Peking, two weeks in Nanking, three weeks in Hangchow and two in Fukien, so back to here to stay I think."

The Young Marshal had returned to China from his European tour by this time and Kjus thought he would be working for him again. On 13 May 1934, Kjus wrote from Nanchang after a flight to Hankow and Changsha, but the Goose was now on course for Shanghai because she needed an engine change—all three engines! He then describes a visit of "Commander" Frank Hawks who was in Nanchang to demonstrate a Curtiss Condor transport. "He was putting on a show here for the General. Everything was going fine until he tried to land without putting his wheels down. He broke a few propellers and messed up the underside of the ship. Frank said later he was just proving the ruggedness of the Condor."

On 1 July 1934 Kjus wrote from Hangchow to say that he had been in Shanghai two weeks changing engines on the Goose, and that the plane then flew back to Nanchang with "General Chiang Kai-shek and the Madame." A few days later they had also flown to Nanking and back to Hangchow where the Generalissimo was to inspect the flying school at Shien Chiao.

Carl Kjus was still in China on 19 September 1934 when there was talk about getting a new airplane, but his next letter is written from San Francisco on 11 February 1935 when he explains to his family that he has not been feeling well for the last five months. On 13 January 1935 Mrs. Harry G. Smith, wife of the new pilot on the Goose, had written the Kjus family in Maywood to explain that Carl was seriously ill, and was being sent home on a ship accompanied by a full-time nurse. He had contracted a virus of some kind which depleted his strength. On 10 May 1935, his old friend, Perry Hutton, wrote to the family to tell them that Carl had died in San Franciso.

The fate of the Young Marshal, Kjus' boss, is also sad. When he saw, in 1931, that his own days were numbered in Manchuria, he had moved his Manchurian troops into North China and set up headquarters in Peking. After his return from Europe he was ordered by Chiang Kai-shek to move his troops west to Sian to assume command as "pacification commissioner," which meant that he was expected to fight the Communists who had reached Shensi on the Long March in 1935. But the Manchurian troops thought they should be fighting the Japanese to recover their homeland in the Northeast instead of the Reds, and the Young Marshal fell into a conspiracy promoted largely by Communists and subordinate generals under his own command. Finally, Chiang Kai-shek, impatient with Chang's inability to carry out orders, flew to Sian, probably in a new Douglas DC-2, piloted by Julius A. Barr, to learn what was causing the delay, at which time he was "kidnapped," 12 December 1936. It seems more appropriate to say that he was taken prisoner and held under house arrest but contemporary texts used the term "kidnap." Army commanders of the Manchurian troops had taken the Generalissimo in an effort, they thought, to force him to abandon his anti-Communist campaign and join a "united front" with the Communists against Japan although the Chinese Red Army had never before engaged the Japanese in any major campaign.[5]

But the Young Marshal lacked the ruthlessness of his father and hesitated to harm the Generalissimo. Mao Tsetung wanted to kill him but was probably restrained by Stalin.

No one ever knew for certain what kind of an arrangement the Gimo made for his own release, but he could not have acquiesced in much, since the Young Marshal agreed to accept responsibility for his detention even though his top aides were also to blame. The Young Marshal flew to Nanking in his own plane, probably a Boeing 237. On Christmas Eve when the news of the Gimo's release was flashed over the news wires and radio, an entire city exploded with firecrackers and relief. But the Chinese could never forgive the Young Marshal. There was no court martial in Nanking; he was simply placed under house arrest and he remained under house arrest until the city was evacuated in November 1938, when he was moved to Hankow and later to Chungking where he remained until after World War II. When the Nationalists moved to Taiwan in 1949, Chang Hsueh-liang, no longer the "Young" Marshal, moved with them, still under house arrest, no doubt the longest house arrest in contemporary Chinese history. In 1983 he was living on Taiwan under relaxed detention, toying with old memories, probably of a noisy Tin Goose.

Junkers JU-52 in Tientsin, 1934.

IX

Air Force Day in Shanghai

Northrup 2E bomber with Chinese Air Force markings under left wing. This may have been one of ships which bombed Shanghai on 14 August 1937. In background stands Curtiss Shrike. Photo probably taken at Shien Chiao air base, 1939.

On 7 July 1937, now referred to in Chinese texts as the "Double Seventh," fighting broke out once again between Chinese and Japanese ground forces, this time at Marco Polo Bridge on the south side of Peking. The fighting spread through north China and by 13 August hostilities had broken out in Shanghai, again in the northern suburb of Chapei where the fighting occurred in 1932.

Lying at anchor on the Whangpoo River near the Japanese consulate was the cruiser *Idzumo,* an old three stacker, flagship of the Japanese 3rd Fleet.

The day after fighting broke out between Chinese and Japanese ground forces, the Chinese Air Force on 14 August took to the air to bomb *Idzumo.* But instead of hitting her, the inexperienced pilots dropped their bombs off the Customs Jetty and struck the Cathay Hotel and the roof of the Palace Hotel, both at the corner of Nanking Road and the Bund. The street was crowded with refugees who were trying to escape from the suburbs of Hongkew into the safety of the International Settlement. There were more than 150 dead and hundreds of wounded.

Later in the afternoon of the same day, another Chinese bomber flew over the city, at about 5000 feet, and released two bombs over the most heavily congested traffic circle in the city, Avenue Edward VII and Race Course Road. The result: more than 3000 killed and wounded.

The day was called "Bloody Saturday."

A week later on 23 August, a bomb dropped on the Sincere Department store on Nanking Road and 140 people were killed and wounded. The airplane which dropped the bomb was flying at altitude and was not seen but there could be no doubt about the origin of the bomb fragments. They were Chinese.

On 31 August, the last of the series of blunders occurred when a flight of Chinese bombers flew over the Dollar liner, *President Hoover* which was approaching the mouth of the Yangtze River en route to Shanghai. Apparently without authorization, a trigger-happy pilot broke formation and went down to attack the ship even though the U.S. flag was flying astern and flags were painted on decks. Bombs were dropped, one hitting the stern. Seven crew members were wounded, two seriously, and one died. Passengers were stunned, and O.K. Yui, mayor of Great Shanghai, apologized to the U.S. Consul-General.

Today, the Republic of China celebrates 14 August as "Air Force Day." Although a disaster, it marked the first time that the Chinese Air Force had carried the attack to the enemy in the China Air Space.

There were four English-language dailies, in addition to several Chinese dailies published in Shanghai at this time. These newspapers had their own staff reporters on the street, but were also subscribers to both foreign and domestic wire services such as the English service, *Reuters,* and *Domei,* the Japanese news agency, *Havas,* French, *Tass,* Russian, and *Transocean,* German. All were involved in the coverage of the land and air battles which surged around the International Settlements. But with all their news gathering facilities, it was difficult to get accurate figures, for example, on the number of aircraft destroyed by either Chinese or Japanese airmen. In war each side exaggerates the losses inflicted on its opponents while minimizing its own.

Headlines in the *China Press* and the other city newspapers described the battles being fought but the lead story was often not the fighting on land but the Japanese air raids which were being carried out against cities all over China. *Domei* said that sixty-two cities and towns in eight Chinese provinces had come under air attack.[1] Nanking, then the national capital of China, and only a short distance by air from Shanghai, became the primary target for aerial attack by the Japanese, and wave after wave of high level bombers, dive bombers and fighters plunged the city into great fires and destruction. On 22 September *Reuters* reported that more than fifty enemy aircraft had been seen over the city in the first wave of that day, and others followed.

In these first days, the Chinese airmen rose to defend their country, and at first they met with considerable success. Their spirited resistance was rewarded by Mme. Chiang Kai-shek who arranged to have "comfort baskets" passed around to all members of the local squadrons in Nanking. The baskets contained tidbits of food, tea leaves, cigarettes and gramaphone records, although she was not there in person to pass out the goodies. Foreign newsmen who attended this ceremony also had an opportunity to interview Captain Lo Yi-chien, a 23-year old native of Szechwan who, caught in a squeeze play between nine enemy bombers, shot his way out of the box and in the process allegedly shot down four of the bombers. There did not appear to be any witnesses to these aerial victories from his own squadron.

One of the first in a series of outrages committed by the Japanese airmen in China occurred on the afternoon of 20 August when ten bombers carried out an attack on South Station in Nantao, a suburb of Shanghai on the south side. The railway from Hangchow terminates here and it was, in a sense, a military target, but the commander who gave the order to destroy it probably did not realize that it was also jammed with refugees trying to escape from the city. A single reconnaisance plane could have established this for the squadron commander. Before the train could back into the station platform, heavy engine noise could be heard overhead and in the next instant bombs exploded on the station blowing fifty people to pieces and wounding 400.

It was in this raid on Nantao station that H.S. ("Newsreel") Wong, a Chinese cameraman working for Hearst, took his immortal picture of the crying child seated on the station platform amid the ruins. "Newsreel" came to the offices of the *China Press* early the next morning with these pictures and described what happened.[2]

Sungkiang, a railway junction southwest of Shanghai, was bombed on 8 September when 300 refugees, mostly from Shanghai, were killed and 400 wounded. On 24 October the Japanese airmen returned to Sungkiang and repeated the carnage of the earlier raid, but this time bombing the city center, government buildings, schools and foreign mission properties, the latter with the United States flag prominently displayed on campus. The tactics used here were followed in many other cities. After the railway station had been destroyed, the airmen returned to bomb government buildings, university centers, hospitals and foreign missions. University centers were especially targeted by the Japanese because they blamed students for much of the agitation to boycott Japanese goods after the Japanese invaded Manchuria in 1931.

By mid-October, The Chinese Air Force, what there was left of it, had been forced to withdraw to bases in the interior, and most enemy aircraft brought down after that date were shot down by anti-aircraft guns. On 8 October the *China Press* ran a photograph of sixteen Japanese airmen who had been taken prisoner after they bailed out or crash-landed in east China.[3] But compared to the overall number of Japanese aircraft engaged, sixteen airmen were a small price to pay for success.

Admiral Mitsumasa Yonai, Japanese Navy Minister, told a

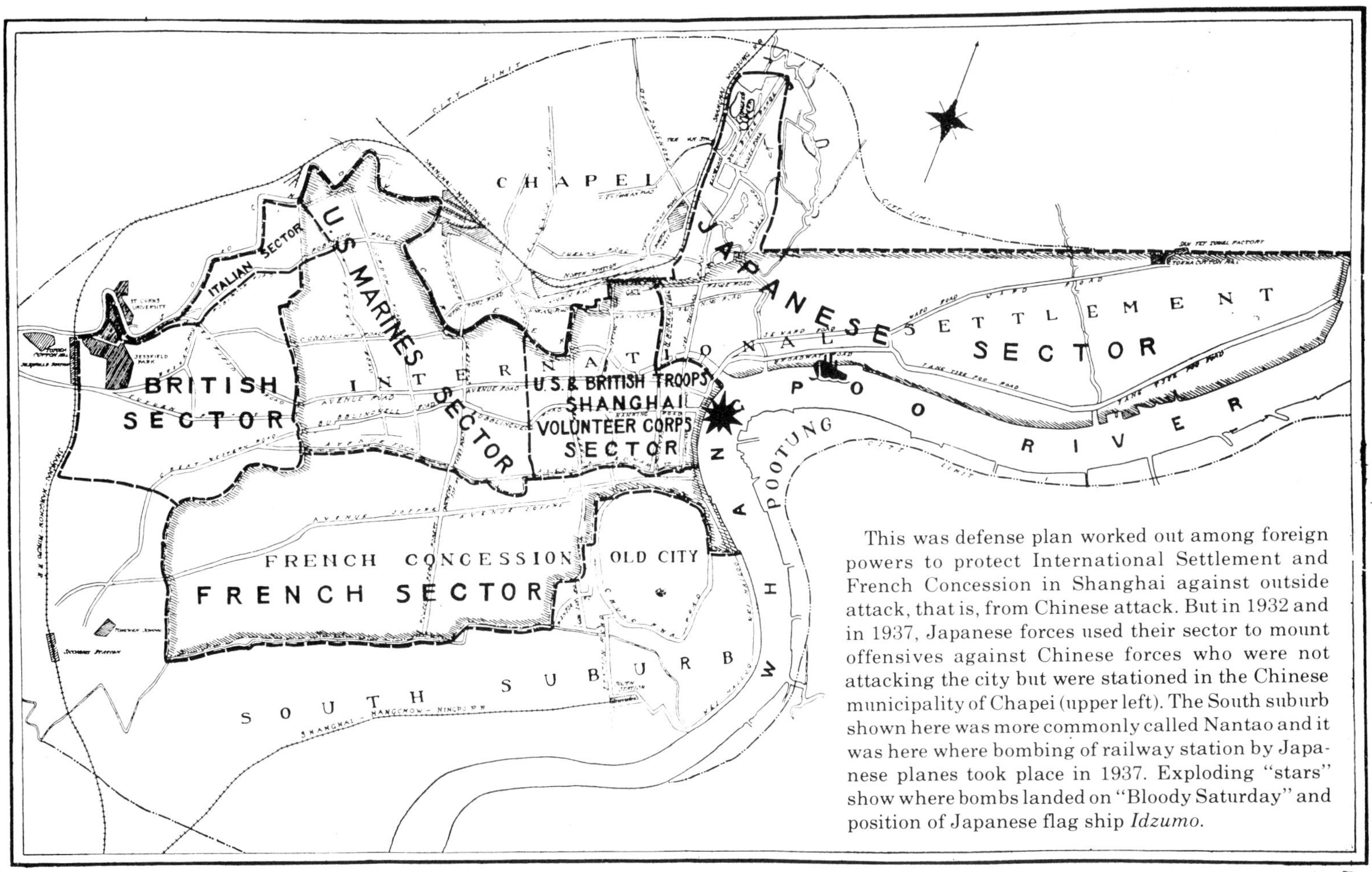

This was defense plan worked out among foreign powers to protect International Settlement and French Concession in Shanghai against outside attack, that is, from Chinese attack. But in 1932 and in 1937, Japanese forces used their sector to mount offensives against Chinese forces who were not attacking the city but were stationed in the Chinese municipality of Chapei (upper left). The South suburb shown here was more commonly called Nantao and it was here where bombing of railway station by Japanese planes took place in 1937. Exploding "stars" show where bombs landed on "Bloody Saturday" and position of Japanese flag ship *Idzumo*.

During fighting between Chinese and Japanese forces in Shanghai, 1937, U.S.S. *Augusta*, a cruiser of Asiatic fleet, moved up Whangpoo River and anchored at No. 1 berth across from Customs Jetty, a birth for decades reserved for British men-of-war but now deserted because of other demands upon the English fleet. Aircraft on *Augusta* are Curtiss SOC-2s, known as "Seagulls." They were from CV8-4 squadron and were put on board in 1936. They were launched from a pair of catapults placed on top of tower located between forward funnel and hangar. Aircraft landed on water and were retrieved by crane. In 1943, *Augusta* stood off Newfoundland Banks where she served as rendezvous for Winston Churchill and President Roosevelt. Above photo taken by author c. 16 August 1937.

meeting of the Imperial Cabinet held in Tokyo on 17 August that Japanese planes had already shot down or otherwise destroyed eighty-seven Chinese aircraft, and bombed or partially destroyed sixteen hangars. The figure of eighty-seven aircraft is not out of line because many of these were probably trainers—Stearmans, Jennys, Moths—located at different fields parked in bamboo nose hangars or waiting for an engine overhaul.

In the first weeks of the air battles, Japanese airmen were being outfought by the Chinese, as already indicated. The Japanese were using the Mitsubishi 96 carrier fighter, a biplane which could not compete with the Curtiss Hawk III, a monoplane.

They needed a better fighter. Rene J. Francillion describes the situation more fully:

> In the early stages of the conflict, when the only fighters available in quantity to the Japanese were the obsolete carrier-based Nakajimi A2NIs and A4NIs, losses reached a dangerous level, but with the arrival at Shanghai of the A5M2-equipped 12th and 13th Kokutais, the Japanese gained complete air superiority (and) with the accelerated tempo of the air operations in China, the development of the Navy Type 96 Carrier fighter was actively pursued, and the A5M2b Model 2-2 was placed in production. Having an enclosed cockpit, the A5M2b was powered by a 640 np Kotubiki 3 driving a three-blade propellor . . . . The A5Ms maintained their superiority over the much faster Polikarpov E-16s used by the Chinese. At the same time the Japanese fighter proved that despite its lightweight structure it was able to absorb considerable battle damage, as demonstrated by Petty Officer Kashimura who brought his A5M2 back to its base after losing one third of the port wing in a ramming attack by a Chinese fighter . . . . Delivered to combat units in China in 1938, the A5M4s increased markedly the area of Japanese air superiority . . . .[4]

Occasionally attacks were launched by Chinese bombers against targets in Yangtzepoo, a suburb of Shanghai occupied by both Chinese and Japanese business firms, but in the weeks to come the daylight bombing

The Shanghai foreign community was dominated by British traditions. One of these was the post board, advertising highlights of a new edition of a newspaper. This poster was written by the author when he served as night editor of the *China Press*. The date is 20 September 1937, at the height of the Chinese-Japanese air battles over Nanking.

Japanese dive bomber (upper right) comes down with torpedos on target along railway in Jessfield sector, west side of Shanghai, August 1937. Railway went to Hangchow.

was replaced by night raids, and foreign property as well as Chinese suffered more than the damage done to enemy warehouses and cotton mills.

Most of the Chinese missions run at night were little more than nuisance raids of one or two aircraft. The approach of one plane, in the dead of night, seemed futile. It buzzed through the night unseen, like a mosquito in the dark, its approach greeted with thundering anti-aircraft guns from the Japanese Fleet, search lights stabbing the sky and the night blazing with tracers. Sailors on the U.S.S. *Augusta,* flagship of the Asiatic Fleet which was anchored at the No. 1 buoy on the Whangpoo River, quickly came to recognize the lone rider in the sky. They called him the "Hoboken Barber."

Probably the last raid over Yangtzepoo by the Chinese Air Force came on the night of 29 October. It was greeted with the usual display of fireworks and blind shooting from the enemy batteries.

Perhaps the single, most successful day the Chinese Air Force had against the Japanese was on 16 August, just two days after the first Japanese air raid on Nanking. *Central News Agency* in Nanking sent in this report:

> "Chinese combating planes in a thrilling air duel with Japanese bombers shortly before noon today, scored another victory against the enemy by bringing down two of the six invading machines in their third unsuccessful air attempt on the Capital in two days.
>
> "Six Japanese bombers took part in this morning's attack on Nanking from the air. Immediately upon the sounding of the alarm, seven Chinese pursuit planes went up [into] the air and combed the surrounding districts. The invaders never succeeded in filtering through the Chinese planes which rose to encounter them. After a thrilling battle midair, the Japanese bombers took flight to Yangchow north of Chinkiang across the Yangtze River.
>
> "During the encounter two of the enemy planes were shot down by the Chinese Air Force. The Chinese pursuit planes, having driven the enemy planes away, returned to Nanking at noon. One of the seven planes was slightly damaged on the right wing.
>
> "When the alarm was sounded of another Japanese air raid on the capital, high government officials were attending a memorial meeting in the National Government. The service was never once interrupted by the news of the raid. Mr. Lin Sen, chairman of the National government, continued his address unperturbed, while an audience of over 400 paid undivided attention to the aged chief executive.
>
> "After the frustration of two enemy air attacks on Nanking yesterday and today, Nanking was now fully confident of its ability in air defense, as reports continued to arrive here of Japanese planes being shot down by Chinese defenders in many cities along the lower Yangtze.
>
> "In addition to the three Japanese planes shot down in Chungyang and two at Yangchow it is definitely learned that the enemy had lost two machines in Shanghai, one in Kashing and one in Chekiang, all during the morning.
>
> "Sixteen Japanese bombers were shot down by Chinese air defenses forces yesterday, the price Japan paid for a series of raids on Nanking, Nanchang, Hangchow [and other cities].
>
> "In defending China's territory against the Japanese aerial attack, five Chinese pursuit planes were also shot down by Japanese planes, all in Chekiang [province] while a sixth was damaged at Nanking.

Curtiss Hawk II of the Chinese Air Forces crash lands in dry rice paddy while pilot examines damage.

Douglas O-2MC, attached to Chinese Air Force. Ideographs on rear of fuselage read *"Ch'i-pa"* (No. 7). Here is another picture showing early version of C.A.F. logo, that is, white sun in blue field, but here the sun's points do not touch outer rim of field.

The Curtiss Hawk II with fixed landing gear and Chinese Air Force markings, probably at Shien Chiao, c. 1934.

"The largest fleet of enemy planes ever to attack a Chinese city visited Hangchow at 5 o'clock yesterday morning as 23 attacking planes rained bombs in the suburbs of the Lakeside city. [The dispatch should have explained that this suburb was probably the airfield at Shien Chiao, the Chinese Air Force training center since 1932.]

"The attacking squadron of Japanese planes, which took off from an airplane carrier anchored off the Chekiang coast, was met in the air by a smaller fleet of Chinese pursuit planes.

"In the course of this engagement, the first of two during the day, five Japanese planes were brought down, while two Chinese machines were also wrecked.

"The second raid on Hangchow [read Shien Chiao] took place shortly after noon as the Japanese planes again attacked the city, but once more met with stiff resistance from Chinese planes. Three of the invading planes were shot down and an equal number of Chinese defense machines suffered the same fate.

"In between the two raids on Hangchow, a squadron of nine bombers subjected Tsaongchiang about 55 miles to the east of Hangchow, to a severe aerial bombardment shortly after 6 o'clock. A Chinese squadron, also of nine planes, engaged the attacking machines in a grueling battle, resulting in one Japanese plane being shot down.

"A fleet of fourteen Japanese machines, composed of eight pursuit planes and six bombers dropped eighteen bombs on the city [Nanchang] and machine gunned the airdrome and the East Lake area. After an engagement lasting nearly an hour the Chinese defense planes drove off the attacking squadron."[5]

The day after the above report appeared, a follow-up story from *Central News Agency* reported that the Ministry of War had officially confirmed that more than thirty Japanese planes had been shot down in the last three days, although without indicating how many were bombers or pursuit. This record was never duplicated by the Chinese. There were days when their luck was running better than others. For example, on 19 September, *Reuters* reported from Nanking, quoting Chinese sources, that six Japanese aircraft were shot down by pursuit planes, one by anti-aircraft in the morning and two by anti-aircraft in the afternoon. The Chinese lost three planes to enemy fighters in the morning and one in the afternoon.

On 20 September, however, Captain Lo Yi-chin (mentioned earlier) and Liu Tsui-kang, Tung Ching-hsiang and Yuan Po-kang shot down five enemy planes. Both Lo and Liu allegedly had four Japanese victories since 14 August.

Meanwhile, Flight Captain Jen Yung-kou and Pilot Officer Liang Hung-yung, who were flying one of the bombers in the attack on *Idzumo* on 14 August, were both killed in action. Jen died in his plane, and Liang, who was wounded, managed to land his aircraft but died in a hospital. They had been attacked by a Japanese float plane while en route back to base, probably at Shien Chiao.

By 19 October, Japanese bombers and fighters had achieved air superiority over Shanghai, Nanking and Hangchow.

This replica of an aerial bomb stood in a traffic circle in Nanking to encourage support and interest in Chinese aviation. Sign at left warns tourists this is no place to sit or sleep. Author photo summer of 1935.

Curtiss BT-32 Condor probably at Nanchang, 1933, ready for demonstration flight by Frank Hawks who may be standing under nose of ship, and, in white dress, Mrs. Hawks. Demonstration was a failure because Hawks failed to lower his gear, but he managed to bring ship in with no serious damage.

In foreground, an Italian Savoia-Marchetti S.M.72. Next airplane, 905, an Italian Caproni CA-101 or CA-111. Photo taken probably at Nanchang, c. 1936.

A bomb crater left by Japanese bombers, probably at Nanking, 1937. In background a Breguet 27-5 bomber and a Northrop 2EC light bomber.

Service work in progress on Curtiss Hawk H-75M fighter in revetment of sand bags, probably Nanking, c. 1937

General Wang Hsu-ming, better known as "Tiger Wang," was in first class of air cadets to be graduated from Whampoa Military Academy, 1925. He later served as chief of the Chinese Air Force. He is shown here at a garden party at Taipei, summer of 1983.

Chinese gunner getting instructions from foreign advisor (left) in Douglas O2M.

Location of this picture in China uncertain. Festive colors suggest it was taken in Nanking and these bombs were exhibited to impress spectators. Bombs at far left are 1000 pound armor piercing, and two in forefront are general purpose (GAP).

X

Mission Accomplished

In early May 1938, two Martin bombers of the Chinese Air Force flew to Japan and, after dropping propaganda leaflets over cities of Kyushu in western Japan, returned to base in China. It was one of the longest and most daring missions ever undertaken by a flight of two planes flown by Chinese crews. It was never duplicated. A report of the mission was written in Chinese by debriefing officers at the time but never published in English until after the war when an English translation was released in Chungking in September 1945, probably by the Ministry of Information, a month or so after the end of World War II.[1]

The literal translation follows:

On 19 May 1938 two Chinese Air Force Martin planes from Hankow Air Base flew to Japan to scatter leaflets for the purpose of "call up the self consciousness" of Japanese civilians. Early in March, CAF Headquarters had worked out an Air Expeditionary Plan which was soon put in action. However, the CAF lacked training in navigation and the CAF air-ground radio communication was still at her cradle stage.

For long range operations, many things pertaining to navigation, operations command, meteorlogy and base to base liaison are dependent on radio communication. On account of the aforesaid factors, immediate installations of radio sets both on ground and the Martin-planes were affected. Furthermore, the said planes were respectively equipped with direction-finder, short wave receiver and short wave transmitter. The various ground stations were equipped with both long wave and short wave radio sets.

Of the ground communication, the main line was from Hankow to Ningpo via Nanchang and Chuchow, and the auxiliary line was from Changsha to Lishui via Wenchow. These two ground radio communication lines were well connected to each other.

Along the line from Hankow to sea coast was established a ground-air station. All these ground communication facilities would enable the planes to operate under any weather condition and simultaneously accurate liaison could be made between the Expeditionary Force, CAF Headquarters and the various air bases. In addition, the various long wave

Captain Hsu Huang-sheng, (holding folio), led two-plane flight of Martin bombers from Ningpo in Chekiang Province to cities of western Japan on 19-20 May 1938 to drop "peace" leaflets, the first military aircraft to enter the Japan Air Space. His wing man was Lt. Tung Yen-po, second from right. Other crew members not identified.

and short wave broadcast stations were so prearranged as to strengthen the effective range of direction finding of the said planes for emergency use.

Being well equipped with various communication facilities, the Expeditionary Force waited for a favorable opportunity to carry out the mission. In the month of May, the daily weather changes so suddenly that within half an hour it is often the case that a fine day becomes overcast or vice verse. As to the weather conditions in enemy territory we could do nothing except to put on record every day the daily weather report of various countries in East Asia for reference only.

Before and after the 15 of May, there was not a day good enough for the long range flight and the moon was waning night by night. At that time Captain Hsu Huan-sheng, commanding officer of the Expeditionary Force had already been in Ningpo where he waited for further instructions. On 19 May 1938 at 14:00 hours Captain Hsu reported to CAF Headquarters at Hankow by a cable stating that the weather in Ningpo became fine and asking for instructions.

At 15:23 the two Martins took off from Hankow and flew to the advanced air base at Ningpo via Nanchang and Chuchow. Enroute to Ningpo the planes began to use the air-ground communication system by which the various stations acquired the signal of the said planes and informed various airfields and air defense organizations of the signal to take necessary precautions. At 17:55 the said planes landed on Lo-she Airfield, Ningpo, where all necessary preparations were completed at 23:30 and then the planes took off from Ningpo, the first plane piloted by Captain Hsu and the second one by 1st-Lieutenant Tung Yen-po, vice squadron-leader of the 19th Squadron under the 8th Group. They actually took off from Ningpo at 23:48.

Both planes carried with them only pamphlets and leaflets to be scattered over Japan for the purpose of "calling up the self-consciousness" of the innocent Japanese civilians, whom the CAF would not kill with bombs. The destinations of the Expeditionary Force were Nagasaki, Fukuoka, Kurume and so forth. The secondary mission of the force was to observe Japanese naval bases and airfields.

On 20 May at 00:42 the moon was covered by clouds and the planes had to fly in darkness for one hour and fifty minutes. When they flew in heavy clouds over Tinghai, two anchored enemy battle ships following their rear searched with searchlights but failed to locate them. At 02:40 the Japanese sea coast was in sight. The planes flew over Nagasaki at an altitude of 3500 meters at 02:45. The city did not black out. After the planes whirling over the city several minutes one flare bomb was dropped down and the lights on the ground were out immediately. By this time the planes, having scattered leaflets, flew toward Fukuoka.

At 03:25 the planes at an altitude of 3500 meters reached Fukuoka where the seaplane and land-based plane aerodromes, battle ships in the harbour, various factories etc. were all clearly visible. Leaflets were scattered and flare bombs were dropped from the planes when whirling over the city before the lights were out. From Fukuoka the planes began to return. At 03:55 when they flew over Kurume, Saga, etc. they could not see the lights any more. On their way back, there were occasionally star-like lights which were put out instantantly.

From 02:50 to 03:32 the two planes, flying separately, scattered leaflets all over Kyushu. During their flight over Kyushu they did not encounter any AA gun fire or interceptor.

After four o'clock the planes departed from enemy territory. Enroute back to China, the weather became inclement and the two planes lost touch with each other owing to heavy clouds. They could neither see nor communicate with each other because the wave sound of radio on both planes was too weak to be heard. By the use of ground radio stations they began to get direction respectively and communication with each other.

At 04:52 and 05:50 the radio stations at Changsha and Hankow began to broadcast successively. At 06:15 the Martin No. 1403 could see the China sea coast. At 06:16 the Martin No. 1404 flew near the sea coast. Meantime, the plane No. 1403 could not find her direction due to the weakness of the broadcast and then the radio station at Chuchow issued long waves and asked the Air Defense Headquarters to locate the said plane.

At 07:12 both planes reached Sanmen Wan where there were anchored several enemy battle ships which fired at the planes with AA guns. Fortunately no damage was done to them on account of long distance. At 07:42 it was reported that five minutes ago there was a roar of heavy bombers over Linhai and that the roar was toward southwest. At 08:48 the plane No. 1404 flew over Ningpo and landed on Yushan airfield. At 09:24 the plane No. 1403 flew over and through Linhai and landed on the Nanchang airfield. Both planes after refueling, took off respectively but met over the vicinity of Hankow at 11:13 and landed on the Hankow air base.

Thousands of Chinese troops were ferried by U.S. planes between training areas for fighting fronts from 1942 to 1945. Here a C-46 for the Air Transport Command, probably at Kunming appears to be loading up.

Soviet Aid To China

One phase of the history of aviation in China which should not be neglected is the role played by the Soviet Air Force and Soviet military equipment in the years from 1937 to 1939. This was three years before the First American Volunteer Group (Flying Tigers) arrived in China.

Two trade agreements were made between China and the USSR in 1937-38 by which China agreed to exchange metals and commodities for Soviet airplanes, artillery and tanks, probably involving a cash value of about 100 million gold dollars. But Stalin was anxious to help China at this time in order to divert Japanese attention away from Siberia, and he traded more than airplanes and weapons; he sent pilots to fly the airplanes and engage the Japanese in air combat, and he sent instructors to teach the Chinese pilots how to fly the Soviet E-15s and E-16s. The aircraft were trucked across the desert from Soviet railheads in central Asia and assembled in China at Lanchow and Pingliang, both in south Kansu Province.

Japanese authorities were aware of the Soviet intention to send aid to the Chinese Air Force. *Osaka Mainichi* reported that "nine high speed Soviet-made battle planes will soon be imported by Chinese military authorities in Shanghai. It is said that these Soviet-made machines had been assembled on August 19 at a certain place and would (be) brought to Shanghai via Lanchow and Hsian."[1]

The translation of this story from Japanese into English may have suffered somewhat, since it is doubtful whether nine airplanes could be assembled in one day. But the nine planes were, no doubt, an advance echelon.

The *New York Times* reported on 24 November 1937, that "100 Russian planes arrived in China."

F. Tillman Durdin *New York Times* correspondent, reported from Nanking that Soviet planes went into action "for the first time yesterday as part of the Chinese Air Force in a battle with Japanese airmen over Nanking at 1 p.m."[2] He did not say Soviet pilots were flying these aircraft, but they were.

Reuters news agency reported from Nanking on the same day that "aerial combats over Nanking yesterday afternoon marked the first appearance of Soviet planes (which were) ultra fast machines not hitherto used by the Chinese Air Force (and) Soviet instructors have been training Chinese pilots." *United Press* from Nanking on the same day reported that "roughly 20 Soviet planes, purchased by China and not donated, arrived recently at an aerodrome somewhere near Nanking (and) several of these planes particiapted in yesterday's defense of Nanking against the Japanese."

Chennault remembers an air battle that took place over Nanking, probably in early December, between Japanese and Soviet pilots:

> "Nine Japanese biplane fighters came over Nanking to drop surrender leaflets one afternoon. Five Russians and one Chinese pilot took them on. The Japs sent one Russian down in flames and the Chinese pilot scuttled for safety, leaving four Russians against nine Japs. The remaining Russians fought off the Japs for twenty minutes in a masterful exhibition of defensive tactics. Neither side claimed a victim and the main fight was broken off to watch an epic encounter between a Russian and a Jap who had separated from the melee and staged a duel I will never forget. The average air battle reaches a decision in a matter of seconds or at the most a few minutes, but this dogfight went on for thirty long, grueling minutes with the lives of both pilots hanging on every movement of their sticks. The Jap could turn inside the Russian, and they began with tail-chasing circles that brought the Jap closer and closer to firing position. With the Jap's first shots the Russian broke away in a dive and the Jap followed him down. In the middle of his screaming dive, the Russian did a snap roll that shook off the Jap and gave him a quick shot. They both ran out of ammunition and broke off.[3]

In an article written for *Collier's,* Elwyn Gibbon, a member of the first volunteers group in China known as the 14th Volunteer Bombardment Squadron, was on the same field at Hankow in 1938 when the Soviet contingent was there and he writes:

> The Russians have done a lot to help the Chinese keep ships in the air. They have sent bombers, and pilots—big, well-disciplined fellows who never get more than twenty feet from their planes all day and who eye us curiously when we play baseball to pass the time away.
>
> They have given the regular Chinese air force teeth too, by ferrying down a lot of Russian models of the Boeing P-16, tiny little fighters that have more power than the American job and are better than anything Japan has used in this war. The Chinese air force was almost demoralized when the new airplanes arrived. Half of

Polikarpov E-16 single seat fighter in foreground, and Douglas DC-2 at right, probably Hankow, 1937-38.

Right: Polikarpov E-16s. The first and second aircraft are E-16 UTIs, a two-seater fighter. Photo taken somewhere in China in 1937-38. Muddy field suggests Hankow. Below: A single seat, Polikarpov E-15 demonstration plane brought to China by Soviet air mission in 1937. The Soviet Union delivered ninety-three E-15bis, sixty-two E-16s, and sixty-two SB-2 bombers in 1937, in addition to replacements in 1938 and 1939. The P-7180 number on tail is Chinese Air Force serial number.

Soviet airman in front of two-engine Tupolev bomber. Airplane seems to be parked in revetment built up of hay or straw. Photo taken 1937-38. Location in China uncertain.

Madame Chiang Kai-shek's aviators had been killed in the early weeks of fighting because of inferior equipment, and the raw pilots who replaced them were convinced that they had no chance against the Japanese.

They accepted the new ships without enthusiasm. They were ready to die, but not to win. Then a squadron of Japanese pursuit ships attacked a Chinese formation of Russian planes. The Chinese looped—its the one maneuver a lot of them know—and found themselves on the tails of the Japanese ships. They shot down twelve of the fifteen Japanese and lost but two ships themselves.

They ran around the field like crazy men when they got back . . . .[4]

Another sidelight of the Soviet air units in China in 1937-39 is provided in the *U.S. Strategic Bombing Survey* which quotes a Japanese informant as follows:

Lanchow was an important point in communication between China and Russia. It was estimated that close to 100 planes (the majority of them made in Russia) were regularly stationed at airfields in the vicinity of this town. The main types of planes were the E-15 and E-16 and many Russians on air duty participated in combat. It was thought that the cutting or threatening of the so-called "Red Route" was of great importance not only in the destruction of the enemy air force in the area but also it would deal a telling blow strategically and politically to Chiang Kai-shek's regime in Chungking.

At the beginning of February [1938], Japanese air attack units were moved to an airfield in Shansi Province and prepared to attack Lanchow and Chengtu. On 12, 20, and 22 February the 12th, 60th and 98th Air Regiments (approximately 30 planes) cooperated in attacks against Lanchow. One hundred enemy planes, both in the air and on the ground, were reported damaged during these attacks. At this time it appeared as though the morale of the enemy air force was high and air battles were often carried on for a long period of time. As both the 12th and 58th Air Regiments suffered heavy casualties the attacks were suspended and the air units ordered to assemble in central China to prepare for future operations.[5]

The actual number of Soviet airmen who participated in the mission to China will probably never be known, nor are there figures on the number of casualties they suffered. There was a cemetery for the Soviet dead near Nanchang Air Base but the headstones were vandalized in the 1960s by the *Hung-wei-ping* (Red Guards) in the Proletarian Great Cultural Revolution.

Only fragmentary accounts have appeared in Western journals on the Soviet presence in China in the 1937-39 period, and thirty years after the fact, a General A. Blagoveschshenskii, who was a pilot in China, wrote an article for *Kraznaya Zvezda* (Red Star). Although brief, it may be the first official account of the Soviet involvement in China at that time. The article, as it appears below, is a translation made into English by an anonymous translator, probably on the staff of *Red Star,* and the title of the piece is "Thirty Years Ago and Today." A copy of the translation follows:

On 7 July 1937, the whole world heard of the Lukouchiao railway station, southwest of Peking. Here, during night maneuvers, Japanese troops attacked a bridge protected by a Chinese unit and the long-prepared "great war" against China began.

The hot-point of the war in the east could not remain unremarked by the Soviet people. The people of China are near and dear to us. The feeling for their struggle was as if in the blood of the Soviet man—an internationalist by his whole nature. It was natural that the Soviet government should then go willingly to the aid of China. In March and July 1938, the USSR placed two loans of 50,000,000 dollars for the purchase of military equipment. Among the credits, our country sent China airplanes, tanks, artillery and other weapons, fuel, ammunition and transport.

Soon after the beginning of the war in China there arrived

A Tupolev SB-2, a twin-engine bomber used by Soviet air mission to China in 1937-39 period.

Soviet military advisors and volunteer airmen. The volunteer fliers protected the Chinese cities of Kanchow, Chuntsin (?), Liuchow, Chengtu, Hsian, Hankow, Chungking, Lanchow and other places from the flight of Japanese bombers. They did not aid China by their personal participation alone; Chinese pilots came to the aerodromes for instruction in those wartime days. Our equipment and the Russian language were unknown to them, but this was no hinderance. The daily war flights and the day-long work on the aerodrome united everybody. They were welded into one endeavor to repulse the attack of the invaders.

A fighter *gruppa* [group] of over 120 aeroplanes was set up as early as the beginning of 1938. In it were included a *zvano* [flight] of Chinese pilots and *otdel'naya eskedril'ya* [partial squadron] of ten Chinese fliers from Canton. Soon afterwards there was established a bomber *gruppa* of SBs commanded by Timofey Khryukin.

The 1938 period of the war was characterized by the change from defense tactics to active offensive operations. The SBs of the bombardment *gruppa* attacked enemy transport proceeding on the Yangtze River. Under arduous conditions it carried out its blows against the invading forces, and gave great support to the army of Chu Teh. An SB *gruppa* under Fedor Polynin bombed an enemy base on Taiwan.

At the time Soviet fights inflicted heartening losses on the Japanese aviators in battles around Hanchung and Hankow. For example, a group of 36 enemy bombers was almost completely wiped out in one battle over Hankow. There was no air raid alarm for the inhabitants for a month after this. The whole evening and night after this memorable battle, delegations of young people, weeping with joy, came to the hotel where we lived, with greetings and gifts and to shake our hands. Many had themselves seen the battle, which took place at 5000 m. (meters) altitude quite near the city.

The treacherous actions of pro-Japanese elements also concerned us. Before one military flight at night, an attempt was made on the life of bomber commander Kikelinski. Fortunately, he was only slightly wounded. In June 1938 the author of this article fell, with his aeroplane, into a corner of a flooded rice field because a traitor's hand placed sugar in the petrol at Nanchang aerodrome.

In the city of Nanchang there is a cemetery where the remains of Soviet fliers lie. The oldest of them, Sergei Smirnov, was 27. His mother, wife, and little daughter were left in Moscow. There also were buried the Komsomol fliers Kostyn Zaboluev, Sasha Orakhov, Ivan Potapov, and others, whose relatives felt their loss deeply. But they knew that their Chinese friends would honor their memory, and guard and tend their graves, but recently it became known that "Red Guards" desecrated the graves of the Soviet fliers. Why was there such a sacrilege? And why, in Socialist China, has the foul wind of enmity and hatred been raised against the Soviet people?

The noble and self-denying aid of the Soviet people and our country is acknowledged in its time by those who now do all they can to eradicate its recollection from the memory of the Chinese people.

For example, at the VII session of the Chinese Communist Party Mao Tse-tung said, "After the war against the Japanese invaders began in 1937, the Soviet Union was once again first in coming to the aid of China in its struggle against the aggressors. The Chinese people give their thanks to the Soviet Government and nation for all this aid." Many leading Chinese communists then highly praised the role of the USSR in the repulse of the Japanese aggressors by the Chinese people. In one of his declarations, Mao Tse-tung stressed that it was the Soviet Union alone that aided China with military aviation and *materiel.* [sic]

In 1951 it fell to me, for the second time, to work out some complicated tasks with Chinese friends. I remember meeting a 39-year old general who began his fighting as a 16-year old lad in the army of Chu Teh. He very much wanted to study in Moscow. "Friendship for ever!" he cried. These words were on the lips of all Chinese comrades we knew. Recalling this, it is with bitterness that I think of the spread of anti-Soviet hysteria in China, and the adventurist course of the Mao Tse-tung group. Soviet people always think of the Chinese nation as a friend and a brother.

And we, the Soviet people, believe that a day will come when the present pernicious course of the Peking leadership will be changed. This will be completely in agreement with the interests of peaceful socialist friendship and the anti-imperialist struggle, and above all, in the interests of China herself.[6]

(End of translation.)

The author of the article says that the Soviets had 120 "fighter aircraft" in China. What he probably means is that there were 120 aircraft, fighters *and* bombers, since the word "fighter" can be construed as any aircraft engaged in combat. Chennault, in *Way of a Fighter,* says the Soviets had four fighter and two bomber squadrons "completely staffed and equipped to fight in China."[7] General Blagoveshchenskii calls them *gruppa* which, of course, can be translated as "Group," a designation for three or four squadrons in the U.S. Air Force. Thus, the figure of 120 Soviet aircraft in China seems fairly accurate.

Peter Mow, trained to fly in the Soviet Union, worked closely with the Russians, as well as Chennault, although the Russians took orders from no one but their own officers. The commander, according to Chennault, was a General Asanov who is not mentioned in the *Red Star* article. Chennault remembered meeting the Russians this way:

> Professionally we got on very well. Socially the Russians kept to themselves . . . . Off duty the Russians enjoyed themselves riotously . . . . On duty (they) enforced discipline. In contrast to the American custom of standing alert duty lounging or playing poker in an alert shack, the Russians sat stiffly all day long in their cramped cockpits. They parked their planes in a circle around the field. When the air-raid warning sounded, they took off madly in all directions. I never saw a Russian

Front and rear, single seat Polikarpov E-16 fighters, and in center, E-15-Bis, probably repair shop at Hankow or Lanchow, 1938.

scramble that failed to produce at least one collision . . . . Their combat behavior was unpredictable. Some days they just decided to not fight and scuttled off full throttle over the tree tops. When they decided to fight they did so with the team work and tenacity of ants, swarming over the Japs and overwhelming them with sheer determination. Like the Japanese, the Russians had an overfondness for rigid air discipline that often backfired when something happened to their leader.[8]

General Blagoveshchenskii says in one battle over Hankow, Soviet airmen knocked down thirty-six enemy bombers. Chennault, in *Way of a Fighter,* gives the same figure which is either a remarkable coincidence or some one in Moscow read the same book.

Saburo Sakai, one of Japan's top aces, was on the field at Hankow that day and in his ghost-written book, *Samurai*, he says there were 200 Japanese navy and army aircraft parked wing to wing which were destroyed or damaged . . . "the destruction wrought by only twelve enemy [Russian] planes was incredible . . . [9]

This was the last great strike by Soviet aircraft. The Soviet squadrons were using Polikarpov E-15 biplanes for fighters, an improved version of the Boeing P-12 with double power and four machine guns syncronized to fire through the props. The other fighter was the E-16, a short stubby monoplane, all engine which looked something like Jimmy Doolittle's "Gee Bee" racer. The Soviet bombers were twin engined Tupolev SB-2s and 3s, which were lightly armed and carried a poor payload of bombs.

Mention is made in the *Red Star* article to the fact that the Soviet squadron was the first to run a mission against Japanese targets on Formosa. The target was an airplane assembly plant at Taihoku, but there does not appear to be any record of damage assessment.

General Blagoveshchenskii describes how pleased he and his comrades were to aid Chu Teh, commander of the Red Army. This is revisionist history. No air support was ever given to the Red Army, often called the *pa-lu* (Eighth Route Army). The Soviets went to China to aid the Chinese Air Force of Chiang Kai-shek. Moreover, the airfields listed by the general were all under the control of Chiang Kai-shek, not the *pa-lu*.

After the loss of Hankow to the Japanese ground forces on 25-26 October 1938, an unspecified number of Soviet aircraft flew south to Hengyang and Nanchang to regroup. They had to disburse soon because the Japanese were after them.

It is uncertain whether Mao Tse-tung ever said that the Soviet Union was the only nation to come to the aid of China with military aviation and supplies. If he did he was no doubt referring to the period between 1937 and 1941, because Mao was fully aware of the American Volunteer Group and later the 14th Air Force. However, Mao took undeserved credit for driving the Japanese out of China and never in anything he ever said or wrote did he admit there was an American in China, although his own troops were often rescuing downed American air crews. Furthermore, an American liaison team, The Dixie Mission, was stationed at Yenan, the Red headquarters, the last year of war.

The Soviets turned over most of the aircraft they brought along to China to the Chinese Air Force before leaving in the late summer of 1939 after the German invasion of Poland. It is impossible at this point in time to determine how many of the airplanes the Chinese paid for, and how much the Soviets were charging the Chinese for the airplanes their own pilots flew. The Soviets did not *"come to the aid of China."* The original trade agreements were just that, one partner furnishing military aid in exchange for minerals, farm produce, pig bristles and tung oil.

In addition to airplanes, the Soviets brought tanks and artillery to China, and advisors and mechanics to teach the Chinese how to use these weapons. Albert Wong, a mechanic with the Chinese Air Force, recalls seeing Soviet tanks north of Hengyang in the summer of 1939. Artillery officers were attached to several war area headquarters of the Chinese army and it is generally conceded that Russian advisors, based on Yolushan Mountain, showed the Chinese how to use their artillery and drive the Japanese back in their first attempt to take Changhsa in 1939.

Most of the Soviet advisors were recalled early in 1944. Three advisors stationed in Changsha disappeared over the Christmas holidays in 1943.[10]

Soviet SB-2s parked on line at Hankow, 1938.

XII

Combat Over Wu-Han

Chinese Air Force pilots at Enshih walking from field to alert shack for debriefing. All wear Nationalist flag on flight suit. Enshih was also headquarters of Chinese Army 6th War Area. Photo by author taken early December 1943.

There are three large cities in central China on the Yangtze River called Hankow, Wuchang and Hanyang, the first city lying on the north bank of the river, and the latter on the south bank, but collectively known to the Chinese as "Wu-Han."

In the period under discussion, Hankow was the most strategically located air base in China; it lies about half way between the Great Wall on the north and Gulf of Tonkin on the south, and half way between Shanghai on the east and Chungking in the west.

After the fall of Nanking in December 1937, some of China's fighter squadrons, which had survived the air battles over Nanking and Shanghai, had moved up the Yangtze river to Hankow. Here they were joined by the newly-organized Soviet Russian squadrons flying mostly E-15 and E-16 fighters. In the text that follows, mention is made several times to the "Volunteer Squadron," and this refers to the Russian unit or *gruppa* (group), not to the 14th Volunteer Bombardment Squadron which, by comparison to the Russian presence, was minimal.

Most source materials agree that the 14th Volunteer Bombardment Squadron was disbanded on 22 March 1938, although it appears that a handful of pilots from the 14th remained in China after the unit was disbanded in March, but no longer an offensive outfit. The Russians were there and the fact that two of the big air battles described below took place *after* the 14th Volunteers were disbanded, suggests that the 14th did not participate. And there is little reason to doubt who the translator means in the script below by "some friendly planes." He refers to the Russians, not to the Americans and Europeans.

Chinese pilots, as the narrative below reveals, recognized the Japanese carrier fighter Type 96 (code name "Claude") as their most formidable adversary. To offset this advantage, the Chinese, who had lost many of their Hawk IIIs over Nanking, were now being checked out by the Russians in the E-15s and E-16s, both fighters, the former a biplane.

The battle of 31 May described below was probably the last in which Chinese pilots gave their utmost to stem the tide of Japanese air superiority. Many of the best Chinese pilots were dead or in a hospital by this time. What was left of the Chinese Air Force had to retire to the mountains of the west and camouflage the few airplanes it had left, never again to reach the peak of excellence achieved in the three air battles described here.

There were no more great air battles over Hankow after 31 May 1938 mainly because after that date Japanese long-range bombers made the fields at Wu-Han untenable. The bombers were probably the Mitsubishi G3M2 (code name "Nell"). Many of these were shot down by the Chinese and Russian pilots because the airplane lacked suitable defensive armament at the time. Nevertheless, these losses were made up from their own factories. The Chinese were wholly dependent on foreign replacements. Chinese courage could match Japanese courage but not Japanese science or aviation technology.

In the early days of Shanghai and Nanking fighting, Chinese fighter pilots were having a field day against the Japanese biplanes taking off from the carriers *Akaga* and *Hosho;* anchored in the mouth of the Yangtze River. It took only a few days for the Japanese high command to realize that it was being out-gunned and outfought in an obsolete airplane, and they sent at once for a new type, the A5M (code name "Claude"), Type 96, and that ended the edge which the Chinese had enjoyed for a few days.

The English text that follows is a translation made from the original document by Chinese translators:[1]

FEBRUARY 18th AIR COMBAT

Reports came in on 18 February 1938 at 1240 hours that a group of 26 enemy fighters escorting 12 bombers were advancing towards Wu-Han. Upon receipt of this information, 29 E-15s and E-16s, the total strength of the CAF 4th Group stationed at Wu-Han, led by Group Commander Captain Li Kwei-tan, took off to intercept the raiders. The following results were achieved:

Lt. Tung Ming-teh, the squadron leader of the 21st Squadron, took off from Hankow at 1310 hours with 10 E-16s led by him in V formation at a height of 3,500 meters. They spotted 10 odd enemy planes at northwest of Hankow in V formation at a height of about 4,000 meters. Air battles ensued with the result that 4 enemy fighters were brought down, one by Lt. Liu Chieh-sheng, the second by Lt. Tung Ming-teh, the squadron leader of the 21st Squadron, Lt. Yang Ku-fan, Lt. Liu Chieh-sheng and Lt. Liu Chung-wu, the deputy squadron leader fo the 23rd Squandron jointly; the third by Lt. Li Wen-chiang, Lt. Yang Wu-fan, Lt. Wang Teh-chien and Lt. Han Sheng jointly, and the fourth one by Lt. Wang Yuan-po, Lt. Kung Yeh-ti and Lt. Wang Teh-chien jointly.

The 22nd Squadron, led by Captain Li Kwei-tan, the group commander with 11 E-15s in V formation took off from Hankow airfield at 1300 hours. When they got to a height of 3,000 meters, they spotted 12 enemy planes on the southwest also in V formation about over 4,000 meters high. The first 6 enemy planes in the formation attacked the first 6 planes in our own formation and the other 6 enemy planes attacked on our flank. After the formation was broken off, individual dog-fight followed. As a result, Lt. Liu Tse-han, the squadron leader, shot down one enemy fighter but owing to engine trouble he bailed out and safely landed. The machine of Lt. Tseng Shao-yu, the deputy commander, was hit and was force-landed, but the pilot was unhurt.

In the dog fight with enemy fighters, Lt. Chang Kwang-ming, pilot officer's plane, was hit by 29 bullets but he returned to his base safely. Lt. Feng Ju-ho, pilot officer, shot down two enemy fighters while Lt. Chang Ming-sheng, pilot officer, shot down one. After having shot one enemy fighter, Lt. Wu Ting-cheng, discovered that one of our own fighters was attacked by enemy planes and he went to the help. Unfortunately the plane in front of him made a sudden turn. The collision crashed his plane, but he bailed out unhurt.

Captain Lu Tsi-shun, squadron leader of 23rd Squadron, led 8 E-15s and took off from Siaokan at 1245 hours for Hankow to meet the attack. They flew in V formation at a height of 3,000 meters. When they were on the approach of the Hankow airfield on the north, they spotted that 5 or 6 enemy fighters were fighting hot against several of our own planes of the 22nd Squadron. He joined in the fighting. As the result of this battle, Lt. Liu Chung-wu, the deputy squadron leader, and Lt. Hsin Shou-shun, pilot officer, both shot down one enemy fighter each. Lt. Wang Yu-kun (bottom of original page mutilated) . . . .

In this battle Captain Li Kwei-tan, the group commander, Captain Lu Tai-hsun, the squadron commander, Lt. Pa Tsing-chen, pilot officer, Lt. Wang I, pilot officer, and Lt. Li Peng-chiang, pilot officer, were killed and Lt. Wang Yu-kun, flighter leader, was wounded.

APRIL 29th AIR COMBAT

On 29 April 1938 reports came in one after the other that at 1310 hours 39 enemy planes were over Lukiang; at 1315 hours they were over Tungcheng and at 1400 hours they had passed Chishui, and one after the other our fighters took off. Our planes assembled in the designated air area according to the Wu-Han air defense plan and patrolled around waiting for the arrival of these enemy planes. At 1430 hours the enemy planes closed in and an air combat followed. The participants of this battle were the 3rd Group CAF, the 14th Group CAF, and the Volunteer Group [Soviets]. The following results were achieved:

Four E-15s of the 3rd Group, combined with the 17th Squadron of six E-15s in two flights of three each, led by Squadron Leader Lt. Chin Tsch-liu which left Siaokan at 1415 hours, were located at upper rear of the 4th Group at a height of 6,500 meters. They patrolled around, when they were south of Wu-Han, Squadron Leader Lt. Chin discovered that the enemy planes, after bombing Hankow, flew southweastward through Wuchang and then turned northeastward. Our planes pursued after them and when they were near Hwangkang, Lt. Chin's plane and another one piloted by Lt. Chiu Kow were directly over the enemy planes. They dived and fired at the enemy planes from their lower rear. Smoke was seen to have come out from one of the enemy planes on the right flank and the plane tended to fall. As enemy planes had now been far off, our planes returned. On their way back, they encountered three other enemy fighters. They pursued after them but they dived away. It was already 1600 hours, when our planes returned to their base and landed. As the result, Lt. Chu Kia-shun, deputy commander, brought down one enemy bomber which dropped southeast of Wuchang, and Lt. Mo Tah-yen, pilot officer, brought down another bomber which dropped in the vicinity of Hungshan.

On this day, the entire strength of the 4th Group, from the group commander down, was out on the mission. The group commander, Lt. Mao Ying-tzu, led 9 E-15s, with the 2nd flight on the right led by Lt. Liu Tze-han, the squadron leader of the 22nd Squadron, and the 3rd Flight on the left led by Lt. Liu Chung-wu, the squadron leader of the 23rd Squadron. When they climbed to the designated height, they patrolled in the designated air area. When they were over Tsingshan, they caught sight of the smoke from the burst AA gun shell and found that enemy bombers had closed in on Wuchang. They also spotted a group of friendly planes south of Wuchang and lest they would lose the chance to attack on the enemy planes, they immediately flew after them. All of a sudden, they discovered over 20 enemy fighters Type 96 over them, which dived and attacked on them. They returned fire but because of the inferior number of our planes, they were surrounded by enemy planes, over and under, right and left. After dog fights with a number of enemy planes, Lt. Mao Ying-tzu, the group commander, spinned down because of too much pressure applied to the control system, and the (word illegible) was not corrected until the plane was at 1000 meters (bottom of original page mutiliated) . . . effective firing range of Lt. Liu's plane, he gave it a good shooting. The enemy plane dropped down to a place northwest of the Tunghu Lake.

Lt. Liu Chung-wu, squadron leader, was also surrounded by three enemy planes, One of them was shot down when it was making a turn in front of Liu's plane. The other two attempted to escape when the first one was brought down but Lt. Liu flew after them. When they were driven to the south of Wuhu Lake, one of the two enemy planes turned around upon Lt. Liu's plane and this one was set afire and brought down, too. However, Lt. Liu's own plane was also hit by 70 odd bullets. On his way back, he spotted several more enemy planes but as his plane had been damaged and he had been flying at a low altitude, he decided to return to his base.

Lt. Yang Shang-hsien, flight leader of the 22nd Squadron, had a dogfight with a Japanese 96 over Wuchang. He was in the favorable position. When the enemy plane was driven to the west of Ocheng, it was brought down.

Lt. Hsin Shou-shun, pilot officer, was surrounded by four enemy planes south of Hankow. Smell and smoke from the plane made him spin down. When spinning was corrected, smoke became even heavier. He put off switches. At this time, he was again surrounded by those four enemy planes. He spinned again and after he corrected it, one enemy plane dived upon him. He climbed up and counter-attacked

Chinese Air Force P-40N being refueled by 10-gallon tins with aid of Chinese Army. Enshih air base, early December 1943.

his enemy. After his enemy was driven away, he force-landed. Through inspection he found that the No. 5 cylinder of his engine was damaged by enemy fire.

Lt. Tung-Ming-teh, deputy group commander, piloting an E-16, was supposed to join the Volunteer Group in the mission. But, after taking off, he found that the Volunteer Group planes had been far off to the south of Wuchang and so he took under his command the 24th Squadron and the monoplanes of the 4th Group, patrolling around the airfield. Lt. Tung also ordered Lt. Li Keh-yuan, squadron leader of the 24th Squadron, to take 5 E-16s and to go to the upmost air zone over the monoplanes. He noticed our AA gun fire over Hanyang and all the planes under his command rushed to the enemy bombers, soon participated by the Volunteer Group. One enemy bomber was hit and burst into fire over Liangtsechu Lake when attacked by both the Volunteer Group and Lt. Tung's planes. Another enemy bomber attempted to fly away at a low altitude, but it was followed by Lt. Tung's planes as well as some friendly planes. When it was driven to vicinity of Hwangkang, it was hit and set afire and finally it came down on the lakeside.

Lt. Li Keh-yuan, squadron leader, and Lt. Tung King-chiang, pilot officer, spotted an enemy bomber south of Wuchang after its bombs were released. They fired at the enemy plane from lower rear but failed to hit it. When they had hardly started a second run, two enemy fighters came upon Lt. Li's plane from the rear. Lt. Li managed to fly away from the enemy fighters and go after the enemy bomber until it was out of sight.

Having already seen that one biplane of ours was attacked (bottom of page mutilated) . . .

After taking off, the Volunteer Group planes went far away from Wuchang but soon they came back over Wuchang and participated in the fighting taking place between enemy planes and those of the 3rd and the 4th Groups. A total of 6 enemy bombers and 7 fighters were shot down in this battle. According to the report submitted by the Volunteer Group, when enemy bombers broke in over Wuchang and the fierce fighting just started, one of the Volunteer Group planes was hit and damaged by AA gun fire from Wuchang and came to ground in Wuchang. The pilot bailed out and got hurt. However, other planes, by taking shelter amidst the smoke from burst AA gun shells, could find out the position of enemy planes easily and give them immediate attacks.

According to the Wu-Han air defense plan, enemy planes could be intercepted and beaten off before breaking in over Wu-Han. On that day, enemy planes did break in through the patrolling area as designated in the plan, because it was the first air combat the Volunteer Group members had ever had and they were not familiar with the terrain of the Wu-Han area. For this reason they did not patrol in the area as designated in the plan and enemy planes were not intercepted and beaten off until after they had released their bombs.

As the result of the day's battle, 11 enemy fighters and 10 bombers, totalling 21 were brought down while we lost 12 planes including three repairable. Only ten of our pilots were wounded including Major Li Hwai-ming, pilot officer of the 23rd Squadron, who was reported missing and whose dead body was found floating on the Yangtze in the vicinity of Tsingshan about a month later.

MAY 31st AIR COMBAT

On 31 May 1938, reports were received that 18 enemy bombers were spotted over Tungcheng at 1051 hours; 36 bombers were spotted over Lukiang at 1145 horus and 46 enemy planes were spotted at Lotouou at 1145 hours. About noon, 39 enemy fighters appeared in the vicinity of Tsingshan.

We had 49 fighters stationed at Hankow including 4 E-15s of the 3rd Group, 8 E-15s and 6 E-16s of the 4th Group, and 21 E-15s and 10 E-16s of the Volunteer Group. On receipt of the above mentioned reports, all these planes took off at 1150 hours in V formation, patrolling in Hankow-Tsingshan area at different heights with a view to intercepting the invading planes. After taking off, they climbed up while circling around the airfield.

At 1207 hours, the first unit of our planes clashed with 18 enemy planes at an altitude of 4,500 meters. The enemy planes, all Type 96, came in V formation at three different altitudes. The topmost formation consisted of 9 planes at an altitude of 6,000 meters; the middle formation of 12 planes at an altitude of 5,000 meters and the lower formation of 18 planes at an altitude of 4,500 meters.

Immediately after the clash, 6 E-15s of ours came for reinforcement and at the same time the middle formation of 12 enemy planes at the altitude of 5,000 meters poured in. While the enemy planes were fighting against our E-15s (nine in number) and E-16s (? in number), their strength was further increased by airplanes from the topmost formation of nine planes (bottom of page mutilated) . . . .

The eight planes led by Lt. Cheng Siao-yu and four planes of the 3rd Group fell into formation over the airfield and after circling around the field once, they rushed to the dogfight which was just going on in the northeast between enemy and our own planes. However, as the fighting was some distance away, the enemy planes had run away when Lt. Cheng's planes got there.

As the result of this combat, Lt. Lo Ying-teh shot down one enemy plane on the south bank of Erh-hu Lake; Lt. Liu Chieh-sheng shot down another one which came down about 20 kilometers north of the airfield whose pilot bailed out and ran away; Lt. Han Sheng and Lt. Wang Yuan-po's planes fired at a third enemy plane with five bursts, which was burst into flames and landed in the vicinity of Tienhsin Island. The Volunteer Group brought down six enemy planes. Among them, one landed at Liehkow whose pilot was killed and another one was burst into flames in air and landed about 30 li from Liehkow, whose pilot bailed out. Because of loss of coverage, the enemy bombers circled over Changlinkan and Tuanfeng and did not get to their bombing objectives. They returned to their base together with their run-away fighters.

Lt. Chang Siao-hsien, flight leader of the 21st Squadron, was killed in this battle and his plane, an E-16, No. 2107, landed at Taihsiashan and crashed. The E-15 No. 90 of the Volunteer Group landed in the vicinity of Hengtien. The pilot safely bailed out but the plane was destroyed."

(End of translation)

There was another Chinese fighter outfit on the field at Hankow in 1938 which was flying British-made Gloster Gladiators. No mention is made to these aircraft in the narrative just quoted, but they could have been on the field after the last big air battle of 31 May, that is, in June-July 1938. A story in the *Chicago Tribune* reported that British manufacturers were building aircraft for "China's air service," and that the Chinese had ordered "the latest types of fighting planes," including a "batch of new Gladiators, single seat, 250 miles per hour fighter," at a cost of US $1,250,900.[2]

The same story said the aircraft would be shipped to China in the near future. Japanese agents were aware of their arrival in Hongkong. *Domei* from Tokyo reported that forty to fifty Glosters had arrived in Hongkong.[3] Actually there were thirty-six fixed gear biplanes, with 830 hp mercury radial engines. All were erected in Canton where four were promptly lost in taxiing accidents and several more on take-offs. The Gladiator was obsolete before it reached China.

In her book, *Lady and the Tigers,* Olga Greenlaw recalls seeing a "squadron" of these Glosters in Hengyang. One of the pilots was a Captain Clifford Louie, better known as "Long-legged Louie," an American of Cantonese descent born in Seattle. He was a student cadet at Shien Chiao in 1934 and learned to fly under Harvey Greenlaw. Mrs. Greenlaw was in Hengyang at this time and she remembers the story of the Glosters this way:

> Louis had come down from Hankow where his outfit, the Gypsy Squadron, a group of Cantonese fliers flying Glocester [sic] Gladiators, had been badly mauled attempting to keep Jap bombers away from the city. There had been one of the most spectacular air battles of the war, with seven of the Gypsies—all that were left of the original squadron, and thirty-five Russian E-15s and 16s against a hundred Japanese. Of the Gypsies, Louis alone escaped unscathed. The squadron had been temporarily disbanded for lack of equipment as the fifty Gladiators with which they started had been used up and no replacements were available . . . .[4]

Art Chen, another American-born pilot of Chinese descent, went to China to fly for the Chinese Air Force not long after war broke out in Shanghai in 1937. Chennault, in his *Way of a Fighter,* gives unstinted praise to this Oregon-born volunteer whom he credits with ramming a Japanese fighter and downing it over Hankow in 1938.[5] Forced to bail out, Chen was machine-gunned by the enemy on the way down but escaped death. Later, Chen himself was shot down, his airplane on fire, after an air battle over Nanning. Fearing that if he bailed out too soon he would land near the Japanese-controlled Nanning air base, he delayed his jump but at the cost of severe burns. He spent several months in a hospital recovering and was finally invalided home. After the war, he returned to China to fly for Central Air Transport Corporation, a Chinese air line, but with the Communist take-over of the mainland in 1949, he returned to Portland where he still lives.

There is little information available in English sources on the activities of the Chinese Air Force after Hankow. What was left of it, was dispersed to several bases in south and west China. A clue is given to activities of the CAF in late 1942 into 1943 in a document which is genuine but whose source is uncertain. The original document is not sharp enough for photographic reproduction and it is here retyped from the original. This report suggests that most missions by the CAF in this period were run against enemy reconnaisance planes and remote airfields. Most of the CAF aircraft listed here were now obsolete, even the P-40E which was probably red-lined by the 14th Air Force and repaired by the CAF. The retyped document follows:[6]

Above: These three P-47-Ds were first "Thunderbolts" to arrive in China, 20 April 1944, and are parked on the line at Kunming air base. Chinese officers (right) are anxious to see them because they were earmarked for Chinese Air Force. But Chinese pilots had not been checked out in P-47s and most of the early planes were lost in accidents. Later P-47s were absorbed into Chinese-American Composite Wing squadrons.

Below: Chinese Air Force P-4ONs on hardstand, Enshih air base, east of Chungking in Hupeh Province. One plane at right is taking off. This photo taken early December 1943. In latter part of 1944 base was occupied by a fighter squadron of C.A.C.W.

CHINESE AIR ATTACKS — BY DATE

Date	Cable	Chinese Planes Participating & Bombs	Description & Results
1942			
Oct 24		2 P-43's.	From Nancheng. Planes were in search of enemy recce planes, and shot down one.
Oct 25		6 P-66's.	From Peishiyi. Planes were in search of enemy recce planes. None seen.
Oct 26		6 P-66's.	From Pieshiyi. Planes were in search of enemy recce planes. None seen.
Oct 27		*4 P-66's.*	From Peishiyi. Planes were in search of enemy recce planes. None seen.
Oct 27		*9 A-29's. 12 P-43's.*	From Wenkiang & Taipingsu. Airdrome & military establishments at Yuncheng, Shansi Province, bombed. Airdrome & barracks damaged, 1 plane destroyed on ground
Nov 2		*4 A-29's.*	From Liangshan. Airfields and Custom House Wharf at Hankow

		bombed. The Yu Hua and Han Hua cotton mills at Wuchang and storehouses nearby and at Hsukiapeng fired. Over 600 enemy killed or wounded and barracks demolished at Yukiahu, Shanu, and Fengkinglu. 3 seaplanes destroyed, certain storehouses fired at Tansuidze, Hankow. 5 planes destroyed at Wong-Kia-Den. Great fire at Wharf. Jap AA shot down one of their own planes.
Nov 10	*4 P-66's.*	From Peishiyi. 4 P-66's. Planes were in search of enemy recce. None seen.
Nov 27	*6 A-29's.* *5 SSB III.*	From Liangshan. Enemy vessels and ferrying equipment bombed at Shasi, Shayang, & Tatukow. Airdrome & Storehouses at Shadi damaged. Hundreds of houses in city & number of military establishements demolished. 1 A-29 and 3 SB III's forced landed due bad weather
Jan 1/43	*3 P-40E's.*	From Kweilin. Planes were in search of enemy recce. None seen.
Jan 10	*5 P-40E's.* *10 P-43's.*	From Liangshan. Enemy airdrome at Kinmeng attacked, storehouses, barracks and 5 planes on ground destroyed. Several enemy trucks destroyed on Kiangling — Kinment — Tanyang highways.

The fragmentary "air report" noted above, suggests how limited this and other Chinese Air Force units were in their operations from 1942 to 1945. The Chinese-American Composite Wing had largely supplanted the Chinese Air Force. Nevertheless, order was still being observed, at least for administrative purposes by the creation of five "route air forces" in west China. An organization chart supplied by General Hsieh Mang at one point in time divided the CAF into five commands as follows: Chungking, 1st Route Air Force, commanded by Liu Mu-chuen; Kueilin, 2nd Route Air Force, commanded by Hsieh Mang; Chengtu, 3rd Route Air Force, commanded by Chou Chih-jou; Lanchow, 4th Route Air Force, commanded by Shih Kuang-fang, and Kunming, 5th Route Air Force, commanded by Yen Yu-chung. Chou Chih-jou was recognized as senior officer and commander of all CAF operations.[7]

After WWII ended in 1945, it was understood that the United States would help train and equip a full military organization for China in all branches of the service. The ground work for this was begun almost at once with the establishment, in Nanking, of an Army Advisory Group. Nanking had been the capital of China under the Nationalists before it was occupied in December 1937 by Japanese forces.

. Since the United States Air Force was at this time in process of becoming an independent branch of the military service and no longer the "Army Air Corps," an "Air Squadron" was set up as part of the Army Advisory Group in Nanking. This was headed by John Paul McConnel who also recruited personnel for the squadron. It was his responsibility to write mission papers as well as organization and function papers for the Group. As for the Chinese Air Force, the United States Air Force sent advisory teams of about ten men each to each tactical unit still operational, and advisors were also sent to a flying school to be re established at Shien Chiao near Hangchow.

William P. Hopson, the last to command the 308th Bomb Group of the 14th Air Force in China, was assigned in July 1946 to the Army Advisory Group. This is how he remembers the situation:[8]

"I returned to China in July 1946 to head the B-24 team at Shanghai. The U.S. had agreed to furnish the equipment for and to help train eight and one-third groups of aircraft consisting of one B-24 group, one B-25 group, four fighter groups, two air transport groups (C-46s), and one photo reconnaisance squadron. I didn't know where the aircraft came from unless from reserves in India. We flew our B-24s home after the war. The teams in Nanking, mine consisting of ten people, included operations, communications, armament, supply and maintenance. We were busy teaching tactics and all the support undertakings.

"The mission headed by General Marshall to patch up peace between the Nationalist and Communists in December 1945 had failed. When he returned to the United States, he became Secretary of State and issued an order to the advisory teams in China to no longer provide tactical advice to their Chinese counterparts although they could give logistical advice to help preserve the tremendous amount of equipment they had received.

"In March 1947, I left the bomb group assignment in Shanghai and went to Nanking as assistant chief of staff for operations of the Air Division. In this capacity I visited most all of the Chinese tactical units to form an opinion as to how well our advisory teams were carrying out their missions and how combat ready the Chinese units on the ground and in the air were. It was during these visits that Chinese officers in command, and in operations positions, told me what a heinous offense it would be in the eyes of the Chinese Air Force headquarters if any pilot lost an airplane, and how tired of facing danger their crews had become after the long years of fighting Japs and Communists. By this time the war between the Nationalists and Communists on the ground was in full bloom and judging from the results I think that the Nationalist ground forces were also tired and disenchanted.

"When the war was going badly for the Nationalists, Marshall relented a little, agreeing that limited combat advice could be given to the Chinese Supreme Command. There were officers designated for meeting with the Supreme command once a week. I was the air member. It was not very successful.

"It was my impression that most of the Nationalists, both army and air force, no longer had the heart to continue the civil war, while the Communists did. So the Communists won the war and our U.S. advisors were forced down to a single echelon headquartered in Shanghai. I left 16 January 1949.

"As to turning over our aircraft, the Nationalists had units designated to accept all aircraft given to them and they had a few pilots, and were training more. Most of these units flew to Taiwan. As far as I know the only operational aircraft the Nationalists had at the end of the war were the Chinese-American Composite Wing. The strafing of Communist troops by planes of the wing, according to their own squadron commander, was not done below 5000 feet."

XIII

The Bombing of Chungking

In late summer of 1938, it was clear that Japanese forces wanted to occupy the city of Hankow in central China more than any other city in China. Their advance was slowed down when Chinese forces blew up dykes along the Yellow River and caused vast floods, and Japanese ground forces had to call on their air force to drop supplies or evacuate stranded troops. But the campaign went on, doggedly, and by mid-October the city was occupied, and a few days later, the city of Ichang, farther up the Yangtze River.

With Hankow air base under control, Japanese airmen could easily reach Chungking with their bombers, and if necessary, with their fighter escort. But the latter were unneeded because by this time the Chinese Air Force could offer only token resistance.

However, long before Hankow was occupied, Japanese long-range bombers were probing the Chungking Air Space, and one of the first missions against the city was run on 5 May 1938 when a flight of six or more aircraft probably flew from Nanking or even Formosa to attack the city.

Although Chungking is usually covered by a heavy fog in the winter months, the weather front sometimes breaks, and this happened in January 1939 when Japanese airmen took advantage of it to hit the city, again without escort. The *New York Times* described the raid as "the worst bombing the new China capital had yet experienced . . . .[1]

Twenty-seven bombers flew over the city in flights of three, boxed in fairly closely, still uncertain of Chinese interceptors. Only one Chinese pursuit plane challenged them and allegedly shot down one bomber, but this was a poor percentage considering the number of enemy aircraft that went on bombing the city, unmolested. Chinese anti-aircraft gunners kept the incoming aircraft above 15,000 feet, and the Japanese airmen quickly learned that at 16,000 feet they were relatively safe. The gunners could not reach them.

After the January raid, the weather closed in for the winter and it was probably not before 3 May 1939 that the next raid hit the city. In fact, the second major attack followed on 4 May, with twenty-seven bombers participating in almost identical formations as those on 15 January. Although heavier air raids followed, these two in eary May were the most devastating because the city was unprepared for them. Not less than 10,000 people were killed and wounded in the two days. Two years later, on the anniversary of the first May day raids, the Japanese bombed the city and only five people died. Shelters had been built into the rocky hillsides and a comprehensive system of air raid drills were enforced by air wardens, police and soldiers.

Carl Crow, an American who ran an advertising agency in Shanghai in the 1930s, had moved to Chungking and witnessed the raids of 3 and 4 May 1939. He later described what he saw:

> The sounds of the anti-aircraft guns and the exploding bombs now became so incessant that I could not distinguish one from the other. What added to the confusion but blended all the sounds was the fact that we were enclosed in high hills on all sides, so that a single rifle shot echoed and re-echoed in fine imitation of the rattle of the machine guns . . . . Several hundred people who were in Central Park were blown to bits. They evidently thought no bombs would be dropped there as it was an open spot with nothing but flowers and trees. There were also heavy casualties among the refugees from previous air raids who were being cared for by the French Catholic priests . . . .[2]

In the spring, summer and fall of 1940, Chungking shared sorrows with the city of London as the most bombed city in the world. There was no heavy industry here, only small factories and cottage industries. There was not even a railroad station. There was a small arsenal making rifles, mortars and ammunition out in the hills but not in the city. There was no room for an airplane to land here, but a short strip had been built on an island of sand, actually a shifting sand bar, along the north bank of the Yangtze River directly below the cliffs upstream from the city. Called Chiulungpo (Old Dragon Bar), it was barely long enough to land a DC-3 but airmen could never depend on it because it was often under water. The main airbase at Peishiyi lay north of Chiulungpo, or west of Chungking across the mountains.

From a primitive system of air raid precautions, Chungking developed a highly efficient Air Raid Warning Net. Portable radio transmitters and telephone operators kept watch on incoming enemy aircraft, and when word reached Chungking that enemy bombers had taken off from Hankow, a "one ball" alert was raised. The ball was a round wicker basket mounted on a pole where people could see it. When the incoming aircraft were half way, a "two ball" alert went up, which meant that pedestrians and householders must hurry to reach their assigned air raid shelter or cave. A few minutes before engine noise was heard, a "three ball" alert went up and that was the final warning. By that time the streets of the city were cleared of civilians and everyone was underground.

In 1940 the Joint Board of Air Raid Relief had 7,800 members working in 260 units, all volunteers without pay. This new board followed a reorganization of the A.R.R. after the disastrous raids of 3 and 4 May 1939. When the bombing "season" approached in 1940, the Chinese were ready. It was called the "bombing season" because shortly after the first of May, the winter fog lifts over Chungking and stays away most of the time until October when it seeps back, cold and clammy.

More than 100 air raid wardens were killed in the line of duty but the organization never lacked for volunteers. It was a badge of courage to belong. And like London, Chungking had its own "suicide squads" who went about after every raid digging out duds and defusing them. In the summer of 1940 forty-one demolition and eighteen incendiary bombs were found, one buried thirty feet into the ground.

Life for the more than one million people of Chungking was difficult even in normal times in the war, but with the threat of constant air raids, it became almost unbearable. The Chinese population could do nothing to divert these attacks. Lacking any power to strike back, they could only wait in their shelters for the blows to fall, praying that their

own homes would be spared.

Despite the lack of military targets in the city, the Japanese sought to destroy it, and had they used incendiaries more than demolitions, or all-purpose bombs, they might have succeeded. But in one of the strategic blunders of the war, they failed to realize this. It was also a blessing for China.

From the evidence of testimony heard in interviews with former Japanese airmen, it seems that they rationalized their terror bombing of Chungking on the assumption that it would end the war, but behind this outward symbol of good will lurked the original intention of the Imperial General Headquarters to "punish China" for refusing to cooperate in the "Greater East Asia Co-Prosperity Sphere," and for refusing to recognize Japanese claims to Manchuria in the newly-created state of "Manchukuo."

Hollington K. Tong, former editor of the *China Press* (Shanghai) and later director of the Ministry of Information in Chungking, lived through the air siege of the city, and admitted after the war that many Chinese leaders wondered how much longer they could tolerate the bombings which were almost too much for the human condition to bear.[3] But Chiang Kai-shek refused to surrender despite offers from the Japanese to "make a deal" and end the war.

Mme. Chiang Kai-shek inspects damage after Japanese air raid on Chungking, China's war time capital in World War II. Women accompanying her are probably members of relief organizations. Photo taken c. 1942.

XIV

The 14th Volunteer Bombardment Squadron

Unidentified Americans studying application forms to join 14th Volunteer Bombardment Squadron. Many were called but few were chosen.

When the war began between Japan and China in July 1937, Japanese air power suddenly become a major factor in the success of its military operations in China. It was not long before there were rumors of an international squadron of foreign pilots which was being organized to bolster the fledgling Chinese Air Force in its fight against Japan. *Domei,* the semi-official news agency in Japan, carried a story from Tokyo about a protest that the Japanese government was going to make to the U.S. State Department about plans for "crack American aviators" to join the Chinese Air Force. The story said that Russell Hearn, a former advisor to Chang Tso-lin, was planning to recruit 182 American pilots "for action in China."[1]

It is not entirely certain whether the United States Secretary of State, Cordell Hull, saw the story, nevertheless, he told reporters that the law against the enlistment of Americans for foreign armies would be applied to any situation, and reminded all prospective pilots that the penalty for disobeying the law was a $2000 fine and up to three years in prison. The law was never enforced, but the Japanese foreign minister, Koki Hirota, assured the Diet in Tokyo that the U.S. was guarding against enlistment of American aviators in the Chinese Air Force.

Despite these disclaimers, a few weeks after the *Domei* story appeared, a group of foreign pilots was being organized to fight for China, although instead of 182 "crack American aviators," there were only four or five, none of whom had ever had a crack at anything. The others, numbering half a dozen or so, were Europeans who did not have any reason to fear the extension of U.S. law to China.

This first volunteer squadron of foreign pilots was variously known as the 14th International Volunteer Squadron or as the 14th Volunteer Bombardment Squadron. It was organized in the late fall of 1937, and had scarcely run a mission before it had to move up the Yangtze River to Hankow to escape the wrath of Japanese carrier fighters.

The idea of using foreign pilots to fight a war in another land was not entirely new to aviation. The Civil War in Spain of 1936-37 attracted a number of foreign mercenaries to a Soviet-sponsored volunteer air force engaged in fighting against the Spanish leader, Franco, although Spain was also a testing ground for the latest tactics in aviation and aircraft not only by the Soviets, but the Germans and Italians.

To recruit pilots for a foreign squadron in China, the Chinese government ordered its consular representatives around the world to look for applicants, discreetly, because it would not be good publicity to offend pacifists in the United States. As a result, a number of applicants were accepted and shipped off to China by sea, although Julius Barr and Royal Leonard, who had been flying in China for several years for commercial airlines or as personal pilots to Chiang Kai-shek, managed to enlist several pilots, including two from the United States and one from Manila.

Claire L. Chennault, an aviation advisor to Chiang Kai-shek, was assigned by the Chinese to take command of the volunteer squadron although he was never entirely sure of his own status; in a letter he wrote to one of the pilots after the squadron was deactivated 22 March 1938, he signed himself "aviation technical advisor," not commanding officer, and in none of of his letters or official correspondence does he use any title or prefix of rank, usually signing his initials only.

The squadron was more of a token force than a threat to the Japanese. After the misfits who drifted in from all parts of the world had been weeded out, there were only a dozen or

A. Boulingre of France, member of 14th Volunteer Bombardment Squadron probably at Hankow, 1938.

Art Chen, Chinese-American from Portland, Oregon, went to China in 1937 to fly for the Chinese Air Force. In air combat over Hankow in 1938 he made ace, but was twice shot down, the last time in flames and was in hospital for many months.

so qualified combat pilots in the entire organization.

Chennault arranged to have Sebie Smith and Rolfe Watson transferred from Shien Chiao to serve with the squadron, Smith as a liaison officer in charge of maintenance, and Watson as armorer. Dudley Long, a mechanic, was also attached to the squadron, in addition to two so-called gunners. Tommy Allen and Herbert Walker. Allen, like some others, arrived in China without a pilot's license and when Billy MacDonald checked him out, Allen confessed he had never flown an airplane before. But Nanking was falling. Everyone was moving up the river to Hankow, and Chennault asked Sebie Smith to bring Allen along and make a gunner out of him.

Pilots who came from Europe and Indo-China to join the squadron included four Frenchmen, William La Bussiere, Marcel Floueir, W. La Roce, A. Boulingre and Jan Roeffear, a Belgian. La Bussiere had fought in a Russian squadron in Spain in 1937. La Roche cracked up three times in China and walked away without serious injury and decided to return to France.

One of the French pilots died on his first attempt to fly. Sebie Smith remembers how it happened.

> He came in with full flaps down, landing gear down, the carburetor heat on and prop in high pitch. He almost fell on me trying to go around and after overshooting the field, he stalled out winged over the edge of the field, and spun in.[2]

Chennault made Vincent Schmidt, a native of Mineola, New York, and veteran of World War I, squadron commander. He had also flown in Spain.

In addition to Schmidt there were four other Americans, namely, Elwyn H. Gibbon, George H. Weigle, Lyman Woelpel, and James W.M. Alison, the last two being commercial pilots from Seattle. Gibbon was the only army-trained pilot, and he brought George Weigle with him from the Philippines, a man described by Sebie Smith as a "good cow pasture pilot" but with no previous time in fighters.

A Chinese-American, Le-ong Lee, from California, may have flown with the squadron. But there were others and though little is known about them, they are mentioned in a letter by Julius Barr to Major King Kia-S'ze, probably base commander at Hankow airport.[3]

Barr had been flight-checking the recruits and he tells Major King (with a copy to Chennault) that he had just checked out nine pilots in the Douglas DC-2 and this the way he judged them.

> James Henrickson can fly, but not very well. He is not qualified to start right out in a plane but will need some dual instruction. He has had no military training and so is of little value as a pilot.
>
> Martial Laroche (sic) passed all right on the Douglas, but I have recommended that he fly only medium weight planes or pursuit.
>
> Marcel Florein (sic) is better than LaRoche and with a little dual instruction could be used for a bomber pilot. I have not inspected the qualifications of these two men because that was done in Nanking and they were authorized a check flight.
>
> R.G. Whitehead passed all right. With him also I do not know his previous experience, for he was examined at Nanking, but he flew the Douglas O.K. and I'm sure he will be all right for pursuit work. Perhaps with a few hours dual in multimotors he will be all right for a bomber.
>
> Elwyn Gibbon is an excellent pilot. A valuable man.
>
> George Weigle is also a very good pilot.
>
> Archie Randolph can fly but not very well. He has no military training, so should not have come to China.
>
> Harry Stokes is a good pilot, but he has had no military training. His ability is good enough that I think he can be of some use to Nanking with a little training.
>
> Harold Welch (co-pilot on the Sikorsky) is a very good

A French member of 14th Volunteer Bombardment Squadron, unidentified, probably at Hankow, 1938. Shield on cap is upraised wings of Phoenix, logo of Chinese Air Force.

Martin 139 bomber (Martin B-10) waiting for armorers to load bomb seen in foreground. White uniforms are Chinese Navy. Photo probably taken at Shien Chiao in 1937.

Martin 139 Bomber (export model of B-10 bomber) of the 14th Volunteer Squadron, No. 1403.

Vultee V-II attack bomber taking off from wet field at Hankow air base, winter of 1937-38.

Pilot Elwyn H. Gibbon before his discharge from U.S. Army Air Corps.

A wrecked Vultee being taken to Central Aircraft Manufacturing Company repair shop in Hankow, winter of 1938. Mervyn ("Murf") Garrold inspects damage.

pilot as well as a mechanic. With a little training he can fly any plane as well as he did the Douglas. However, he has had no military training, so Nanking may not use him right now.

According to my recent instructions from Nanking it seems that Henrickson, Randolph and Stokes should be rejected and sent back. Mr. Vincent Schmidt is trying to get them kept here as co-pilots and gunners in his foreign squadron. He will talk to you about that at an early date.

It seems clear that Barr was seriously trying to weed out the misfits. Since Henrickson, Randolph and Stokes are not heard of again, they were probably some of the people who were returned to the United States.

Gibbon checked out James Allison in a Northrop 2E on 21 February. He said Allison flew well but paid little attention to instructions. He also embarrassed the entire squadron by his escapades and brawls with the Chinese.

Gibbon was apparently put on the payroll on 3 December 1937 in Nanking. He was hired for $1000 a month, that is, Chinese dollars which, at roughly three to one at the time, would mean that he was making little more than $300 a month. He probably got food and quarters. He flew from Nanking to Hankow on 6 December, where he assisted Barr in the last flight checks.

When Barr wrote his report to Major King, the struggle for Nanking was not over, and that is why he uses the expression "of some use to Nanking," etc.

Aviation Technical Advisor Chennault did not have a staff trained to debrief his crews. As a result most reports of operations come from personal letters and files kept by the pilots. One of these is preserved in a collection held by the San Diego Aero Space Museum. From this it is learned that on 27 February 1938 Gibbon was on a mission to the Yellow River, with two elements of Vultees No. 29 and 32, and Northrop 1411 with Vultees 1421, 1426, and No. 30 in the second element. Pilots assigned to the mission were Gibbon, Boulingre, LeBrussiere, Weigle and Leon, but Gibbon fails to include the name of the sixth pilot while the name Leon is also not identified. It could be a nickname. This is the way Gibbon wrote his own mission report on the basis of the 27 February mission:

Crossed Yellow River at Yin She (east of Loyang at 3:03. Began bombing small village containing troops 5 miles S.W. of Bee Yang at 3:08 from 1500 feet. Reached Bee Yang at 3:12 and found an estimated 75 to 100 convoy trucks and a large force of troops quartered there. About 45 bombs were dropped in Bee Yang and numerous fires were started. Recrossed Yellow River east of Gee Siu and saw approximately 15 small boats at that point and from indications of the ground personnel it was the enemy attempting to cross the river. The leading element circled the spot allowing the machine-gunners to administer a severe strafing which undoubtedly resulted in many casualties. All ships returned to Hankow without damage to planes or personnel.[4]

Lacking knowledge of the proper spelling for Chinese place names, Gibbon spelled them the way they sounded. It seems that the raid was a success.

On 23 February 1938 several Soviet SB-2s bombed an aircraft plant at Taihoku in northern Formosa (Taiwan). Chennault says he was along briefing this mission. He fails to explain any details although it is generally accepted as a fact that Soviet pilots were flying the airplanes, probably with a Chinese co-pilot and mixed Russian and Chinese crews.

The raid on Formosa caused friction between Chinese Air Force pilots and European and American pilots of the 14th Squadron. To the Chinese it was one more example of the arrogance often displayed by the majority of foreign pilots, the most outspoken of whom was the squadron leader, Vincent Schmidt.

Sebie Smith, who was on the Hankow field at the time, recalls the incident on the Formosa raid. Said he: "Foreign pilots got a lot of credit in newspaper articles about having bombed Formosa (but) that was just a bunch of hokum. They never did go to Formosa."[5] Smith also remembers Vincent Smith: "In spite of his love for fighting wars, he was kind of an excitable type. He came to the field one day and looked up and saw a bunch of wild geese and he thought it was a Japanese formation, and he got very excited until finally he was convinced it was a flock of geese."

Chennault may not have held the 14th Squadron in very high esteem. With the exception of Gibbon, most of the others, at least the Americans, were former commercial pilots and, to an old army man like Chennault, they were civilians. In the letter to Gibbon quoted below, dated 30 April, the "friends" to whom Chennault refers, are Soviet Russian pilots who were still flying in China after the 14th Squadron broke up. This is what Chennault told Gibbon:

We had a bigger dogfight yesterday than on February 18. The Japs finally fell into the trap we've been baiting so long. We got 8 bombers and 13 pursuits, according to a check this a.m. Lost nine (planes) of which 2 jumped safely and 2 others had forced landings away from the field. About 30 of our "friends" from Nanchang were *visiting* up here yesterday and they really poured it on the Japs. We had a total of about 60 pursuit (in the air) against 12 (Japanese) bombers and 25 pursuit.[6]

As to the condition of the field at Hankow that winter, Sebie Smith has this to say:

The one runway at Hankow was a little over 10,000 feet, a mixture of clay and gravel and sand . . . quite wide, especially the shoulders which had enough gravel or hardpack to taxi the airplane up on it and not bog down. As a result we parked the airplanes all along the side of it, facing the runway, so that they wouldn't bog down in the mush out on the field. We had a lot of snow that seemed to stay there. It was cold and disagreeable, with water several inches deep. We would have to taxi the airplane all through that mush.[7]

The entire squadron was often grounded by bad weather, and on many days only part of the squadron was up to strength owing to lack of parts. Without hangars to park the aircraft, the engines were hard to start in the morning. Smith devised a method of keeping them pre-heated with small stoves and blanket hoods over the cowling.

Pilot Gibbon probably had more flight time than any other member of the organization and after the unit dispersed he wrote to Chennault and asked for confirmation of his flight time.

Chennault replied:

Vultee	250 hours
Hawk	25 hours
Curtiss CR	25 hours
Northrup	15 hours
Martin	350 hours
Fleet	15 hours
Douglas	5 hours
	685 hours

In a private letter sent to a friend in Canada by La Bussiere, who is probably the only surviving member of the 14th Squadron, he says that in the five months the 14th was in action, he flew 116 sorties, including twenty-eight bombing missions and fifteen night missions.[8]

Most of the targets these pilots were attacking were in north China, and a mission from Hankow could be 80 to 1500 miles round trip. This could mean five to nine hours in the air at speeds of 160 to 210 mph. No fighter escort was provided. It seems the squadron was lucky in this respect. Most of the damage it suffered was from Japanese fighters over Hankow, or from bad landings on a treacherous field.

The big dogfight mentioned earlier by Chennault in the letter to Gibbon took place over Hankow more than a month after the 14th Squadron was deactivated. Elements of the Chinese Air Force and the Soviet squadrons were given credit for the success of this fight. But there does not seem to be any evidence that any Japanese fighter or bomber was shot down by any member of the 14th Squadron. However, it should be borne in mind that the volunteers were flying mostly bombers, not pursuits.

Gibbon on his return to the States, later drifted from one flying job to another and ended up in India in 1942 where he was a test pilot for an aircraft company and was killed in a crash.

Gibbon's friend and wing man, George Weigle, remained in China and found employment as a test pilot with the Pawley organization (CAMCO).

Robert A. Fausel, a Curtiss airplane division test pilot, brought two deomonstration planes, a Hawk 75Q and a CW-21 interceptor, to Burma, for the Chinese Air Force to consider. The planes were shipped in crates to Rangoon and from there they were carried by steamboat up the Irrawaddy River to Bhamo where Fausel was met by Charles Hunter, factory foreman for CAMCO in Loiwing. The crates were trucked to Loiwing and assembled under bamboo mat sheds with bamboo hoists supporting block & tackle.

Fausel test flew the Hawk on 11 February 1939 and two days later flew the ship to Kunming, and on 17 February to Chengtu. He later went back to Loiwing for the CW-21 interceptor.

In Chengtu, Rolfe Watson, an armament specialist, made the modifications necessary for the installation of two Danish-manufactured Madsen cannon under the wings, and in his diary, dated 6 March, Watson notes that he "fired cannons of Hawk 75 first time in history . . . "

On 19 March, Weigle, flying the Hawk and Fausel the CW-21, followed Royal Leonard in a DC-2 to Chungking with Pawley, Arch McEwen, Bruce Leighton, a Pawley vice-president, and two Madsen technicians from Denmark as passengers. Weigle later flew the Hawk back to Chengtu. On 4 May Fausel called Weigle on the phone and told him to bring the Hawk back to Chungking, unaware that the Japanese were planning a big air raid on the same day. Weigle's instructions were to land at Chiulungpo, the sandbar strip lying on the north bank of the Yangtze River below the cliffs of Chungking, but when he arrived over the city, the smoke-dust from the bombing raid was so great it completely obscured the air strip at Chiulungpo, and he flew on to the nearest field at Kwangyuenpa, a Chinese fighter base about fifteen miles down river. The next morning he was preparing to return to Chiulungpo and while attempting too steep a climb, he crashed on take-off and was killed.

Fausel flew down in the CW-21 to inspect the wreckage. The Madsen cannon had not been fired, contrary to an assumption by Chennault that Weigle "shot down four Japanese bombers" (with the Madsen cannon).[10]

The U.S. Navy gunboat, *Tutuila,* stationed at Chungking, went down to Kwangyuenpa and brought back Weigle's body. Fausel flew the CW-21 back to Chiulungpo and helped Leighton arrange for the funeral services. Internment followed at the International Cemetery in Chungking.[11]

George Weigle, former member of the 14th Volunteer Bombardment Squadron, stands behind wing of Hawk 75Q, probably last photo taken before his death in crash on 5 May 1939. He was flying as test pilot for Pawley organization (CAMCO).

George Weigle and Elwyn H. ("the Mad Irishman") Gibbon, probably in Hankow, winter of 1938. Weigle wears rescue patch issued to foreign pilots by the Aero Commission to alert Chinese civilians to give them aid and to protect them in the event they were forced down. There is no reference to any reward on this patch. This photo appears to be the first evidence of patch worn by foreign airmen. The wording was revised later to make clear that the airman was an American.

XV

The First American Volunteer Group (Flying Tigers)

Claire L. Chennault had been in China nearly three years when Generalissimo Chaing Kai-shek asked him to explore the possibility of recruiting a group of American airmen to come to China to fight. The rewards would be good pay, glory, or death. Of course there were no real fighters pilots in the U.S. Army Air Corps yet because there had been no war to make them into fighters. But there were pilots who had been trained in simulated exercises to engage a would-be foe either from land bases or from flat tops.

The Generalissimo was fully aware that Chennault had some experience in handling foreign pilots and though the few months Chennault had commanded the 14th Volunteer Squadron at Hankow were not necessarily auspicious, neither had he much material to work with. But the willingness of foreigners to come to China to fight for pay, albeit only about twelve of them, gave the Generalissimo reason to believe that others might be willing to come, and this time, they would be all Americans.

Squadron Leader Arvid Olson confers with Claire L. Chennault who wears old blouse with Sam Brown belt and "overseas" cap. Olson wears rescue patch which was successor to patch used in 1938 by 14th Volunteer Bombardment Squadron" in Hankow. Patch seen here was issued by Aero Commission and has the word "American" on it which the earlier patch did not. Photo probably at Toungoo.

Chennault, the man the Gimo chose to command this miniature air force, had already witnessed numerous Japanese air strikes on Nanking, Hankow and Chungking. He had watched dog fights between Chinese and Japanese pilots over Nanking, and between Russian and Japanese pilots over Hankow and, having written a treatise on *The Role of Defensive Pursuit* a few years earlier while at Maxwell Field, he was constantly comparing what he saw in China with what he believed to be the best in pursuit tactics. He was in the right place at the right time with the expertise to beat the Japanese. He was at the moment probably the most qualified airman in the Army Air Force (retired), east or west of the International Date Line.

After Hankow fell in October 1938, Chennault, together with his aide and interpreter, P.Y.Shu, and Wang the chauffeur, drove southwest to an unfinished airfield called Chihkiang in west Hunan Province, and here he spent a couple of months trying to teach a few cadets how to fly a Hawk III, of which there were only three on the field. All were lost in crashes and there were no replacements. Chennault was then ordered to Kunming where he arrived on New Year's Eve in time to enjoy a party put on by friends and flight instructors, including Billy MacDonald, at the Kunming aviation school.[1]

Kunming Air Base, lying 6000 feet above sea level, is not the best place for a flying school, but it has one of the most healthy climates in China, a landscape made beautiful by tall eucalyptus trees which line the roadways and canals, and a sun which shines most of the year. The city north of the airfield, Kunming, is the capital of Yunnan Province, a province which marches west through the mountains to the east line of Burma.

Although there had been a flying school in Kunming for more than a decade, it had never played a significant role in aviation. But after war broke out between Japan and China in July 1937, it became an important training center for air cadets. The government realized after the Aviation School at Shien Chiao came under attack that it had to make every effort to keep the training program going elsewhere, and ordered the student cadets and instructors to move to Kunming. Thus when Aviation Technical Advisor Chennault—now up-graded to "Colonel" in the Chinese Air Force—arrived in Kunming, the school was in the midst of a training program directed by Wang Shu-ming ("Tiger Wang"), assisted by Billy MacDonald and several instructors, mostly Americans.

Kunming is connected with the port city of Hanoi in northern Vietnam by a narrow gauge railway. After the Japanese occupation of Nanking in 1937, the easiest route to Kunming for any traveler or business executive, was via Hanoi.

Panel of P-40B with last minute instructions for modifications before packing at Curtiss plant, dated 10 March 1941, serial number 14033.

The Burma Road which ran to the west of Kunming into the mountains was completed up to a railhead at Lashio by 1938, but it was still easier to reach Kunming via Hanoi. The rail route continued to be used until the summer of 1940 when Japanese forces occupied Hanoi and what was then French Indo-China.

With the closing of Hanoi to traffic, China became dependent entirely on the Burma Road. Supplies were offloaded from ocean vessels in the port of Rangoon in lower Burma and from Rangoon shipped north by railroad to Lashio and from their by truck over the mountains into China. Thus, almost overnight, Rangoon became a key link in the supply line to China.

Meanwhile, in 1939, Japanese aircraft began to bomb Kunming and other cities in western Yunnan in an attempt to halt the flow of supplies over the Burma Road. At this time there were no observation posts or telephone hookups to warn the base commander in Kunming of incoming aircraft. Thus, when John M. Williams, a communications expert who had served under Chennault at Maxwell Field, arrived in Kunming in 1939, Chennault at once ordered him to set up an air raid warning net covering the entire province of Yunnan.[2] The reason for the net was clear in Chennault's mind because he had written in *The Role of Defensive Pursuit* about the need for such a net in the United States, and now he had a chance to apply his theory to China, a nation actually under enemy attack.[3]

It took several months to organize the net but eventually it became an effective weapon in the hands of the Chinese spotters to warn the American squadrons of incoming aircraft, usually described on the telephone as "heavy engine noise" whether it came from a lone Japanese reconnaissance plane at 23,000 feet or a flight of twin-engine bombers at 10,000 feet.

Chennault spent several months with the cadet school in Kunming and then in the late fall of 1940, he was summoned to Chungking for a meeting with the Generalissimo that was to change his life. From an obscure "aviation technical advisor," he would, within three years, become a war hero and have a portrait painting on the cover of *Time*.[4]

To reach the United States now, he could not take the train to Hanoi. He had to fly west to India across the Hump

Unloading P-40 aircraft crate from truck at Rangoon airport. Crate was slid down railroad rails to ground and then slid on bottom of crate to assembly area.

Close-up of cabin open and closed on P-40B. Photo taken 26 February 1941 at San Francisco before packing. Serial number 140066.

or, as some combat cargo crews preferred, "the rock pile."

In Washington, D.C., he worked in an office provided by China Defense Supplies, a purchasing agency of the Chinese government headed by Soong Tze-ven, better know as T.V. Soong, brother of Madame Chiang Kai-shek.

Through the winter and spring of 1941, Soong and Chennault labored to establish, or to organize a small air force of American pilots. But the entire project hinged on a sensitive issue: neutrality. Since 1937 the United States, under the Neutrality Act, could not sell arms or munitions to nations at war with one another, and that included China.

Japanese propagandists made much of the Neutrality Act to needle United States foreign policy away from any move that might be unfriendly to Japan or friendly to China, such as organizing a volunteer group of airmen. No protests or *de-marches* were ever lodged by the Japanese ambassador to the Soviet Union against the several squadrons of Soviet fighters and bombers which engaged Japanese airmen in China from 1937-1939. There were no pacifists in the Soviet Union. But in the United States pacifist sentiment was strong and it was also dangerous for any Congressman to suggest the United States should assist either China or England despite efforts both nations were making to defend themselves against air forces which had suddenly become superior to their own.

Finally, on 15 April 1941 President Roosevelt, convinced that something had to be done to keep China from surrendering to the Japanese, signed a secret executive order which allowed officers and enlisted men in the armed forces to resign their commissions without prejudice to their careers, and to join a volunteer air force to "defend the Burma Road" as it was put euphemistically. The volunteer pilots would be paid \$500 to \$750 per month, enlisted personnel less. These men were volunteering to put their lives on the line, and a number of them gave their lives to fulfill their contract.

When William Pawley, president of Central Aircraft Manufacturing Company, learned of these plans, he offered the use of his corporate name as a cover for recruiting pilots, and he also offered to send some of his own people to help in the recruitment program. The Chinese also made an arrangment with Pawley to pay the salaries and arrange for steamship tickets for the volunteers.

The recruiting was done in a clandestine manner and the entire operation went off without mention in the *New York Times* or a radio newscast from Gabriel Heater.

The recruiting officers went to air bases and naval air stations across the country. One officer went aboard the carrier *Ranger* in Norfolk and recruited three future aces. All told, about 100 pilots were recruited in addition of 150 or so crew chiefs, engineers, clerks, propellor specialists, armorers, doctors, and two female nurses. The first contingent sailed from San Francisco in July 1941 for Rangoon, Burma. Their passports listed them as business men, mechanics, missionaries, and school teachers.

Through a complicated deal made with the British in Washington, T.V. Soong was able to buy an allotment of P-40s for which the British no longer had any use. The original table of equipment for a Group (of three squadrons) called for 100 P-40Bs and 25 P-40Es, but owing to the great difficulty of servicing the aircraft in Burma, there were seldom more than fifty that were combat ready. Replacement aircraft were brought in later.

The First American Volunteer Group was formally constituted under General Orders No Hi Sy 05987 cut by the Ministry of War in Chungking on 1 August 1941:

> To Colonel C. L. Chennault
>
> 1. The First American Volunteer Group is constituted this date.
> 2. You are requested to organize this group with foreign volunteer personnel now arriving in China for service with the Chinese Air Force. Additional personnel required to complete the organization of this group will be supplied by the Commission on Aeronautical Affairs.
> 3. You will assume command of the Group and render full reports when the organization is effected.[5]

The reference, in the above, to "additional personnel meant Chinese mechanics and engineers. This auxiliary force, commanded by "Tiger Wang," served in Burma, and veterans of the A.V.G. agree that their assistance in the assembly and maintenance of the aircraft was invaluable.

The Pawley organization was supposed to service the aircraft but with a repair shop at Loiwing, 350 miles to the north of Toungoo, where the action took place the first weeks of combat, was logistically not easy.[6]

The P-40s were shipped in crates to Rangoon, offloaded there and assembled by Walter Pentecost, a representative of Curtiss-Wright, and then flown to the training area at Toungoo.

Burma in 1941 was a colony of the British Empire, and British Colonials were not enthusiastic about helping either the Americans or the Chinese. It had been only a short time before the Americans arrived that the British reopened the Burma Road to traffic into China. Under pressure from the Japanese Foreign Office, the road had been closed for several months.

Royal Air Force units stationed in Burma at the time were equipped with Brewster Buffalo fighters, not Spitfires, but the Buffalos were considered adequate to meet any opposition that might arise in this part of the world.

When the United States joined Britain and China in the war against the Axis Powers on 8 December 1941, the Royal Air Force in Burma automatically went into action and also shared in the struggle in which the First American Volunteer Group was about to engage under the Chinese flag. The courage of the British fliers in Burma was never in doubt, and they rose to meet the enemy over Rangoon and Toungoo, on the side of the Americans, but their aircraft were obsolete and most of them were shot down.

Chennault was on hand to greet the first contingent of volunteers who came ashore at Rangoon in July 1941. By November the last group had arrived and was quickly checked out. In the training sessions, Chennault stood before a blackboard, diagraming his theories of air combat like plays on a football clip board. At other times he was out on the line observing simulated attacks between members of the squadrons.[7]

In the beginning, the record of these volunteers was not very impressive and a rash of crack-ups and careless accidents had skeptics among the Royal Air Force, stationed at Rangoon, as well as American ground crews believing these were not the professional killers they had hoped to see. Visiting Chinese Air Force officers from Chungking who came down to Toungoo to observe the progress of the training program were appalled at the rate of attrition. But all these judgments were premature. Chennault knew he was dealing with trained officers, but the fact is, most of them had never flown a P-40 before. Equally important, they had to

Hoisting fuselage into position on P-40 to install wings at Rangoon, Burma, summer of 1941. Flooring is part of old crate. At left, hangar under construction.

learn to work as a team. Finally, they had to learn what Chennault was telling them about the strong points of the Mitsubishi AM5s and *Hayabusas.* He had seen these planes in action over Nanking and Hankow and he imparted this information to his pilots, and then, as if by the wave of a wand, he had a first-class fighter outfit.

The First American Volunteer Group was divided into the 1st, 2nd, and 3rd Squadrons, the 1st called the "Adam and Eves," and 2nd called the "Panda Bears," and the 3rd, "Hell's Angels," all names not meant to frighten anyone. But the mood changed abruptly when the pilots painted the mouth of a shark on the nose of their airplanes. They had seen a magazine article showing German aircraft in North Africa with a similar logo on the nose, but none were as fearsome as the ones on the nose of the P-40, an airplane ideally suited to this symbol. Not only did it have ugly teeth but evil eyes glaring from either side of the cowling.

For the American Volunteer Group, quickly abbreviated to the "A.V.G.," the first brush with the enemy came on 20 December 1941, not in Burma where it might have started, but over Kunming. Ten Japanese bombers out of Hanoi made a surprise attack into the Yunnan Air Space, without escort, and without knowing that they were being tracked on the air raid warning net. As they approached Kunming, the A.V.G., warned in advance, was on top of them and within a few minutes, four of the bombers were shot down with three probables. The others, probably damaged, escaped.

Edward Rector, ex-*Ranger* pilot, went after one of the cripples and tried to catch it but ran out of gas and had to crash-land. He walked away from his airplane and it took a week on foot to return to base even though he was less than fifty miles away.[8] The mountains of southern Yunnan are part of the sub-Himalayan ranges and roads are few, narrow and twisted.

The Japanese raid on Kunming was no dry run. They had been here before, without escort, but that was before Chennault shifted two of his A.V.G. squadrons to the field. In any event, after that surprise intercept by the A.V.G. the Japanese did not return for some time. The main theater of action moved to Burma, especially over Rangoon and Toungoo where Chennault now sent two of his squadrons. And except for nusiance raids, or night raids, Chungking was no longer threatened and people could enjoy a respite from the 1940-1941 summers of constant bombing. This respite continued for the rest of the war except for a few moonlight raids.

Another reason the people of Chungking breathed more easily was because the Japanese were shifting some of their air regiments, *sentai* and *chutai,* out of China into Indo-China and Thailand. The Thais were neutral in the war, but this did not deter the Japanese from occupying the country. Now, with the Burma Road and the Hump air route, Burma, not Chungking, was the primary target for Japanese air strikes, and the place to begin was the Hump air route to cut off China's supply line to her allies in India.

In addition to air units, Japanese ground forces were now engaged in a land campaign to conquer Burma and India and they almost succeeded, actually reaching as far as Imphal and Kohima, northeast of Calcutta, before being turned back by British and Indian troops with the help of the Royal Air Force and squadrons of the U.S. 10th Air Force stationed in India.

In addition, several divisions of Chinese troops under the overall command of Joseph W. Stilwell, and strong guerrilla forces under Henry Merrill, known as "Merrill's Marauders," and the Chindits under Orde Wingate, and the Mars Task Force, counter-attacked the Japanese in northern Burma in the early weeks of 1944 and helped to drive them back to save Chabua and Dinjan, both key bases on the supply route to China. After this final thrust by the enemy, the threat to these bases grew less each month thereafter except for occasional moonlight raids.

By the beginning of 1945 there were few attacks by Japanese aircraft over the Hump, and most losses to the Americans were blamed on weather or pilot error. In the end, the Japanese attempt to halt the flow of supplies into China failed and when that failed, there was no way they could win.

The role played by the three squadrons of the A.V.G. in the overall defeat of the Japanese Air Force cannot be overestimated. In the five months or so of operations, the A.V.G. recorded the highest ratio of "victories" to losses of any aerial combat team in the world. According to official Chinese statistics, confirmed losses to the enemy by the A.V.G. were 193 aircraft shot down, seventy-five on the ground and forty damaged. Unconfirmed included fifty-six probables shot down and eight on the ground for a total of 372 aircraft confirmed and unconfirmed, against twelve losses for the A.V.G. not all by enemy action.[9]

A "victory" in air combat is often a matter of conjecture, for quite often more than one airplane shoots at the same enemy plane. Occasionally, three or four will divide a "victory." In a report submitted by Chennault to the Central Aircraft Manufacturing Company on 25 July 1942, the total amounts of bonus money owing to individual pilots in the A.V.G., the basic figure was US $500 for each enemy plane destroyed in the air or on the ground. One pilot received a bonus of US $133.33, suggesting that a "victory" was carefully determined and equally as carefully divided. Below are the bonus payments made to the leading pilots of the A.V.G. from 20 December 1941 to 6 July 1942, two days after the unit was officially deactivated.[10]

Names	Amount Requested
F.W. Adkins	$500.00
N.R. Bacon	1,750.00
P.R. Bartelt	3,500.00
W.E. Bartling	3,633.33
L.S. Bishop	2,600.00
J.E. Blackburn	1,000.00
H.R. Bolster	500.00
C.R. Bond	4,383.33
G. Boyington	1,750.00
J.G. Bright	3,000.00
R.R. Brouk	1,750.00
C.K. Brown	133.33

Initial servicing on the last P-40 to be assembled in Rangoon, Burma, summer of 1941. Thatched roof built to protect mechanics from sun.

G.T. Burgard	5,392.85
T.J. Cole	500.00
J.D. Cross	133.33
J.J.Dean	1,633.33
P.S. Dupouy	1,750.00
J.T. Donovan	2,000.00
J.W. Farrell	500.00
H.M. Geselbracht	750.00
C.G. Groh	1,000.00
P.J.Green	1,000.00
R.N. Gunvordahl	500.00
R.L. Hastey	500.00
T.C. Haywood	2,541.66
R.I. Hedman	2,416.66
D.L. Hill	5,625.00
F.S. Hodges	500.00
L. Hoffman	133.33
J.R. Howard	3,166.67
K.A. Jernstedt	5,250.00
T.A. Jones	2,000.00
R.B. Keeton	1,250.00
M.W. Kuykendall	500.00
C.H. Laughlin	2,600.00
F.L. Lawlor	4,250.00
R.F. Layher	416.67
E.J. Liebolt	133.33
R.L. Little	5,276.18
G.B. McMillan	2,041.66
W.D. McGarry	5,142.85
K.T. Merritt	500.00
E.I. Mickelson	133.33
R.C. Moss	2,000.00
R.D. Mott	1,000.00
R.H. Neale	7,776.18
J.V.K. Newkirk	5,250.00
C.H. Older	5,041.66
A.E. Olson	500.00
E.F. Overend	2,916.66
J.E. Petach	1,991.67
R.W. Prescott	2,642.85
E.J. Raines	1,600.00
E.F. Rector	3,258.33
W.N. Reed	5,250.00
F.I. Ricketts	600.00
C.J. Rosbert	2,276.18
J.R. Rossi	3,142.85
R.J. Sandell	2,633.33
C.W. Sawyer	1,133.33
F. Schiel	3,500.00
Van Shapard	500.00
E.E. Schilling	375.00
R.H. Smith	2,750.00
R.T. Smith	4,366.66
F.E. Wolf	1,133.33
P. Wright	1,325.00
	Total $146,999.87

Service hangar at Rangoon, Burma, 1941, showing removal of propeller.

The Chinese government also honored forty-nine members of the A.V.G. with one of the grades in the Order of the Cloud and Banner, one of the highest conferred on either Chinese or foreigners.

Early news dispatches to the wire services attributed the spectacular success of the Americans to the "American Volunteer Group" or to the "A.V.G." This designation continued to be used in newspapers from late 1941 into the first months of 1942. Spencer Moosa, *Associated Press* correspondent in Chungking, noted a new expression about the American airmen which his secretary had culled from a Chinese newspaper. It said that the American airmen were so good they were " 'virtually *fei-hu*' Flying Tigers."[11] In his next telegram to New York, Spencer also used the expression and it pin-balled around the world. The first magazine article ever written about the A.V.G. appeared in *China At War* edited and published by the Ministry of Information in Chungking, and the title was "The Flying Tigers."[12]

A.V.G. 3rd Squadron (Hell's Angels) in photo probably at Toungoo, Burma. Chinese white sun in blue field, logo of Chinese Air Force, visible under wing of P-40B. Front row, l. to r., William N. Reed, C.H. ("Link") Laughlin, Edward F. Overend, Thomas C. Haywood, Charles N. Older, Lewis S. Bishop, Benjamin C. Foshee, Robert P. ("Duke") Hedman, John T. Donovan, and Patrick Cavanaugh; back row, left, P.J. Greene, Parker. S. Depouy, Clifford G. Groh, R.W. Adkins, Robert J. Raines, George B. McMillan, Arvid E. Olson, R.T. Smith, Kenneth A. Jernstedt, R.R. Brouk, Erick E. Shilling, and Frederick S. Hodges.

Fuel from a tank truck for P-40B, somewhere in Burma, 1941. The shark nose painted on this P-40 was an unlikely symbol in China, and the idea to use it came not from the Chinese but from the British and the Germans. The Americans saw pictures of the R.A.F. airplanes from the Libyan desert in the *Illustrated Weekly* published in India at the time. And the British may have copied the logo from the Luftwaffe in which some of the Messerschmitt 210 fighters were painted with shark's teeth. But the shape of the nose and nacelle on the P40 allowed greater latitude and a better depiction of a shark's mouth than either the British or German airplanes.

A former universi
dormitory in Ku
ming was convert
to Hostel I for use
A.V.G. personne
1941.

Stinson 105 used f
liaison flights b
tween Rangoon ar
Toungoo by Willia
D. Pawley, Curti
agent in China. Th
picture taken at Ra
goon, summer of 194

Enlisted men a
mechanics for t
A.V.G. relax in b
installed in Hostel
Customer third fro
right wears patch
3rd Squadron know
as "Hell's Angels
Wayne Ricks, a p
pellor specialist, s
second from left.

XVI

China Air Task Force

After the Japanese attacked Pearl Harbor on 7 December 941, the United States declared war on Japan and in the ame breath became an ally of China. One of the first steps ne Commander-in-Chief, President Roosevelt, made, was o upgrade the military presence of the United States in China. A U.S. military advisory group had been stationed in Chungking for several months before Pearl Harbor but now was superseded by a higher echelon of command and the ppointment of a theater commander, Joseph W. Stilwell, a Vest Pointer and former language officer with the American Embassy in Peking.

Clayton Bissell was promoted to brigadier general and nade Stilwell's air member. Chennault was promoted to rigadier general in the United States Army Air Corps, one ay in seniority behind Bissell.

The American Volunteer Group was deactivated at a eremony held on 4 July 1942 at Kunming air base, and on ne same day the China Air Task Force was activated and by nat act, the United States officially entered the China Air pace as a belligerent nation allied with China against apan. For administrative purposes, the China Air Task orce was attached to the 10th Air Force which had its eadquarters and field of operations in India.

After the A.V.G. was deactivated, Bissell contacted the nembers and urged them to join with the China Air Task orce. Instead of wooing the men, he threatened them with nduction into the army when they returned home. However, hey were civilians, and he had no control over them. Many vould have enlisted if they had been given a furlough home, or had not been threatened. It was a costly mistake.[1]

Chennault made a personal plea to the A.V.G. to stay for wo weeks until replacements could arrive, and fifty-five ilots and ground crew extended their contract. Among nem was John Petach, Jr., who went off on a strafing nission over Nanchang on 10 July and was hit by ground re and crashed to his death.

Colonel Alfred Hagenberger, chief of staff to Chennault in ne later part of the war, made a study of the C.A.T.F. and ound it had had fifty-one fighters, most of them P-40s from ne A.V.G. squadrons. In addition, there were seven medium ombers of the 11th Bomb Squadron, and ten P-43s borowed from the Chinese Air Force.

Aside from the A.V.G. personnel who elected to stay, dditional people came from crews of the Doolittle Raiders after the bombing of Tokyo in April. There were also seven pilots and ground crew from R.A.F. who had to leave Burma after the Japanese overran most of the country in the spring of 1942.

The U.S. Army Air Corps, according to Hagenberger, supplied "a dozen pilots" for the China Air Task Force and others drifted in almost at random, including Robert L. Scott, Jr., and Caleb Haynes, who had been ferrying supplies across the Hump. Bruce Holloway and John Alison left the 51st Fighter Group of the 10th Air Force in India to fly with Chennault, and Marian Cooper and Herbert ("Butch") Morgan came all the way from the States to serve on his staff. Clinton D. Vincent preferred flying in China to a desk job in Karachi and asked to be transferred to Kunming, He later became the first commander of the 68th Composite Wing in Kueilin.

The old 1st, 2nd, and 3rd squadrons of the A.V.G. became the 74th, 75th and 76th fighter squadrons in the C.A.T.F. and, together with the 16th Fighter Squadron and 11th Bomb Squadron (M) less one flight of the 349th Bomb Group, became the 23rd Fighter Group. The three former A.V.G. squadrons were led by A.V.G. aces, namely, Frank Schiel, Jr., commanding the 74th, David ("Tex") Hill, the 75th, and Edward Rector, the 76th. Robert Scott who arrived in China in time to run a few missions with the A.V.G.; in the spring of 1942, became the first commander of the 23rd Fighter Group under the China Air Task Force.

The C.A.T.F. was ostensibly an interim organization,. Outnumbered by the Japanese air regiments in China, it managed to stay alive against superior numbers by the foresight and skill of its commander who struck when he was least expected and equally confused the enemy by not striking at all.

Less than a year after it was activated, C.A.T.F. air operations were increased dramatically and the Japanese countered this new force with equally determined opposition as each side struggled to knock the other out of the China Air Space. This increased pressure forced the United States to enlarge its air power in China.

Women workers, carrying their own baskets, flee from danger. Nearby airfield is under attack and black smoke in background suggests aircraft burning as a result of hit. Headgear worn by woman at left suggests this picture was taken at Paoshan or other field in western Yunnan Province.

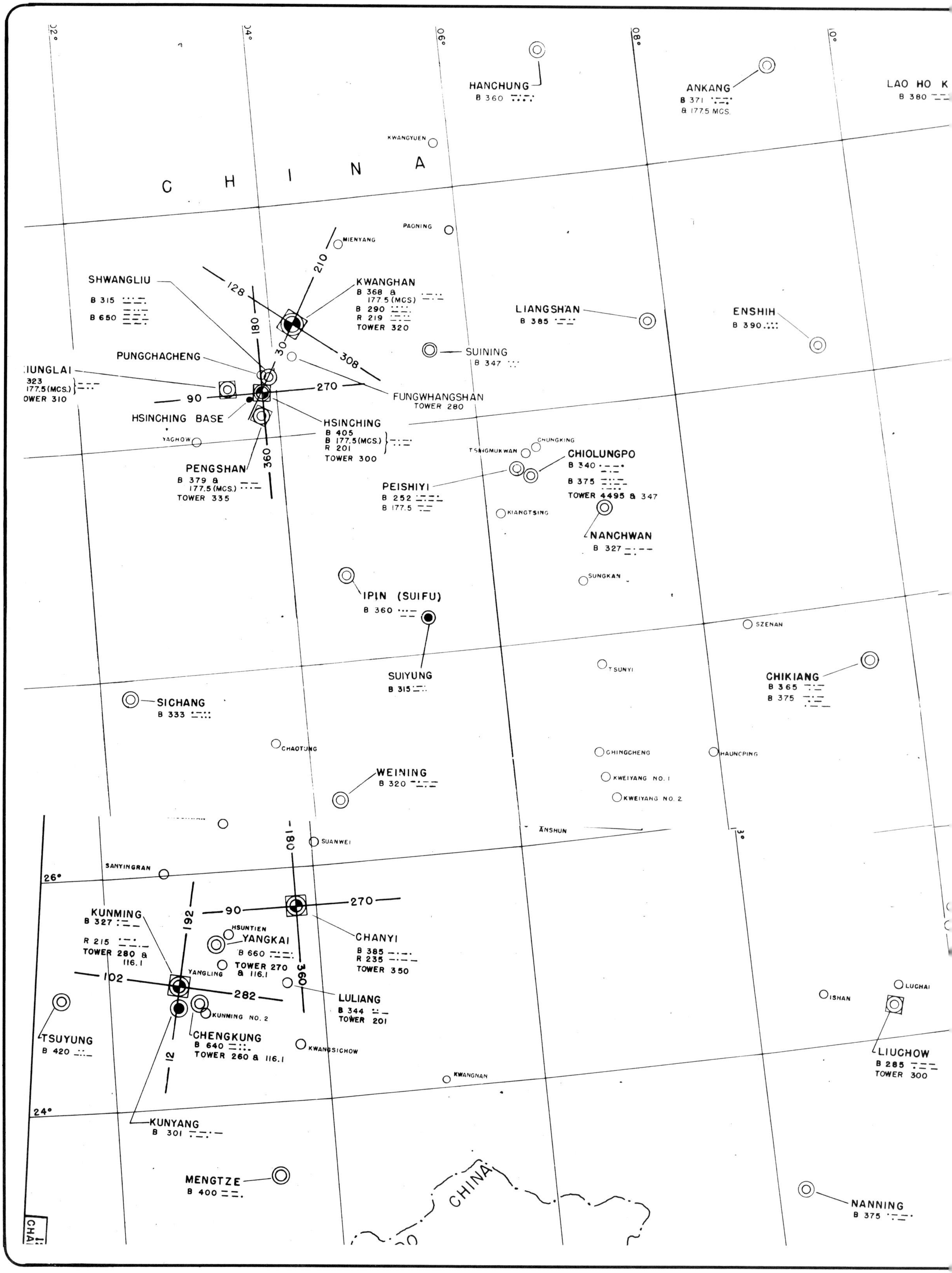

C H I N A
HANCHUNG
B 360
ANKANG
B 371
B 177.5 MCS.
LAO HO K
B 380
KWANGYUEN
PAONING
MIENYANG
SHWANGLIU
B 315
B 650
KWANGHAN
B 368 & 177.5 (MCS)
B 290
R 219
TOWER 320
LIANGSHAN
B 385
ENSHIH
B 390
SUINING
B 347
PUNGCHACHENG
FUNGWHANGSHAN
TOWER 280
HSINCHING BASE
HSINCHING
B 405
B 177.5 (MCS.)
R 201
TOWER 300
YACHOW
PENGSHAN
B 379 & 177.5 (MCS.)
TOWER 335
TSINGMUKWAN
CHUNGKING
CHIOLUNGPO
B 340
B 375
TOWER 4495 & 347
PEISHIYI
B 252
B 177.5
KIANGTSING
NANCHWAN
B 327
SUNGKAN
IPIN (SUIFU)
B 360
SUIYUNG
B 315
SZENAN
TSUNYI
CHIKIANG
B 365
B 375
SICHANG
B 333
CHAOTUNG
WEINING
B 320
CHINGCHENG
HAUNCPING
KWEIYANG NO. 1
KWEIYANG NO. 2
ANSHUN
SUANWEI
SANYINGRAN
26°
KUNMING
B 327
R 215
TOWER 280 & 116.1
HSUNTIEN
YANGKAI
B 660
TOWER 270 & 116.1
YANGLING
CHANYI
B 385
R 235
TOWER 350
LULIANG
B 344
TOWER 201
KUNMING NO. 2
CHENGKUNG
B 640
TOWER 260 & 116.1
KWANGSICHOW
TSUYUNG
B 420
ISHAN
LUCHAI
LIUCHOW
B 285
TOWER 300
KWANGNAN
24°
KUNYANG
B 301
MENGTZE
B 400
CHINA
NANNING
B 375

Main runway at Chabua, upper
sam, 1 November 1944.

Billeting office at Chabua,
ember 1944. Author photo.

Bamboo Music Hall, Chabua,
December 1944.

Opposite page: Map shows air bases most frequently used by China Air Task Force and later by the 14th Air Force from 1941 to 1945.

China Air Task Force "sharks" from 16th Fighter Squadron being fitted with external fuel tanks (foreground) preparatory to long-range mission. Airplane in foreground "owned" by R.M. Mooney. Photo taken 21 October 1942, probably at Kunming air base.

Enlisted men of 11th Bomb Squadron (M) in first group picture taken in Kunming, summer of 1942. 11th Bomb was activated in France in World War I, and "Jigg" patch seen on nose of airplane here, was adopted from "Maggie and Jiggs," a favorite comic strip in the Sunday "funny papers" before and after WWI.

Chinese workmen pulling concrete roller to pack down runway.

Crash landing of C-46 ends in fire ball, 23 November 1944, probably at Kunming. Crew saved.

Cleaning 50 caliber guns 20 March 1945, probably at Kunming.

Crater left by Japanese bomb on Suichuan air base, January 1945.

Dental office mobile, probably at Kunming.

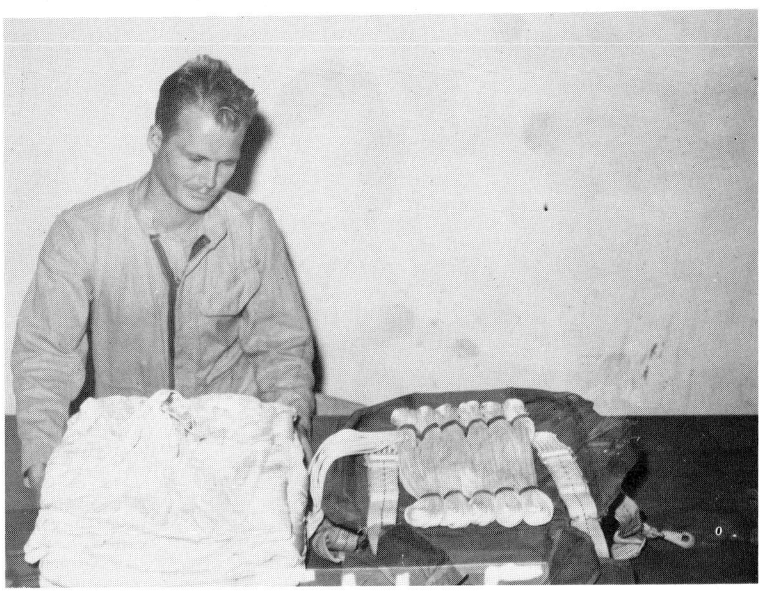

Parachute rigger, November 1943, probably Kunming air base. ("It don't mean a thing. If you don't pull that string.")

Thousands of men, women and children were employed, almost the year around, building airfields for the Americans in the years from 1941 to 1945, and it was done the hard way, as these pictures reveal. Top cover was made by chipping small stones into smaller stones and sharply pointed stones that caused any number of blowouts to tires and fatal accidents. But hauling concrete and concrete mixers and trucks over the Hump into China was never seriously considered because of the logistics involved. People got paid according to the weight of stones in the baskets extended from the "yo-yo" pole across the shoulders of the man at left. Concrete rollers, pulled by scores of men, were used to pack the layers of stones which had to be at least eighteen inches in depth. In lower left, workers move out to field.

'op: Kunming air base, N 2506
10242 as it looked in 1943-44
'ore being lengthened. Slit
ıch runs along field at left for
sonnel to jump into during air
ds. Hangars stand at left, and
nters are parked on the hard-
nd. In upper right is hostel area
l Hostel 7 for transients. Air
ınsport Command and China
tional Aviation Corporation
ight offices extend out from
e of hostel area, with several
nsports parked in front of hos-
area and upper right taxiway.
wer stood in angle where road
ds off at right center. Several
craft are parked nearby but
ver does not show. It was des-
yed in direct hit by enemy
nbers in 1943 and never rebuilt
nough control was maintained
m lower building not visible
e. Lower photo shows field after
as lengthened from 6100 feet to
0 feet. Main runway lay north-
thwest and both pictures point
therly. The field lay at about
00 feet altitude. The lower pic-
e was taken 6 February 1945
er runway extension was
npleted.

Close-up air view of headquarters buildings and medical stations at Yang Tong air base, southwest of Kueilin. Photo was taken 31 August 1943 at 1400 hours (2 p.m.) by 21st Photo Reconnaissance Squadron.

Yang Tong air base, located several miles southwest of City of Kuelin, is shown here with satellite field at lower right. Main runway lay roughly north, and mountain peaks on facing page appear here in the southwest, or at lower left of picture. Photo taken shortly after Japanese bombed field, as noted by pock marks.

Photo shows main runway at Yang Tong and roads leading to headquarters buildings. Slit trench follows edge of field on near side. Aircraft on line appear to be fighters. This photo was shot by 21st PRS 31 August 1943 and the exact position is given at 25°20' N and 10°10 E.

Aerial view looking northeast across "ice cream cone" peaks of Yang Tong air base with main runway visible beyond center peak. Headquarters buildings of 68th Compose Wing nestle under peaks in center. Today, Chinese tour conductors take American tourists on Li River cruise to show them unique formations of these mountains but field shown here has been abandoned.

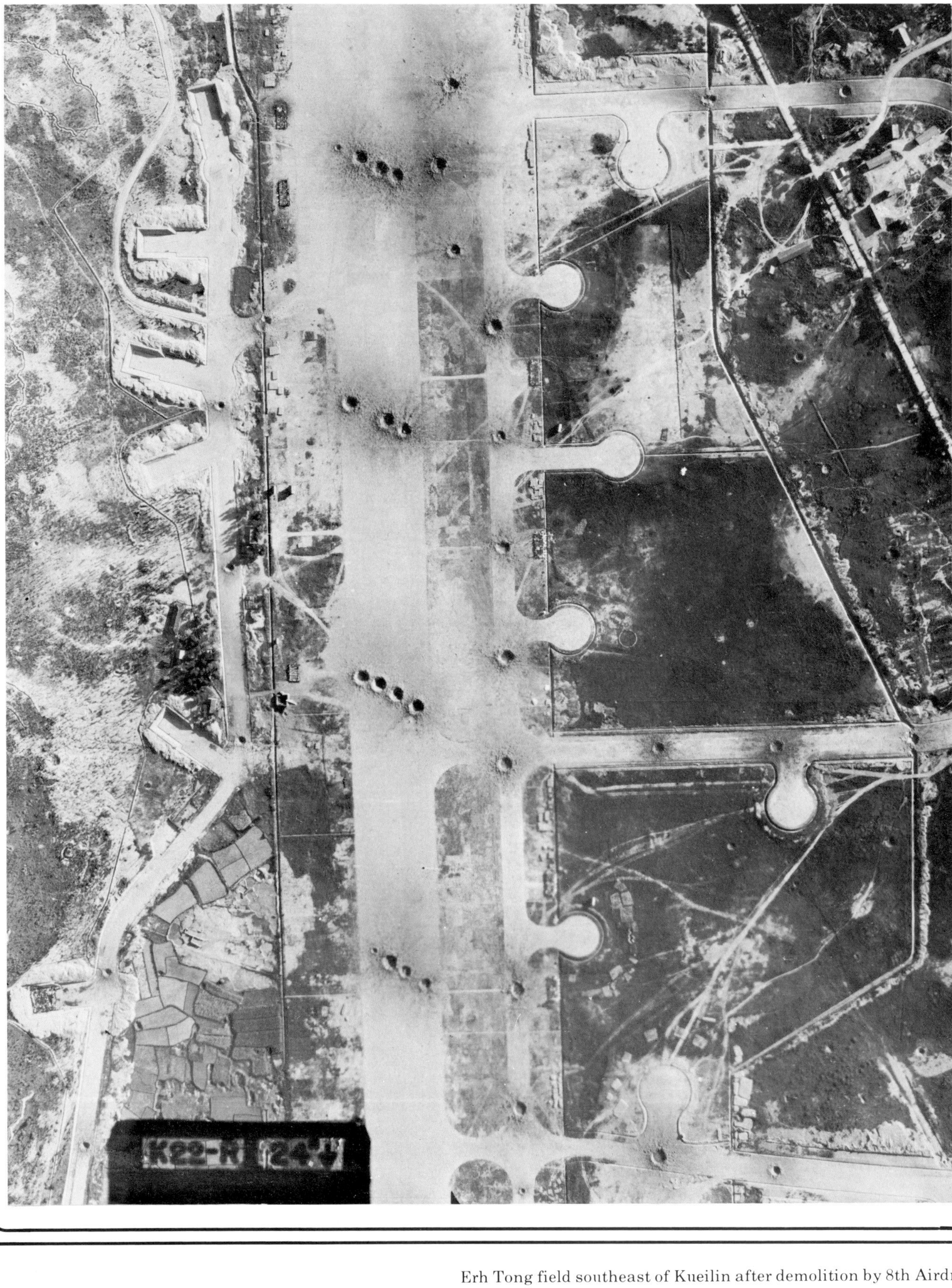

Erh Tong field southeast of Kueilin after demolition by 8th Airdrome Squadron, November 1944.

Lichiachen, a satellite field lying to east of Erh Tong (southeast of Kueilin city), was destroyed by demolitions set by 8th Airdrome Squadron in evacuation of field in November 1944.

Liuchow air base on Loching River in central Kwangsi Province with "ice cream cone" peaks visible at left.

Suichuan air base (left) lay along Kan River in south Kiangsi Province. At left is hostel area, Kan River plainly visible. In lower photo, river runs north and west of field. Aircraft are parked at both ends of field, revetments at left of strip.

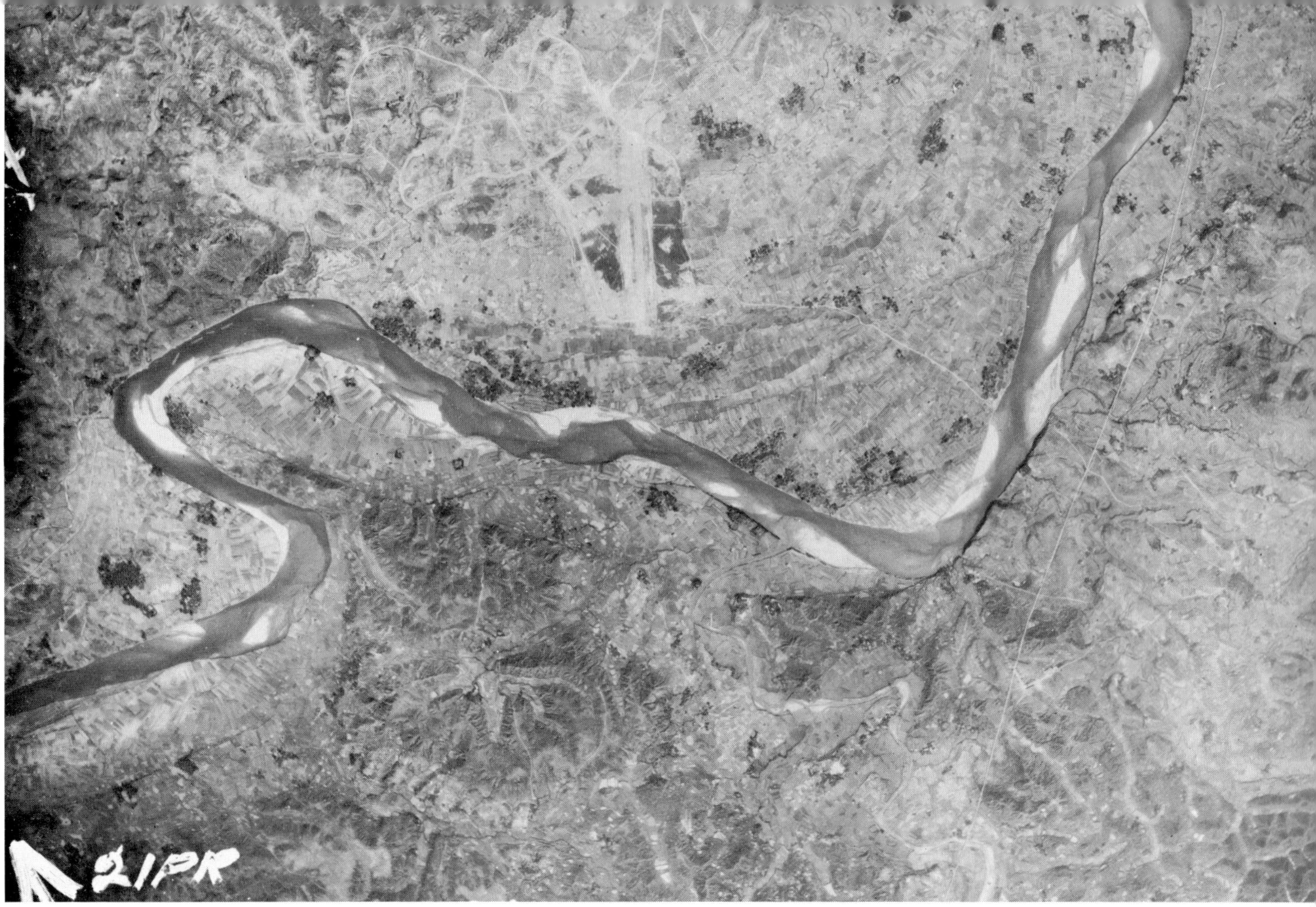

Above: Kanchow air base in south Kiangsi Province with Shang-yu River circling around. Runway barely visible at 25,000 feet. Arrow (lower left) points north. Base was built under supervision of Russian engineers, probably in 1939. This photo taken in 1944, and markings on negative were made by lab technicians of 21st Photo Recon.

Lower photo: Marking at bottom of photo made by lab technicians of 21st Photo Reconnaissance Squadron to identify exact position, date and altitude of camera. Photo shows air base at Lingling taken 11 December 1943. Lingling was base for 74th Fighter Squadron in late 1943 and into fall of 1944. Hsiang River paralleled runway.

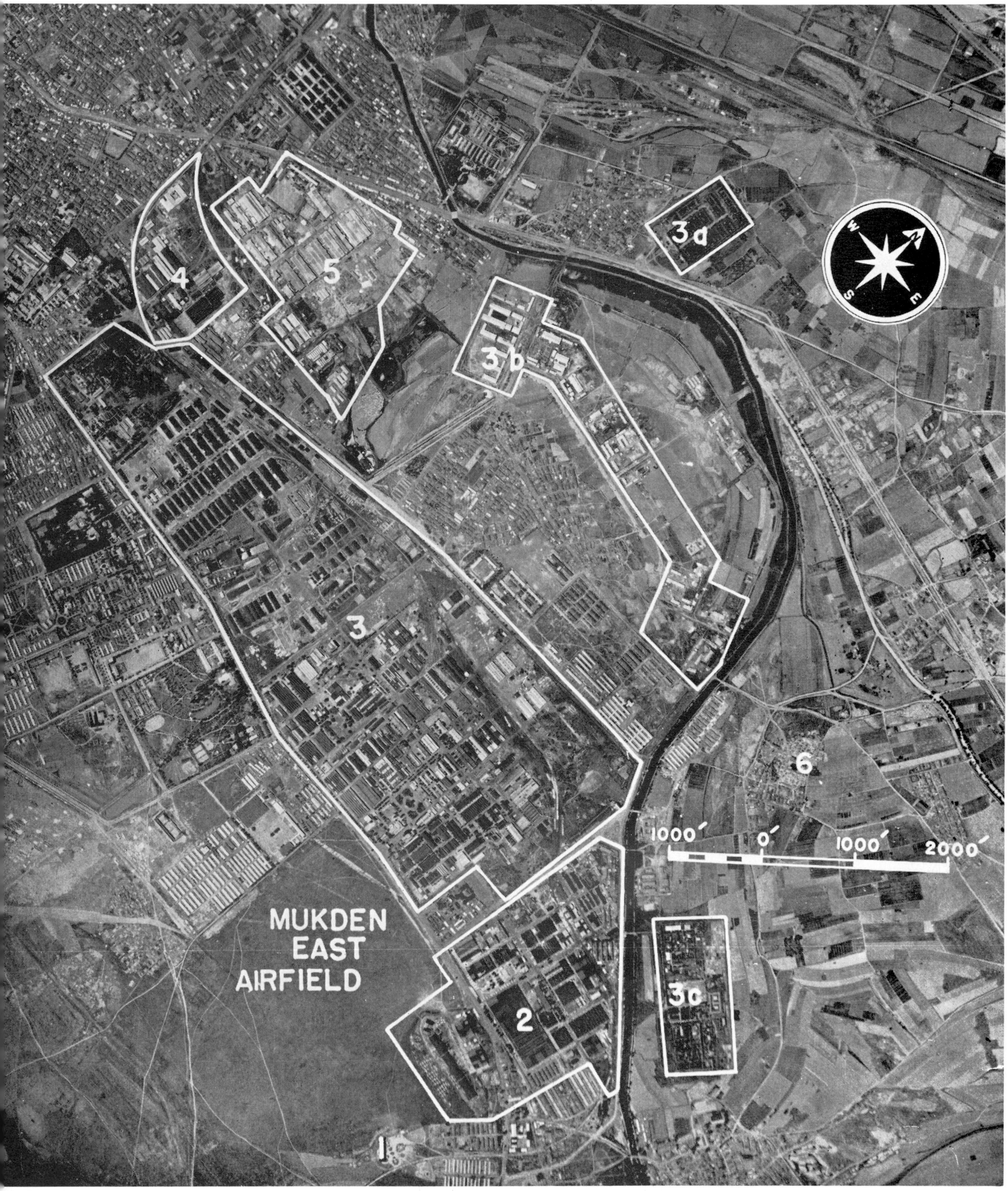

Between 18 September 1931, when Japanese forces occupied Mukden, Manchuria, and 1945, the city became an industrial center for the Japanese war economy. This photo, taken on 14 June 1945, concentrates on east side of city where the East Airfield was located, a field laid out in the early 1920s by Chang Tso-lin, but the air view does not take in the North Mukden field developed by the Japanese. The vital areas in the picture have been interpreted by American photo intelligence experts and, (1) shows anti-aircraft batteries; (2) shows Manchuria Airplane Manufacturing Company's aircraft assembly plant; (3) Mukden arsenal (3a and 3c) explosive storage and probable shell loading, (3b) probable components loading plant; (4) Manchuria Iron Works; (5) machine tool plant, (6) ceramics plant. The number of aircraft counted here, 126 planes, exceeded in number any ever photographed by the 21st Photo Reconnaissance Squadron in Asia.

There were no photographs taken of the 16 B-25Bs which landed in China and Siberia after the Doolittle raid on Tokyo 18 April 1942. All were lost, destroyed or captured. The photo above was taken over the Burma Hump on 17 November 1944 and shows a B-25J from the 22nd Bomb Squadron. Photo was shot by W.H. Van Vleck whose ship was flying in the wing position.

XVII

The Doolittle Raiders

The most publicized air raid in the Pacific theater of operations in World War II was led by Jimmy Doolittle against Japan on 18 April 1942. It was brilliantly conceived and daringly carried out, but it had tragic consequences.

To make the attack, it was necessary to launch the bombers from a point as close to Japan as possible without losing the element of surprise. To do this, sixteen B-25s, medium bombers, were launched from the carrier *Hornet* in the north Pacific about 400 miles east of Tokyo. After bombing military targets in the main cities of Japan, the flight plan was for the bombers to continue on to China and land there. But the mission was planned under such deep secrecy that Doolittle had not even told his allies, the Chinese, that he was coming. It was fairly certain that none of the planes would reach China before dark and the chance that any airfield held by the Chinese would be lit up, without prior notice, was slim. But Doolittle had been in China before, in 1933, demonstrating the Curtiss Hawk II, and he apparently thought the Chinese would not mind this unannounced intrusion into their Air Space.

Since most of the crews were running out of gas by the time they reached the coastline of China, they did not have far to go before they had to bail out, which most of them did in the province of Kiangsi lying to the south of the Yangtze River in central China. One crew crash landed on a beach near Ningpo, south of Shanghai, and one got its bearings completely turned around and landed in Siberia where the Soviets held the men for a year before releasing them even though the United States was an ally of the Soviet Union.

After bailing out in the night, most of the crewmen wrapped themselves in their parachutes and waited for morning. Eventually all were picked up by Chinese guerillas and civilians and forwarded by train and by plane to the west of China, all except the one crew that crashed at Ningpo and another that bailed out near an airfield held by the Japanese at Nancheng in south Kiangsi, where they were captured.

A number of officers, including Doolittle, were flown to Chungking and here on 3 May 1942 they heard congratulatory messages from President Roosevelt, and Generals Marshall and Arnold. The raiders were awarded the Presidential Unit Citation and the Distinguished Flying Cross, and Doolittle was advanced two ranks in grade from light colonel to brigadier-general.

The raid over Tokyo and other cities electrified the American people. Since the Japanese attack on Pearl Harbor, American forces had been on the defensive, holding on to precious pieces of real estate like an island called Guadalcanal, and many people on the West Coast heard they might come under attack by Japanese naval artillery or submarines.

The raid on Japan did not shorten the war but it served to warn Japanese leaders and their people that they were no longer immune to attack from the air themselves. But to the Grand Imperial Headquarters, protectors of the Emperor, the raid was a personal insult. No one can look *down* on the Emperor, the Son of Heaven, and the Air Space over his head is sacred, never to be violated by anyone, foreign or domestic.

Shortly after the raid, with the approval of Imperial Headquarters in Tokyo, the 13th Japanese Army headquarters in Shanghai issued a set of regulations called the "Military Law Concerning Punishment of Enemy Airmen." It was made retroactive in order to bring charges against captured Doolittle fliers. This code legalized the execution of a prisoner of war, and in doing this, Japan violated civilized procedure recognized throughout the world on treatment of prisoners.

Eight officers and men from the two crews who were captured were eventually brought before a military court in Kiangwan Military Prison, north of Shanghai, on 28 August 1942. However, some of the men had already been tried at an earlier hearing held in Tokyo and then returned to Shanghai.

The American airmen, in flying over Tokyo, even though scrupulously avoiding the Imperial Palace, had violated the belief held by most Japanese that no one is supposed to "look down" on the Emperor. Sanctimonious Japanese army leaders saw no contradiction in the fact that their air crews had been bombing Chungking, the capital of China, for three years, long before the Doolittle raiders took off for Tokyo. There was no emperor in Chungking, but there were people.

The trial of the eight Americans was a propaganda device to threaten future American airmen who would have the temerity to invade Japan. The outcome of the trial was a forgone conclusion: all eight men were sentenced to die.

The details of the charges brought against the airmen may be found in a Japanese document, translated into English, presumably by the Judge Advocate's section of MacArthur's headquarters in connection with the numerous war crimes trials held after the end of World War II.[1] Under the heading, *Reasons for the Sentence,* the charges made against the two flight crews, almost identical, are as follows:

> It has been proven to the satisfaction of the Court that in March 1942, the defendants, motivated by their sporting instincts and sense of glory, in response to a call for participants in an air raid, made application to their currently attached unit; that for a period of over a month, they, together with the other trainees, received special combat training; that they embarked from the port of San Francisco on 2 April aboard the aircraft carrier *HORNET*, and on that day were apprised, for the first time by Colonel Doolittle, that they were to raid Japan in five formations of 16 bombers, and were briefed as to the details of the raid and the respective missions of each craft; and that on the morning of 18 April, at about 0800

hours, at a point approximately 35° north latitude, and 153° east longitude, the first plane took off from the flight deck, followed in close order by the other planes, headed for the raid on Japan.

1. The defendants, *Horumaku* [Dean Hallmark] pilot, *Meeda* [Robert J. Beder] co-pilot, and *Nerruson* [Chase Jay Nielson] navigator, occupants of plane No. 6, a North American B-25 medium bomber, arrived over Tokyo at about 1300 hours, suddenly exhibited cowardice when confronted with opposition in the air from the ground, and with intent of cowing, killing and wounding innocent civilians, and wrecking havoc on residences and other living quarters of no military significance whatsoever, together with the other twelve planes did carry on indiscriminate bombing and strafing, thereby causing the death and injury of about 10 civilians and the destruction of numerous residences; following which they attempted to escape to the China mainland. At noon of the same

Photo of five Doolittle fliers who were captured in China on 19-20 April 1942. Two crews of four men each were captured, and on this picture there are five Americans. One died in Nanking prison. Four who were condemned to die were given a life sentence, and three were executed. In front of the Americans stand their judges, the one with braid probably representing army headquarters company.

day, they were forced to land on the water in the vicinity of Ningpo, Chekiang Province, where they were captured by the Japanese army on the 20th.

2. The defendants, *Fuaro* [William G. Farrow], pilot, *Haito* [Robert L. Hite], assistant pilot, *Bha* [George Barr] navigator, *Supatsu* [Harold Spatz], gunner, and *Deshieja* [Jacob D. Deshieza], bombardier, occupants of plane No. 16, a North American B-25 medium bomber, arrived over Nagoya at about 1615 hours, afternoon of 18 April, suddenly exhibited cowardice . . . [here the Japanese repeat the charge in the same words as the first paragraph up to the last sentence]. At 21:40 that evening while over Nanchang, Kiangsi Province, they were forced to parachute because of fuel exhaustion, and were immediately captured by Japanese army units in the vicinity.

In a long, sworn statement made by Captain Hite and Sergeant Deshieza on 11 September 1945, they denied that the Americans had attacked civilians and insisted that only military targets had been hit. As to the other defendants at the trial held at Shanghai, Hite and Deshieza had this to say, in part:

> On 15 October 1942, Hite, Barr, Deshieza, Nielsen and Meder were taken back to the court martial for a second trial which lasted three or four minutes. Hallmark, Farrow and Spatz were not with us and we were informed by Remedios [Louis Dos Remedios, civilian interpreter] that these men had been removed from the prison on 14 October. We did not see these men again, although at a later date various articles of their clothing were distributed to us. At the trial the sentence was read to us in English. We were informed that we had been sentenced to death for bombing and strafing innocent school children, but that through the mercy of His Imperial Majesty, the Emperor, our sentence had been commuted to life imprisonment with special treatment . . . [2]

The reason Hallmark, Farrow and Spatz were never seen again: they were dead. Hite and the others knew about Meder because he had died in prison at Nanking where the men spent several months before the trial began in August. His death was traced to several causes, chiefly malnutrition.

The reason Farrow and Spatz were singled out was that the Japanese accused Spatz, a tail gunner, of machine-gunning a school yard filled with children, and Farrow, being the pilot, was guilty by association. But it remained a mystery why Hallmark in the other crew was singled out for execution.

In a war crimes trial held in Shanghai in December 1945, the Judge Advocate's department of the 14th Air Force brought charges against eight Japanese officers, enlisted men and civilians for the death of the three Doolittle fliers, and this is the way the Japanese defendants described the day of execution:

> We arrived at the cemetary in automobile and we got off. Captain Tatsuta, the branch section chief, asked the prisoners if they wanted to have a last word. (They did not). I was not able to understand them clearly because it was interpreted to the prisoners by an interpreter. After that, the three prisoners were each brought to the respective crosses. They were tied to the cross with a brand new white cloth and likewise a brand new white cloth was draped over their head. The firing squad lined up a certain distance away from them. By that time we had retreated to the background of the scene. Then the firing squad fired. After the shots were fired an army medical officer went and examined the bodies to ascertain that they were dead. After he was satisfied the bodies were released from the crosses and placed in three separate coffins. The three coffins were lined up side by side and at that time all those who were present at the execution participated in a very short ceremony or service.[3]

In cross-examination it was revealed that the cloth around the head of each man had a mark directly between the eyes and over the nose. This was the target for the riflemen to shoot at.

The crosses used were not what might be imagined from Christian tradition. They were short, and the airmen had to kneel down on their knees with outstretched arms while their arms were bound, with cloth, to the cross-arms. John Hendren, chief of the prosecution, explained why the crosses were used. A Japanese officer observer at the trial of the Japanese defendants told Hendren that the commander in charge of the firing detail had assumed that the Americans were Christians who would want to die in the manner of Christ.[4]

The coffins were taken to a Shanghai crematorium. The

ashes were put in boxes and brought back to a waiting room in Kiangwan Military Prison. This room, according to one of the defendants, was not used "from that time on. The ashes were placed on a shrine and incense burners were placed in front of them."

The ashes were found after the war in a Shanghai mortuary along with some clothing of the deceased. Hendren had these items collected and turned over to the army for return to the families of the deceased.

In the trial of the defendants held after the war, the prosecution could not find any document or witness to point a finger at any one officer who gave the order for the execution, and as a result, no one was given the death sentence. The guilty officers were probably casualties of the war themselves by this time. However, the defendants were found guilty, not of murder, but of contributing to the death of the three Americans, and they were sentenced to varying terms of hard labor, none exceeding nine years.

From the dark side of the moon came a more somber, bonechilling revelation in the wake of the Doolittle raid. While the Doolittle crews were being feted in Chungking, thousands of Chinese civilians were being beaten, tortured and killed by Japanese troops for rescuing the Americans.

The American airmen, thankful to their Chinese rescuers, had given them tokens of their appreciation such as fountain pens, sun glasses, knives, in fact anything they still had on them after the bailout. Later, these possessions became talismans not of good luck but of death. The Japanese went from door to door searching for any sign or token, and torturing people to confess what they knew about the Americans, and, in the end, burning their villages.[5]

A Catholic missionary, the Reverend Charles L. Meeus, had a mission station in Kiangsi and fortunately for him he was absent at the time of the massacre but returned shortly after to bury some of his parishioners. He later wrote an article for *China at War*[6] in which he describes some of the horrors of the Japanese massacre. He estimates that not less than 25,000 people lost their lives in his parish alone, not only men and women, but children as well, many slain in cruel and inhuman manner. One family of nine, parishioners of Father Meese, were thrown into a well and all but one drowned.

A young women, a niece of a local magistrate, was burned by cigarette butts over her body, gang raped and then trussed to a pole to stand in the heat of the sun all day.

In Nanchang, ten people were used for a "bullet contest" to see how many bodies one bullet could penetrate. In another instance, eight people were tied to a bamboo pole and thrown from a bridge into a river while soldiers watched their victims thrash and turn until they drowned.

At Ihuang, south of Nanchang, the Japanese learned that one Ma Eng-lin and his wife had actually provided a bed for one of the airmen, Harold F. Watson who was wounded. Ma was wrapped in a blanket, tied to a chair and covered with kerosene, and his wife was made to light the match to her own husband.

No war crimes trial was held to bring Japanese officers back to Kiangsi Province or to face charges of murder in the death of more Chinese civilians than Japanese civilians who died at Hiroshima and Nagasaki.[7]

Aerial close-up of hostel area for transient personnel, Kunming air base. At right are warehouses of Air Freight, China National Aviation Corporation, Air Transport Command and 90th Airdrome Squadron. Dark row at right are trees along road leading north to city of Kunming, about seven miles distance. Aircraft on line mostly C-47's.

XVIII

The 14th Air Force

The China Air Task Force under the command of Brigadier General Claire L. Chennault, with headquarters at Kunming, China, was administered by the over-all authority of the 10th U.S. Air Force headquartered at Hastings Mill (near Calcutta), India. On 10 March 1943 the China Air Task Force was declared a surplus organization and a new military organization called the 14th Air Force, under the command of the now Major General Claire L. Chennault, took its place. Henry F. Strickland of the adjutant general's office in Kunming, was sent to Hastings Mill to handle the details for the transfer of authority.

Special order No. 1,[1] cut in Kunming, 10 March, named the following officers and warrant officers as being relieved from assignment to the 10th Air Force and attached to the 14th Air Force without change of duty or station:

Colonel Clinton D. Vincent, acting chief of staff.
Lt. Col. Herbert Morgan, Jr., acting assistant chief of staff A-3.
Lt. Col. Henry F. Strickland, adjutant general.

Brigadier-General Claire L. Chennault smiles, standing between two of his top men, Clinton D. ("Casey") Vincent (left) and Bruce K. Holloway. Vincent was first chief of staff to Chennault in 14th Air Force activated on 10 March 1943 a few days after this picture was taken in Kunming outside headquarters building. Holloway became C.O. of 76th Fighter Squadron, and later 23rd Group commander. He was later rotated back to the Zone of Interior while Vincent was sent to Kueilin to assume command of 68th Composite Wing. Both Vincent and Holloway became aces before their tour of duty in China was up. Holloway wears patch on jacket showing a tiger tearing the rising sun, a "tryout" patch for 23rd Fighter Group, a logo later changed to the now familiar tiger flying through the sky on a bolt of lightning.

Lt. Col. Jesse C. Williams, acting assistant chief of staff A-2.
Major James M. Brooksband, administrative inspector.
Major John M. Williams, communications officer.
Captain Robert E. Fleming, weather officer.
Captain Edwin Kessler, acting assistant A-4.
Captain Frank R. Otte, acting assistant A-2.
Captain Wilfred J. Smith, acting assistant A-2.
Captain Richard V. Taylor, acting assistant A-2.
Captain Robert A. Waidner, acting chief of staff A-1.
1st Lt. Gerald J. McAllister, public relations.
1st Lt. Morgan E. O'Connor, acting assistant A-2.
1st Lt. John Birch, acting assistant A-2.
Warrant officer (jg) Eldon E. Johnson, acting assistant A-1.

Below on the same order, Major Mark T. Gilkison was appointed acting assistant chief of staff A-4.

Among enlisted men in this first headquarters squadron were Cpl. Howard M. Arnegard, S/Sgt. Robert J. Gillespie, Sgt. Anton M. Vargo, and Cpl. Robert W. McGregor.

Also, by Order No. 1, S/Sgt. George J. Atack of the 74th Fighter Squadron was transferred in grade to the 11th Bombardment Squadron; M/Sgt. Joseph J. Stanovich, of the 16th Fighter Squadron, was relieved of duty and attached to the 74th Fighter Squadron, and Captain James O. Pierce, M.C. and Private Albert E. Rucker of the 1st Troop Carrier Squadron were transferred from Yangkai to New Delhi for duty with the 1st Troop Carrier Squadron.

1st. Lt. John M. Birch, listed above, was a Baptist missionary in east China and when the Doolittle Raiders came down in his area, Chinese partisans asked for his assistance in rescuing the fliers. When Birch arrived in Kunming with a number of the Doolittle crews, Chennault offered him a field commission which he accepted. He spoke fluent Chinese and he was the first man to establish a liaison station for intelligence with the Chinese army headquarters at Changsha.[2] He survived the war but was killed by a Communist patrol in north China after the war. The John Birch Society, an anti-communist organization, is named for him.

Until 5 September 1943, the 14th Air Force was a single command. On that day, two composite wings, the 68th and the 69th, were activated which divided operations roughly between east and west China. The 69th was first commanded by John Kennedy with headquarters at Kunming, and the 68th under Robert L. Scott, Jr., with headquarters at Kueilin Air Base. The latter was also designated the "forward echelon" of the 14th Air Force. On 28 February 1944, Clinton D. ("Casey") Vincent assumed command of the 68th Composite Wing.[3]

According to a report prepared by Alfred Hagenberger, chief of staff to headquarters, 14th Air Force, the 14th began on 10 March 1943 with 170 aircraft, mostly fighters and one squadron of medium bombers. This force was opposed, he says, by approximately 450 enemy aircraft. In December 1943, the strength of the 14th Air Force had risen to 270 aircraft of all types and was opposed by approximately 750 enemy aircraft. In June 1944, the 14th had 498 planes against 1110 for the enemy. Six months later in December 1944 the strength of the 14th had risen to 716 aircraft against 1200 for the enemy, and in June 1945 the number of 14th aircraft had risen only slightly to 723 while the enemy strength had been cut back to approximately 910.

Hagenberger further states that during the war in China 1281 enemy planes were destroyed in aerial combat, 558 probably destroyed, and 655 damaged for a total of 2504. Also, 1,057 enemy planes were destroyed on the ground either by strafing or bombing, and 220 were probably destroyed, and 618 probably damaged for a total of 1895. He estimated that the total number of enemy planes destroyed in the *China-Burma-India* (my italics) theater of operations and probably destroyed or damaged was 4,399.

American losses, Hagenberger states, were 117 shot down and 76 destroyed on the ground for a total of 193, and he estimated that elements of the 14th Air Force, including the Chinese-American Composite Wing, sank or damaged

Making final approach at left is probably a P-51, at Yangkai. Parked on line are C-46s, Yangkai was turnaround and headquarters for Air Transport Command in China.

1,100,000 tons of shipping, destroyed 1,079 locomotives, 4836 trucks and 580 bridges.

This report is confused and no doubt exaggerated. It refers to the China-Burma-India theater which included India where the 10th Air Force was based. Squadrons from the 10th Air Force operated against targets and enemy aircraft over Burma. The 14th Air Force operated against targets in Burma as well as China. Where did the chief of staff draw the line?

The figure of 580 bridges destroyed is highly suspect. For example, based on the number of times the 75th Fighter Squadron, based at Hengyang, attempted to destroy the Puchi railway bridge, south of Hankow, and never succeeded, and considering how many times bombers in north China attacked the main bridge on the Yellow River and failed to disrupt traffic for more than a few hours, there is little reason to accept the accuracy of Hagenberger's figures. In an order of the day cut by Chennault on 1 March 1945, he says 100 bridges were knocked out, and 250 damaged. These figures probably come closer to the truth.[4]

Two types of radio transmitters were used in China in WWII by agents behind the lines and for air-to-ground communications between agents and airplanes. In this picture, probably taken near headquarters, Chennault watches test being made by intelligence spotting team. Radio in foreground, called a V-100, could be operated off normal electric current or manually by hand crank. The other portable transmitter (not seen here) used by agents in China was the ST-RI, slightly smaller, but more efficient than the V-100.

As for destroying 1,079 locomotives it is difficult to believe there were that many locomotives in all of China. Most of the "locomotive-hunting" was done in north China in the last six months of the war. In his order of the day, Chennault, however says 700 locomotives were destroyed and 450 damaged which differentiates between destroyed and damaged, nonetheless these figures are probably exaggerated, too.

The strength of the Japanese Air Force given by Hagenberger is also questionable. Accumulated evidence of Japanese witnesses in the *U.S. Stragegic Bombing Survey* suggests that the Japanese did not have as many planes in China as Hagenberger says.

Both Hagenberger and the Japanese agree on one point: both sides thought it was operating against an enemy with many more planes than its own.

The Chennault order of the day declares that "we have destroyed more than eight enemy planes in air combat for every plane of ours that he has shot down." The ratio of American "victories" to Japanese losses scored by the First American Volunteer Group was about 12-1, but it is extremely doubtful whether an eight to one ratio was maintained after the United States entered the war, that is, after 1942. The fact is, in the first several months of the war, American squadrons were struggling mightly to stay alive against a superior foe who was coming at them from all directions.

In retrospect, therefore, it seems that American intelligence estimates of Japanese air strength in China are exaggerated as are the estimates made by the Japanese on the strength of the American-Chinese squadrons. Neither side was getting accurate information on the other. The Japanese employed spies, Chinese-speaking Koreans or Chinese traitors to infiltrate the air fields of the Americans. Whatever reports these agents sent back or brought back were suffused with guesswork. On several occasions paid agents attempted to sabotage American bases by blinking lights at night to guide Japanese bombers. Veterans of the 5th Fighter Group also remember the officer who was shot by long range rifle fire while sitting in front of an alert shack at Chihkiang. (He survived).

Japanese historians, recording their memories with the U.S. historians after the war, recall that two and three reconnaisance planes were attached to an army corps in the field, or to an air group to spy on Chinese and American air bases. And most American airmen who served in China recall listening to the approach of "Washing Machine Charlie," or seeing "Photo Joe" fly overhead at 23,000 feet, taking pictures or counting aircraft. However, the technology avialable in photo interpretation in the 1940s made it difficult to distinguish between dummy planes and real planes at 23,000 feet. Chinese technicians became masters at creating dummy planes on most fields.

By the beginning of March 1945, it was becoming clear to the Joint Chiefs of Staff in Washington that the Japanese air forces in the Pacific and in China were running out of aircraft. In an entry in the daily journal of the 14th Air Force recorded by William d'Ettore, this comment appears: TOP SEC (top secret) from Hq. AAF (ARMY AIR FORCE) Washington dated 19 March: "Subject: "The Deterioration of the Japanese Air Forces and its Implication on Future Strategy."[5]

The last major effort the Japanese made to neutralize an airfield in China came in the spring of 1945 when their air and ground forces attacked Chihkiang. Aircraft of the U.S. 5th Fighter Group, stationed at Chihkiang, had created havoc to their supply lines running through the Siang River Valley and the Japanese had to try to stop them.

U.S. troop carrier squadrons transported thousand of Chinese troops and mules from Burma and west China (operation "Rooster") to the Chihkiang front. With strong support from both fighters and bombers based at Chihkiang, the weakened Japanese ground and air forces were thrown back and the campaign ended in a stalemate. At one point in the campaign, Chinese troops at a key salient wavered but were revived by the aggressive leadership of an American officer, Woods King, who was with the Zebra Forces, an artillery training command under Stillwell.

While the Japanese supply lines were being interdicted by American submarines and long range bombers in China and the Philippines, the 14th Air Force in China was being

Flight of P-40N Sharks lowering gear, Central China ca. 1944.

Above: Fighter control personnel at Yangkai tower watch as man puts last nail in field elevation poster. Right: control tower at Kunming air base as it looked in 1942, not long after it was built. Tower was destroyed in 1943 by direct hit in enemy raid and never rebuilt.

slowed down not by the enemy as much as by the mountain ranges which march south out of Tibet into Burma. On a more southerly route it was possible to fly from Kunming to Calcutta without oxygen and without scraping the underbelly of an airplane. But for two years the Japanese controlled this lower route and it was necessary for the 14th Air Force to be supplied over routes which rose from 15,000 to 22,000 feet. This was "altitude" in 1944.

The sub-Himalayan mountains of Burma limited combat operations in China because there was no way to transport gasoline into China except over the Hump by air. But the airplanes used gas to get across the Hump and they had to have enough gas to get back into India to pick up more gas.

The Burma Hump fooled everyone, including Henry Arnold, chief of the Army Air Corps. He thought an airplane, like the heavy bomber, B-24, could fly a combat mission after four trips back to India for gas. No one challenged his information until experience proved him wrong. Most B-24 pilots agree that it took nine and ten trips back to India to store up enough gas for one airplane to run one mission against enemy targets in China. It was an expensive operation, for not only was this a waste of time, but also of aircraft. Too many airplanes were lost crossing the Hump either to enemy interceptors or to the weather, or because they were overloaded. By the beginning of 1945, 14th Air Force squadrons were swollen with too many pilots without aircraft and without gas to fly the ones they had. The problem was never solved. A four-inch pipe line was laid from India along the Burma Road and completed as far as Kunming by the summer of 1945, but by then the war was almost over.

The men who flew the Hump to haul personnel and supplies to the squadrons in China were mostly from the Air Transport Command and several combat cargo squadrons. They were fighting a war just as hazardous as the bomber crews and fighter pilots in China. A report in *Time* Magazine states that 879 men died "in the wrecks of 468 different airplanes trying to fly the cloud-hung Himalaya Hump."[6]

An *Associated Press* dispatch from Shanghai referred to the Hump as the "most congested air route in the world" in which "overall Hump activity cost the lives of 910 crewmen and 130 passengers. 694 planes lost and seventy-five rescued."[7]

The real losses in men and airplanes will never be known because this would include losses suffered by commercial airlines, by fighter pilots crossing the Hump in a P-40 for the first time, and three B-25s running into a weather front and crashing on top of each other against a mountain.

On a clear day the mountain ranges of the Hump are an awesome, yet beautiful sight. An observer, if he were permitted to fly in the glass nose of a B-24 bomber, could look down into a corrugated jungle on the Salween River, a river flowing without motion at the bottom of a canyon two miles deep.

The scope of aerial activity in the China Theater was limited both by the vastness of the country and the logistics of the gasoline supply across the Hump. Despite the limited number of aircraft available for combat in China, as compared to the European theater of operations, fighter pilots engaged the enemy at all levels and against all odds. When they engaged the enemy they could actually see him through the plexiglass canopies as he turned or dove to gain advantage or avoid a disadvantage. The aerial flights over China represented the last personalized, eye-ball to eye-ball fighter tactics which grew out of World War I when pilots fought in open cockpits and scored either because they had greater skills or better airplanes.

Today the human factor has been largely replaced by a long-nose airplane which operates like a computer. A Navy "Tomcat," for example, can track five planes simultaneously without even seeing them. The one-on-one engagements disappeared with the introduction of jet aircraft and electronic sophistication. Airplanes are making deflection shots so far from target as to be almost invisible to the enemy under attack. Arcs, miles wide, have replaced sharp turns.

Meanwhile, an insight into one day of aerial activity on one particular front in China appears in an intelligence summary which was sent out to several headquarters simultaneously by a radio control network manned by American personnel. This summary was prepared by the commander of the 75th Fighter Squadron, then based at Hengyang, during the battle waged by Japanese ground forces for possession of the city of Changte in western Hunan Province, November-December 1943. The message is written in the terse style of code clerks but numbers are spelled out lest errors creep in. This message came off a decoding machine (109) and was copied on a typewriter by a code clerk. It is addressed to "Sandy," a code name for a U.S. liaison team stationed a few weeks at Enshih during the Changte campaign.[8] The term "shark" as used here refers to P-40s because they had a shark's nose. "Oscars" are Japanese Zeroes. The abbreviations in the message are: "prn" for paragraph; "prd" for period; "cma" for comma, and "msg" for message.

The two squadrons engaged in this action were the 75th and 76th. Between them they ran sixty sories that day, that is, sixty airplanes were in the air at one time or another.

HEAVY ACTION PRD. FIRST VIDOVITCH WITH EIGHT SHARKS SEVENTY FOURTH OFF ZERO SEVEN ZERO INTERCEPTED BY APPROXIMATELY SIXTEEN OSCAR TYPE WITH WHITE SPINNERS PRD. VIDOVITCH HIT ONE SEEN TO EXPLODE CMA GAGE AND LUNDY ONE EACH DAMAGED PRD GOBTTLE HIT IN WING BUT RETURNED CMA ALL DOWN ZERO NINE ONE FIVE PRD NEXT GROSVENOR OFF WITH EIGHT SEVENTY FIFTH ZERO SEVEN THREE ZERO TO PATROL CHANGTE PRD ONE RETURNED PROP VIBRATION CMA ONE BACK AFTER PATROL WITH BAD ENGINE PRD. PATROLLED FORTY FIVE MINUTES THEN FLEW NORTH EAST STRAFED LAKES PRD TWO HUNDRED BOATS AT LEAST IN NANSHIEN AREA WITH MEN AND GOODS CMA ESTIMATE TWENTY BADLY HIT WITH FORTY CASUALTIES PD. ALLBACK ONE ONE ZERO FIVE PRD THEN MIMMACK WITH EIGHT SHARKS CMA FIVE SEVENTY FOURTH THREE SEVENTY FIFTH PATROLLED CHANGTE FOR ONE HOUR CMA OFF EIGHT THREE ZERO PRD TRAFFIC NOTED ON ROAD NORTHEAST CHANGTE CMA ALL RETURNED SAFE ONE ONE THREE ZERO PRD NEXT RICHARDSON WITH EIGHT SHARKS SEVENTY FIFTH OFF ONE ZERO ZERO ZERO PATROLLED CHANGTE PRD EAST PART OF CITY BURNING PRD HIT SIX LARGE SAMPANS AND BOATS LOADED WITH SUPPLIES NEAR ANSIANG CMA THREE BADLY DAMAGED CMA OVER ONE HUNDRED BOATS SEEN SOUTHEAST HANCHOW APPROACHING CITY PRD ALL BACK ONE TWO FIVE ZERO PRD THEN LUNDY WITH EIGHT SHARKS SEVENTY FOUR OFF ONE ZERO THREE ZERO PATROLLED NEAR CHANGTE THEN STRAFED LAKE AREA BOATS STRAFED BOATS CONTAINING BOXED GOODS PROB-

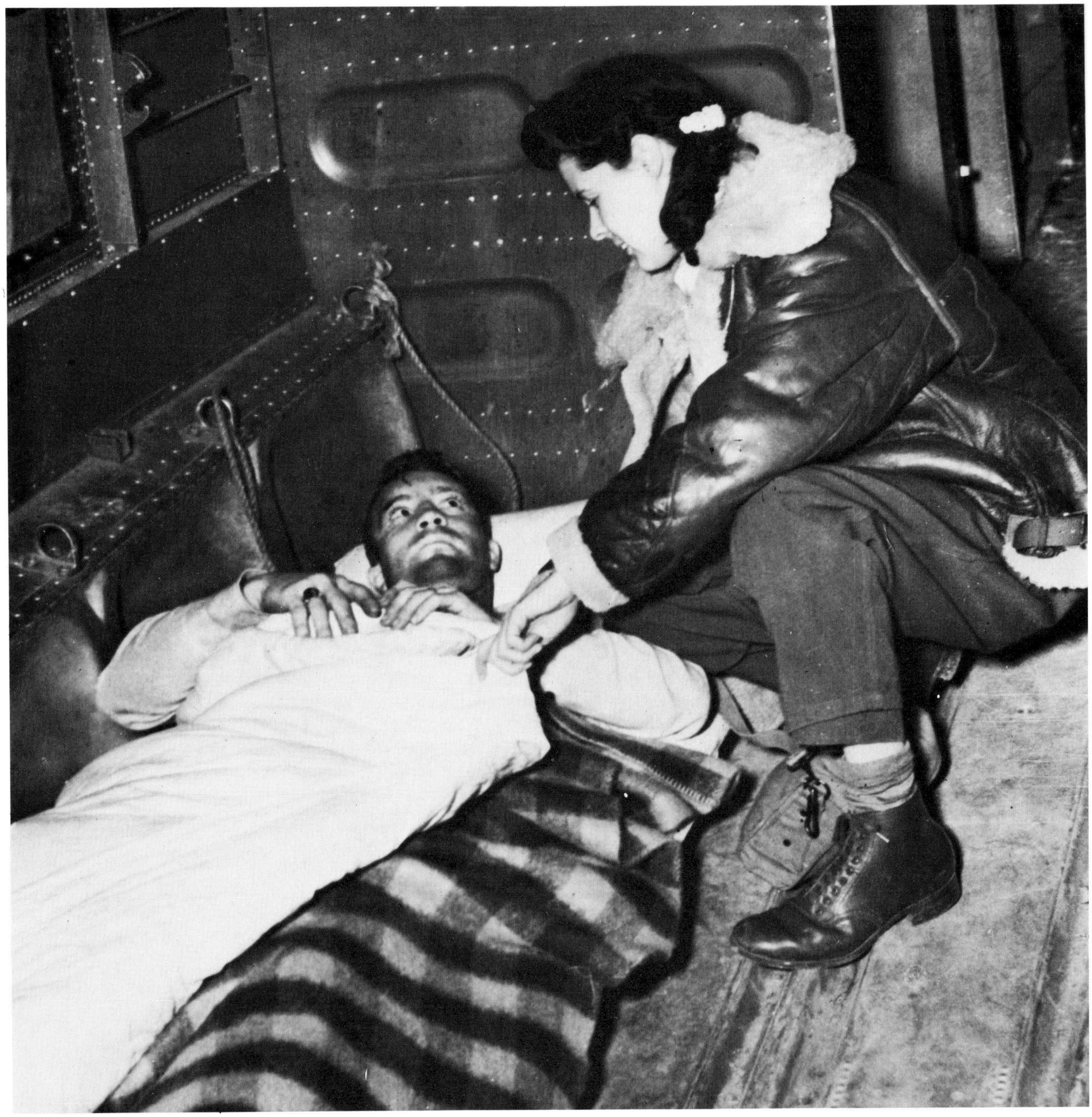

Flight Nurse Rose M. Smith attends patient on board evacuation plane, east China, c. 1944.

ABLY SINKING THREE PRD ALL DOWN ONE THREE THREE ZERO PRD AT ONE TWO ONE FIVE GROSVENOR WITH EIGHT SHARKS SEVENTY FIFTH PATROLLED CHANGTE FOR FORTY FIVE MINUTES THEN STRAFED HOUSES FROM SHIMENCHAO TO TEHSHAN AND HSUTU LAKE PRD INTERCEPTED BY TWENTY OLD TYPE ZEROS PRN MEBBE OSCAR TWOS PRN. FORCED TO RUN BECAUSE OF ZERO ALTITUDE PRD LT. TANNER HAD HYDRAULIC SYSTEM SHOT OUT CMA MADE BELLY LANDING ON FIELD UNHURT PRD THREE CANNON SHELLS IN ELEVATORS OF LT. VURGAROPULAS PLANE BUT MADE SAFE LANDING PRD REST ESCAPED GAINED ALTITUDE AND RETURNED BUT MADE NO CONTACT PRD LT GREEN LOST PRESSURE ON RETURN CMA LANDED WITH PLANE ON FIRE CMA PILOT UNINJURED CMA REST DOWN SAFE ONE SIX ZERO THEN MAURITSON WITH FIVE SHARK SEVENTY FOURTH OFF ONE THREE THREE ZERO INTERCEPTED BY TEN OSCARS TWO FLYING UNDER OVERCAST PRD LT SMITH ONE CONFIRMED SHOT DOWN IN FLAMES CMA MAURITSON ONE PROBABLE MEYER ONE DAMAGED PRD ALL RETURNED SAFELY ONE SIX ZERO ZERO PRD LASTLY COTTON OFF ONE FOUR TWO FIVE WITH SEVEN SHARKS SEVENTY FIFTH TO DROP SUPPLIES TO CHINESE ON TEHSHAN MOUNTAIN PRD SAW FIFTEEN OSCAR TWOS NEAR TEHSHAN SO DIVED TOWARDS MOUNTAIN PRD TWO TANKS BROKE LOOSE CMA ONE SALVOED SHORT OF TARGET CMA FOUR IN TARGET AREA PRD NO CONTACT WITH THE JAPS CMA ALL BACK ONE SIX TWO FIVE PRD SIXTY SORTIES TODAY PRD.[9]

The abbreviated names above are Harland S. Vidovich, William J. Grosvenor, Richard A. Mauritson, Russell F. Mimmack, Elmer W. Richardson, squadron commander, Vern A. Tanner, Glen E. Lundy, Laurence W. Smith, Fred F. Meyer, Thomas Cotton, Nathan E. Green, and James C. Vurgaropulas. The last named was later KIA, and Green was shot down but "walked out" with the help of Chinese guerillas.

The message below is typed on Chinese rice paper by code clerks attached to U.S. radio liaison station, Dog Sugar Eight, at Chinese 9th War Area headquarters at Changsha 27 May 1944, the message is from Philip C. Loofbourrow, C.O. of 75th Fighter Squadron based at Hengyang, ninety miles south of Changsha. The message reads:

> TO DS8 FROM SM5 (HENGYANG) GOT TWO TRUCKS AND SOME CAVALRY BETWEEN YOCHOW AND PUCHI TODAY PRN ROSHOLT FROM LOOFBURROW MSG GEORGE ONE ONE ONE PRN INTERCEPTED BY FIFTEEN TO TWENTY HAMPS OSCARS TOJOS SOUTH OF PUCHI WE GOT ONE HAMP CONFIRMED ONE TOJO DAMAGED PRD ELKER AND GREEN MISSING CMA DANIELS WOUNDED NOT SERIOUS CMA BELLIED IN HERE PLANE A LOSS.[10]

Post script: Oswin H. Elker was shot down north of Changsha and reached Dog Sugar Eight about a week after bail out. Nathan Green went down near Yochow and was more than a month walking out because he was surrounded by enemy troops. He asked, through his Chinese rescuers, for a parachute drop of medical supplies and the message was relayed to 9th War Area headquarters and from there to Rosholt who advised Loofbourrow of the request by Green. A drop was arranged, but when the rescue mission reached the drop zone—indicated by a panel agreed upon—Green was gone. His rescuers had to move him, quickly. Walter Daniels, though wounded, flew again, but was later KIA.

Between March 1943 and the end of the war in August 1945, the 14th Air Force became involved in three major

Factory No. 10, main repair depot for Kunming Air Base. Spare storage sheds at right, empty gas drums used for support functions and platforms. P-51 under repair at left, No. 259-2106950, is probably 51st Fighter Group.

Sunday religious service in chapel at Yangkai shared by all squadrons. Enlarged patch on wall is 373rd Bomb Squadron, based at Yangkai together with 22nd Bomb Squadron (M) and 491st Bomb (M).

For something else to do, these ground crews from 22nd Bomb Squad ron at Yangkai went off base for a Sunday picnic in the hills.

Two jeeps pick up returning crews with 22nd Bomb Squadron at Yangkai.

Six-by-six loads up crew for mission with 22nd Bomb Squadron a Yangkai.

Parachute riggers used long tables to lay out 'chutes for proper folding. Shelves hold spares for 22nd Bomb Squadron.

Chow line at Yangkai for troops of 22nd Bomb Squadron.

and one minor military campaign launched by Japanese expeditionary forces in China, namely, the attack on Changte in November-December 1943, the attack on the Hunan-Kwangsi railway to neutralize U.S. airfields in the summer and fall of 1944, the attack on the airfield at Chihkiang in western Hunan which was launched in the early part of 1945, and the brief campaign for the city and airfield at Laohokou, also in the spring of 1945.

But none of these fields posed a threat to the Japanese homeland. This would come from a complex of fields yet to be built on the Chengtu plain in west China for the long-range aircraft, the B-29s.

Fighter squadrons which saw the most action in China were the 74th, 75th, and 76th, and later the 118th Tactical Reconnaissance Squadron of the 23rd Fighter Group; the 16th and 449th Fighter Squadrons of the 51st Fighter Group; and the 26th Fighter Squadron of the 5th Fighter Group. With the exception of the 26th Fighter Squadron, which defended Chihkiang, the others saw more action than later arrivals because they were first in the China Theater. The 74th, 75th, and 76th were activated out of the three former squadrons of the First American Volunteer Group.

The following Unit Citation was cut on 27 August 1945:

23RD FIGHTER GROUP

The Japanese were sending 70,000 crack ground troops down the Siang River Valley in Hunan Province, China, aiming at the capture of Hengyang, vital communications center and mid-way point in the Japanese strategy to drive an inland corridor across China. A major defense stand by ill-equipped Chinese ground forces was planned at Hengshan, 25 miles north of Hengyang, to attempt to stop the drive. Between 17 June 1944 and 25 June 1944 the 23RD FIGHTER GROUP threw its total effort into the battle. On all but three days during this period the weather was adverse to aerial operations, with an overcast arched over the river valley and resting on the mountains that lined both sides of the valley. Demonstrating extraordinary heroism, the pilots flew nearly half of their missions during this period through this "tunnel" created low above the valley. The valley floor was studded with machine guns, anti-aircraft guns and thousands of rifles in the hands of the troops, forcing the pilots to fly through deadly curtains of machine gun and small arms fire. Despite the extreme hazards, the Group's pilots flew 538 sorties, strafing and bombing the enemy spearhead forces. They killed 1,640 troops and destroyed approximately 780 cavalry and pack horses. Striking at the supply lines immediately behind the front, they destroyed 377 small boats and damaged 372 more; sank fifteen large river vessels 100-or-more feet in length and damaged eight. They destroyed 91 motor trucks and damaged 50. They also sank three and damaged two heavily-armed gunboats that the Japanese had rushed into the area to protect their water supply lines. In addition, they wrought extensive destruction among supplies and equipment in the 100 or more compound storage centers they destroyed and damaged. In four encounters with enemy aircraft, the Group's pilots shot down seven enemy planes, probably destroyed seven more and damaged eight, losing none of their own aircraft. This lone, gallant stand by the 23RD FIGHTER GROUP against 70,000 enemy troops, despite adverse weather and even after the Allied ground defense stand at Hengyang failed to develop, is expressive of an extraordinary heroism, gallantry, determination and esprit de corps in keeping with the highest traditions of the American military service.

BY COMMAND OF MAJOR-GENERAL STONE (Charles B.):

The bomber squadron with the most time in the air was the 11th, which arrived in China in the summer of 1942 before any other bomber unit. Although a medium bomber, the B-25 was also used for strafing enemy boats on rivers and lakes, skip bombing enemy shipping off the coast of China and low-level bombing enemy shipping off the coast of China, and low-level bombing of enemy airfields.

The 341st Bomb Group, also a medium bomb outfit, saw action in China in the last months of the war. The Unit Citation, dated 25 August 1945, is repeated here:

341 BOMBARDMENT GROUP (M)

Between 11 December 1944 and 12 March 1945 this Group waged an extremely successful and highly dangerous bridge-busting campaign along the land corridor then held by the Japanese between north China and the tremendous raw material potential found in her conquests in southern Asia and adjacent islands. To thwart the Japanese plan to capitalize on this land line, the Group was assigned the hazardous task of destroying the numerous steel and concrete bridges on the modern rail lines in French Indo-China. Two primary considerations faced the Group in preparing its method of executing this mission. Supplies, gasoline and bombs were at a premium in air-supplied China, and the Japanese had ringed the bridges with extensive antiaircraft defenses. To meet these considerations, the Group developed and employed its own style of attack, "Glip Bombing," which employed a triple change in bombing level at low altitude as an elusive maneuver over heavily defended, channelled approaches to the targets. This technique yielded such accuracy that the tonnage of bombs expended per bridge destroyed reached a new record low of 7.75 tons per bridge. Despite hazardous conditions of low visibility, rugged terrain and intense, accurate antiaircraft fire under such a large part of these "Glip Bombing" missions were run, the Group destroyed 21 major bridges and damaged 17 in twenty-three missions. Particularly representative of the determination and perseverance of the combat crews are the missions of 27 February 1945 and 5 March 1945. A total of 10 high priority bridges were des-

Evacuating Suichuan, January 1945. Jeep, called "gee-poo-cah" by Chinese, is being shoved up ramp with driver trying to help.

Air evacuation nurse supervised movement of patient into airplane. Note parachute gear strapped at door at right.

troyed and two damaged under enemy fire, with the expenditure of 3.85 bomb tons per bridge. Because the bridges were generally located in gorges and valleys, approach to the targets was restricted to narrow lanes in which the enemy could easily concentrate their defensive fire. Four of the Group's planes were shot down in these attacks and 31 others were damaged. Twenty crew members were killed and twelve were wounded. The cost was not light to the group, but the interdiction of this overland route was imperative. With extraordinary heroism, gallantry, determination and esprit de corps, the Group not only met the dangerous challenge but established a new record in economy of operations in doing so. These achievements of the 341ST BOMBARDMENT GROUP (M) are worthy of the gallant traditions of the American military service.

BY COMMAND OF MAJOR-GENERAL STONE.

Photo probably taken near Chengtu of P-47.

P-47Ds ("Thunderbolts") peel off somewhere in west China, probably 1945.

P-47Ds on line for 81st Fighter Group at Ankang (near Chengtu), spring of 1945.

The 81st Fighter Group landed in Ankang without much advance notice and it had to use this old Buddhist Temple for a headquarters building when the staff began to arrive in February 1945.

Repairs to engine under way, somewhere in China in WWII.

Court martial proceedings probably at Kunming, 1944. Two sergeants at right taking down testimony. M.P. stands behind at rear with shoulder-holstered gun.

Sacks of rice packed inside second sack of burlap are being kicked out of cargo carrier to supply Chinese engineers working on Burma Road. Inner sack would burst on impact without second sack to absorb shock.

Lettering on negative written by lab technicians shows strike by 11th Bomb Squadron, 16 November 1943 on Hongkong.

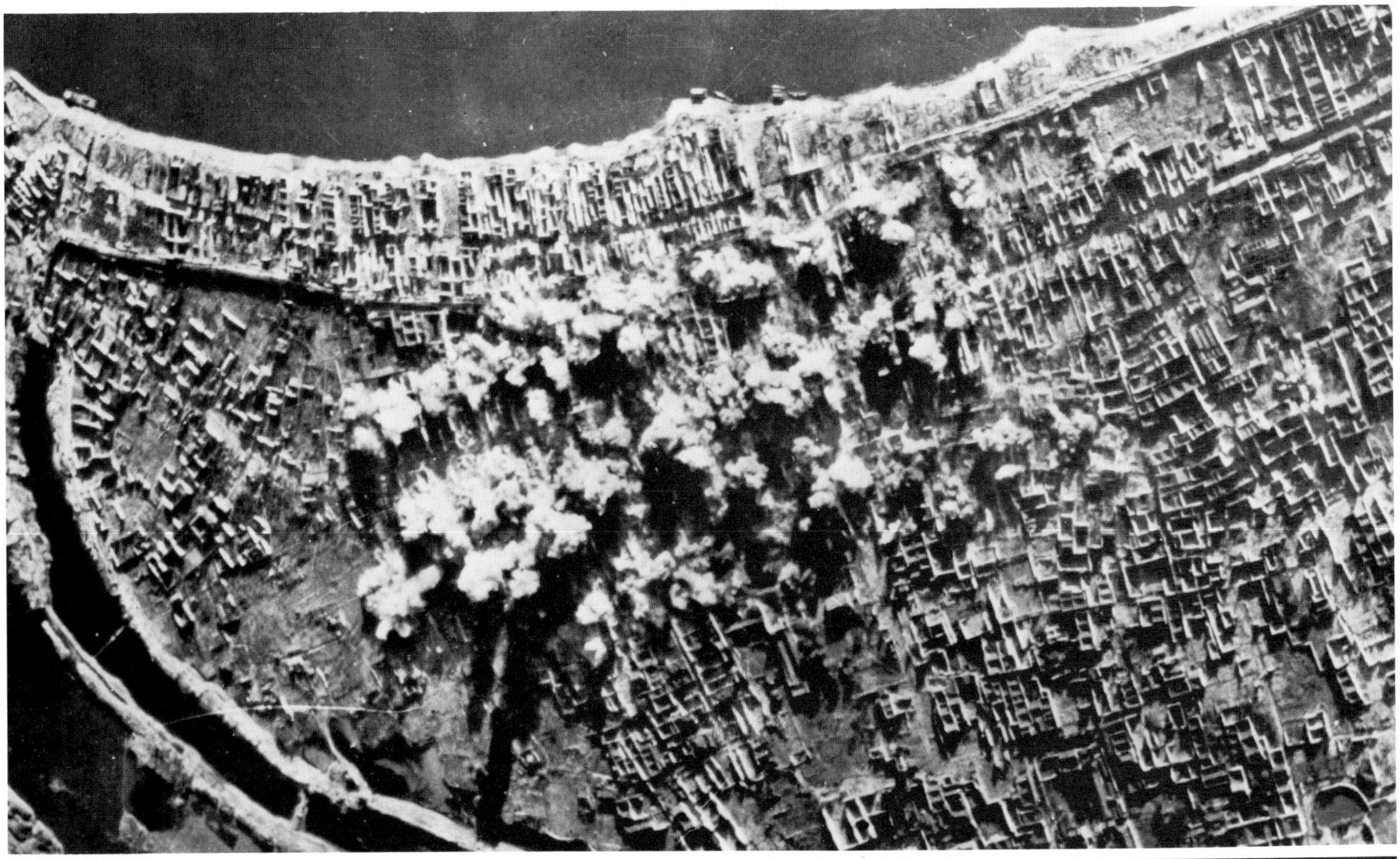

City of Changte on Yuan River in western Human Province comes under attack by U.S. bombers during brief occupation by Japanese forces who were driven out a few days later by Chinese but not before entire city was in ruins, mostly as a result of fires which were started in the fighting with no one to put them out. Battle for Changte by Japanese began in November and continued until third week in December, 1943. Most Chinese houses and stores were made of adobe and the honeycomb resemblance seen here can be attributed to the wood structure burning while leaving the adobe walls and chimneys standing.

Overleaf: painting by Hal Rozema catches moment in time when fourteen B-25s, eight P-38s and seven P-51s are poised for attack on Shinchiku (Formosa) air base, Thanksgiving Day, 25 November 1943.

Number 448 of "Ringer Squadron" (491st (M) Bomb) at moment bomb bays were opened to release demolitions. Terrain suggests North China. The B-25 medium bomber could carry twenty-four one-hundred pound bombs, or eight 300 pounders, or six 500 pounders, or three 1000 pounders. There was also the "blockbuster" but seldom used. Both these photos were probably shot with an Army K-17 camera from the hatch between the bomb bay and the tail section. Some photos were made with a hand held camera known as the K-20 which the tail gunner could use when not in action.

P-38s escort medium bombers of 491st Squadron in attack on Kowloon docks 1944.

Below: Airplane makers in U.S. sent technical representatives to all zones of combat to observe performance of their aircraft under combat conditions and to suggest modifications if necessary. One of these "tech-reps" in China was Eugene Felt who represented Rolls Royce motors made by Packard Motor Company at Detroit.

Army helicopters arrived in China in spring of 1945 and crews had to live in tents. Here No. 345330 is having its rear prop serviced, crew chief standing on up-ended gasoline barrel. "Choppers" were used mainly for rescue work, and Richard Treiber commanded the squadron which was assigned to 14th Air Service Group at Kunming.

P-51Ds from 75th Fighter Squadron on flight over south China, spring of 1945.

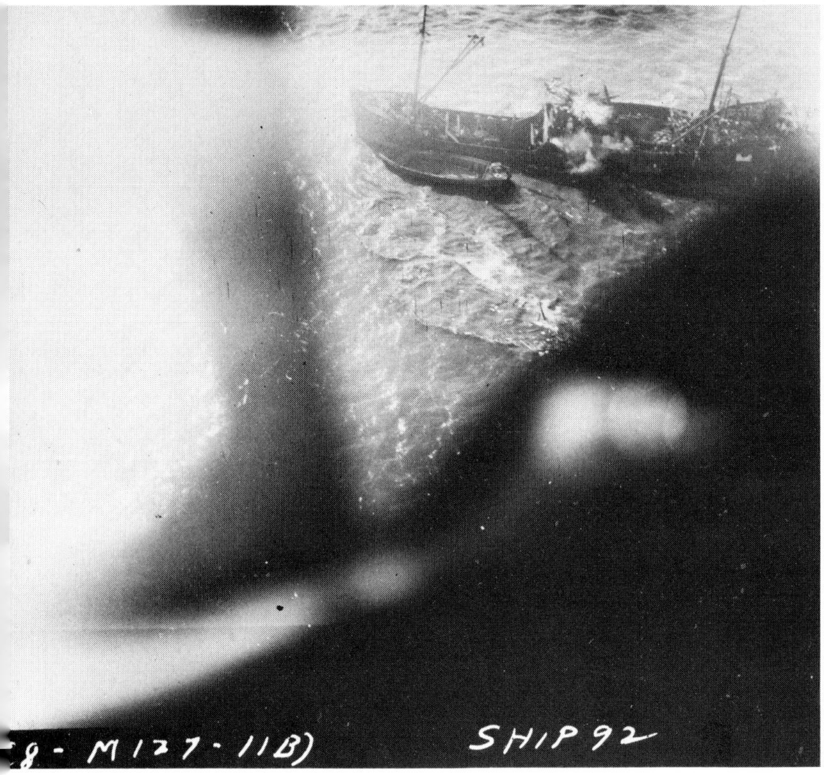

Enemy freighter (top) off China south China coast before mission 157 for 11th Bomb Squadron went into action. Center, close up of freighter seconds before it was hit, and below, explosion astern. Lab technician's notations on bottom of negative.

This airplane, as all can see, has been "red-lined," meaning it can fly but don't ask how far?

Two night fighter squadrons, the 426th and 427th, reached China in the last months of WWII. This P-61A was from the 426th. Background of field suggests picture was taken on one of satellite fields around Chengtu.

P-51Bs on line, August 1944, probably Kunming air base. No guns.

P-38s either 21st or 35th Photo Reconnaissance squadrons. For more than a year, the two squadrons had only one ship each. The ones shown here arrived in China in last months of war.

Below: Plane gets refueled by tanker truck. In **early days, planes got gas** from five-gallon tins carried forward, one by one by platoon of Chinese **workers.**

After a combat mission, it was customary in China Theater to issue a ration of liquor to each member of a returning crew. Here crewmen of the 22nd Bomb Squadron (M) at Yangkai, line up for a shot from a bottle held by the staff sargeant, while officer next to him sees that each glass is filled equally.

Left: Evacuation of Suichuan air base. Here base personnel prepare to board Gooney-bird. Last plane took off on 23 January 1945.

Troop getting ready to leave Liuchow, B-bags filling back end of weapons carrier, November 1944.

American troops being evacuated from Liuchow air base November 1944 by Air Transport Command. Hood on truck painted with shark symbol used on nose of P-40s.

Bucket seats and personal gear of Ground crew of 491st Bomb (M) Squadron en route to Liuchow. After Liuchow was evacuated in November 1944, crews moved to Yangkai.

U.S. flag at half mast on an American air base in China observes passing of President Franklin D. Roosevelt on 18 April 1945. Chinese flags were also flown at half mast in all cities of "Free China" (not under Japanese occupation).

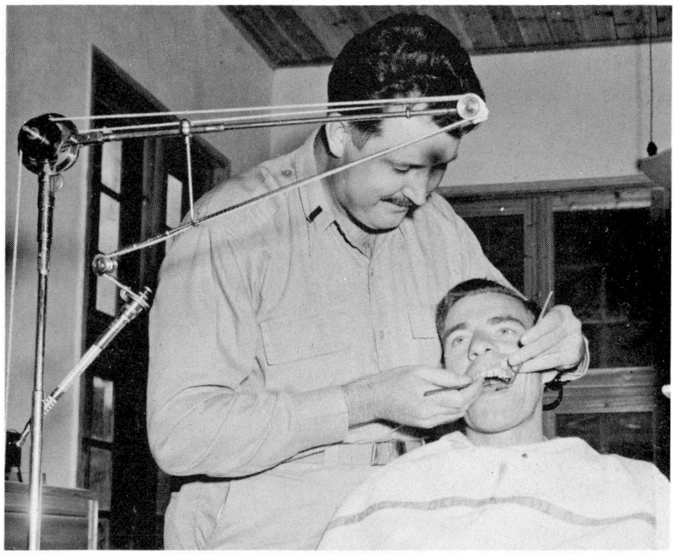

1st Lieutenant dentist probes for infected teeth, foot pedal drill hanging menacingly at left.

Hot Springs Sanitary Hotel, R & R center, near Kunming.

14TH. AIR FORCE

CHINA – WORLD WAR II : 1943-1945

AIR OPERATIONS CHART

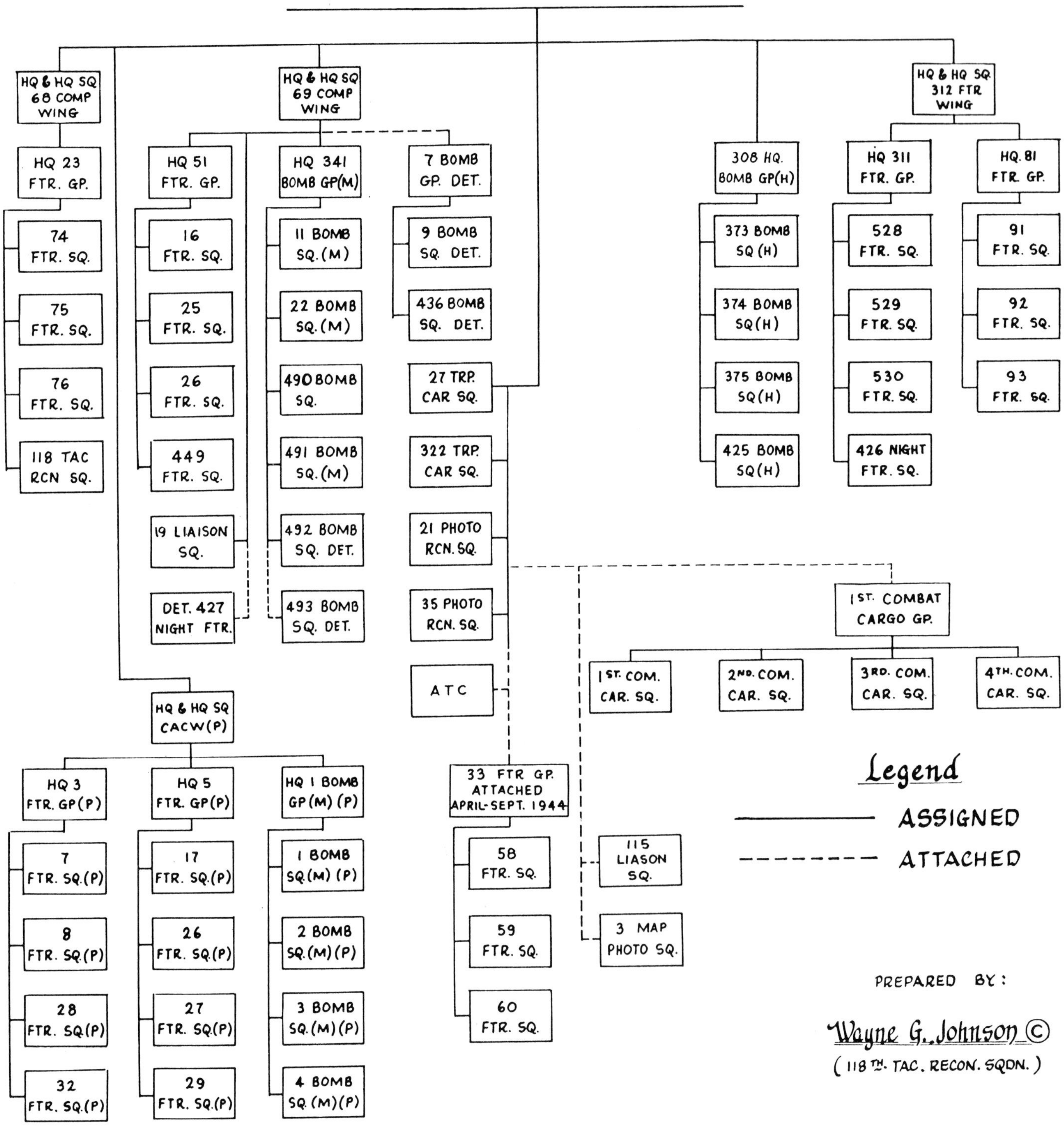

A rare picture, preserved by Eugene H. Beebe, first commander of the 308th Bomb Group in China, shows a two ball alert which signifies that enemy aircraft are within a few miles of field. A one-ball alert meant that enemy aircraft had taken off from some known field, like Hankow or Hanoi, and were heading "our way." A three ball alert warned people that enemy aircraft were almost overhead and to take cover. The air raid warning system, called *ching pao* in Chinese, was not uniform. On some fields, a gong was pounded, and a gong was usually pounded at night when hill signals were not visible. These air raid alert devices were usually perched on the top of strategic hills where they could be seen by the majority of people and, in Chungking, a city of many hills, there were several poles, not trees, from which "baskets" were hung. The two-ball alert seen in the photo above, perched on top of a great tree, was probably taken at Chabua.

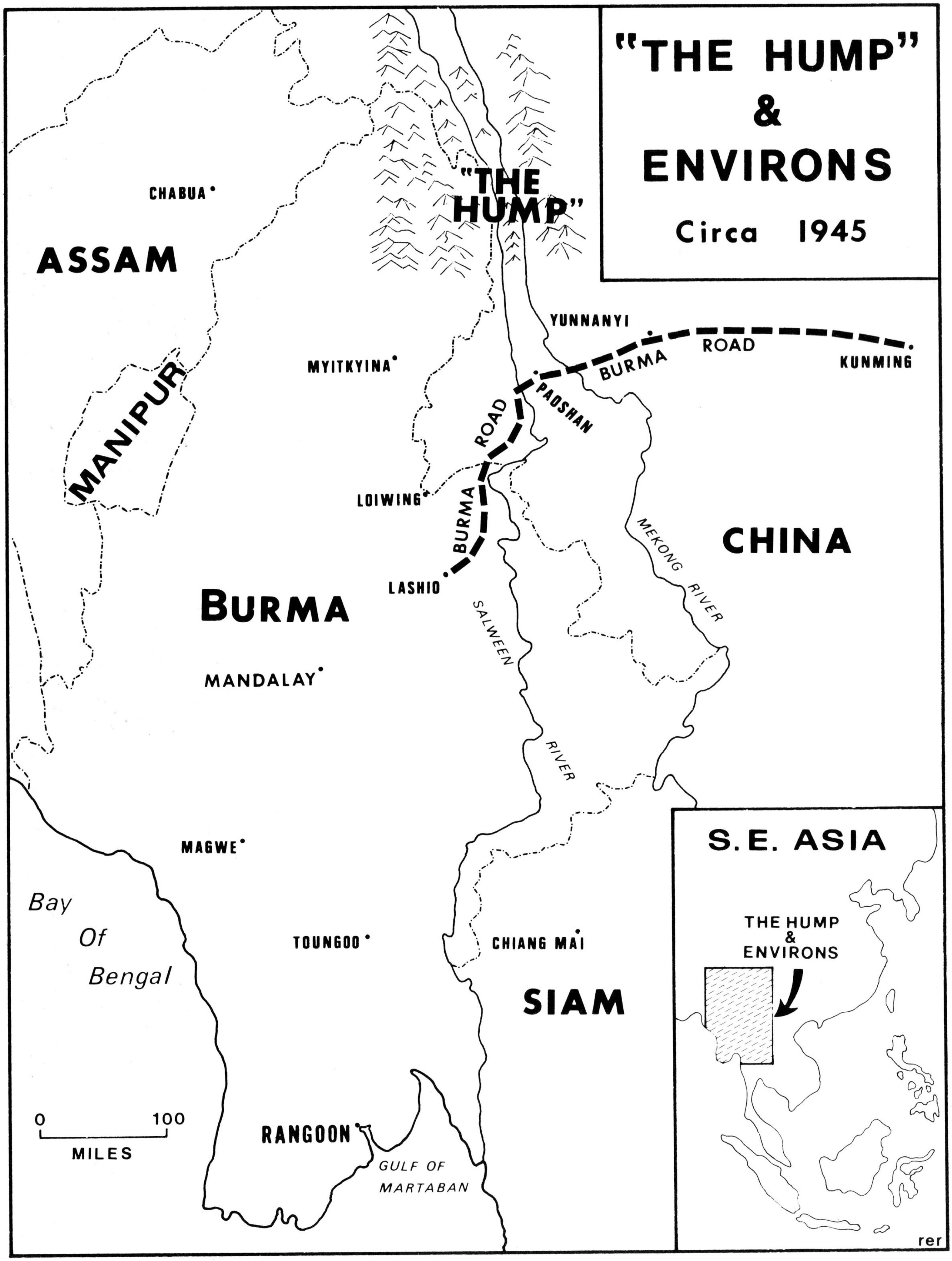

—Map by Randall E. Rohe

XIX

Mission to Shinchiku

By far the single most successful yet audacious mission ever made by squadrons of the 14th Air Force was carried out on Thanksgiving Day, 25 November 1943, against Shinchiku in northwest Formosa, probably the most important enemy air base between Clark Field and Haneda. By staging out of Suichuan, an advance base in east China, the mission was completed in less than five hours without running out of gas.[1]

Clinton D. ("Casey") Vincent, commanding the 68th Composite Wing at Kueilin, flew to Suichuan with David L. Hill, commanding the 23rd Fighter Group, to set up the mission. Hill, better known as "Tex," a former A.V.G. ace who had never lost his way home in two years of combat, was to direct the attack. There were eight P-51As and Bs in his formation, all from the 76th Fighter Squadron, but one was forced to abort shortly after take-off owing to a hydraulic leak.

Spearheading the attack would be eight P-38s from the 449th Fighter Squadron, led by Sam Palmer, followed in flights of two fours by Ryan Moon, Kendal B. Dowis, Robert B. Schultz, Alfred Yorston, Jr., Walter L. Rose, and the squadron commander, George McMillin, also a former A.V.G. ace.

Reconnaisscance photos showed a large concentration of enemy aircraft on the base at Shinchiku, some on runways, some in front of hangars, and along the taxiways of the three main runways. (See air views after strike).

Eight Mitchells of the 11th Bomb Squadron, and six Mitchells from the Chinese-American Composite Wing, were to approach the target, fourteen abreast, dropping fragmentation clusters, the most effective demolition against aircraft on the ground.

The 11th Bomb formation was led by Joseph B. ("Preacher") Wells, flying No. 61 plane, and crew including Cleve L. Bingham, Raymond J. Mazanowski and Joseph O'Connel; ship No. 39 was flown by George T. Grottle and crew including Harold Rochelle, Herbert I. Robinson, and Charles Fox; ship No. 79 was flown by Loren Nickols and crew including John L. Booth, Sanborn V. Howard, and Frank Norton; ship No. 40 was flown by Harold K. Searle, and crew including Stanley Johnson, Paul J. Dickmann, and John Costello; ship No. 55 was flown by John R. Hill, and crew including John Schmidt, Frank H. Gibson, and William Johnson; ship No. 57 was flown by Carl J. LaValle, and crew including Richard L. Schwartz, Gerald O'Day, and Robert T. Hamilton; ship No. 92 was flown by William A. Brenner and crew including Ernest J. Bell, George E. Dykstra, William G. Jones and Carl Cannon.

The six planes from the C.A.C.W. included No. 270 which was flown by Irving L. Branch, actually group commander in C.A.C.W. medium bombers. He was accompanied by (no first name given in mission report) Chang, F. Grover, E.C. Chang and J.C. Sun; ship No. 135 was flown by L. Daniels and his crew included Y.L. Lee, S.C. Chang, C.S. Chen, and C.K. Teng; ship no. 839 was flown by, (no first names given) Harpy, and his crew included (?) Wu, (?) Sheldon, (?) Penny, and (?) Sun; ship No. 136 as flown by (?) Beaucrest and his crew included S.Y. Lo, W.S. Fu, K.C. Chen, C.C. Yang; ship No. 266 was flown by W.P. Carson and his crew included K.C. Wen, H.P. Lee, T.J. Ho, and P.L. Tang; ship No. 208 was flown by S. Churchill and his crew included T.H. Liang, H.H. Chin, H.H. Cheng and P.S. Tu.

A passenger on "Preacher" Wells' lead ship, not listed in the mission report, was Theodore H. ("Teddy") White, correspondent for *Time-Life*.

James M. Williams was the flight leader in Hill's formation of P-51s, and Williams was followed by John S. Stewart, Donald Hedrick, Richard O. Olney, Dale Bell, Robert T. Colbert and Les Manbeck. It was Manbeck who had to abort due to a leak in the hydraulic system.

B-25 from the 11th Bomb Squadron en route to Shinchiku 25 November 1943.

All three squadrons crossed the Straits down on the deck, almost close enough for the spray to cloud up their windshields, in an effort to avoid any possible radar detection.

The main runway at Shinchiku ran roughly north-south with diagonal runways to the east and west pointing to the pivotal north end of the main runway. As soon as the coast was sighted, all aircraft began to ascend from the deck to at least 500 feet altitude, approaching the field from the southwest corner where they banked left to fly north across the target area.

Today forty years after the fact strong beliefs are held about the exact position taken by the two fighter squadrons as they approached the field. The official report has this to say:

> Eight P-38s went into strafe and got those planes in the traffic pattern followed by the B-25s (14) at 1000 feet with fragmentation bombs, followed by our flight of 7 P-51As to straff (sic) the runways . . .

One pilot who participated, said that the three squadrons, that is, the two fighter and one medium bomber, were not much more than 200 feet behind each other, and that the

The entry at the b
tom of this pict
was made by a
technician on the
ginal negative as
was being proces
in a dark room of
21st Photo Reconna
sance Squadron. T
caption, lower le
reading left to rig
gives the direction
the picture (arr
pointing north) a
the file number,
squadron, date, a
tude—29,300 fee
and time picture w
taken—1220 hour
plus longitude a
latitude. If this pict
was taken at 12
hours, it was less th
twelve minutes a
the other one (see
companying pho
was taken by a came
man riding in one
the B-25s who to
his picture at 12
Thus, the higher
ing photo recon s
was over the tar
only minutes after
last airplane in
mission had beg
climbing for return
Suichuan.

B-25s were spread out fourteen abreast as they swept across the field, a point not mentioned in the mission report. A photograph taken by one of the B-25s over target also has the airplane at 400 feet altitude, not 1000.

Thus, the plan was for the P-38s to go in first, followed by the B-25s, with the P-51s coming down to "mop up" anything still not smoking or flying in the traffic pattern that the others had missed.

Hill recalls, according to the mission report, shooting down an olive drab Oscar Mk II which was making a chandelle in front of him while apparently trying to get on the tail of the last bomber. Bell, in the second element leader's position, caught a jet black Zeke in a similar attitude at 400 feet a few seconds after Hill's blast, and both enemy plane went down, according to the mission report.

A Nate came in on Williams from 1 o'clock above, bu Wiliams continued to strafe, and Oleny, on Williams' wing shot a burst at the Nate as it turned on Williams and scared it away.

Among the P-38s, McMillan destroyed one bomber or the ground; Moon confirmed two bombers in the air and two on the ground; Palmer destroyed three fighters and bombers in the air; Schultz got two in the air, and Dowi knocked one out of the sky and blew up three on the ground Yorkston confirmed two in the air, and Rose destroyed two and damaged one in the air.

Probably first photo of air strike on Shinchiku taken by cameraman on board one of B-25s at 1208 hours at 1100 feet. Smoke obscures wrecked Japanese aircraft, but bombers in lower edge of field not yet hit.

Among the fighters of the 76th Fighter Squadron, Williams saw one enemy plane, probably an Oscar, crash and burn at the northwest end of the field, and a short distance farther up the field, another one that went down in flames, definitely a Zeke. Stewart, in the second flight, also saw two planes crash in this area, probably the aircraft claimed by Hill and Bell.

As the first flights approached the target, Hill and Hedrick fired at several Nells on the west corner of the field. Two were confirmed, and a second flight noticed a good fire in this area. Near the center of the field, Williams saw a Sally [bomber] badly damaged with one landing gear gone and a wing down. He passed over this to hit another Sally parked nearby, setting it on fire.

Stewart and Colbert went to the right around a fire near the south corner of the field and each confirmed one Sally parked in front of a hangar. The pilots from the 76th, as a squadron, believe they saw twelve planes burning on the field as they banked left to return home, most of the enemy planes hit by fragmentation clusters dropped by the Mitchells.

The Japanese had a machine-gun in a pit on the southwest corner of the runway and the gunner may have been responsible for the single bullet hole found in one of the P-38s on return to Suichuan.

Intelligence officers, after debriefing, and after studying photos of the strike made during and a few minutes after the strike, said between fifty and sixty aircraft were destroyed in the air and on the ground, plus others damaged. The P-38s took credit for twenty-two destroyed of which twelve were in the air and ten on the ground, according to the report.

The first flight began to return to base at a quarter to three. All aircraft landed safely although a ground loop was avoided when Oleny over-controlled his P-51 and damaged the prop in a kangaroo landing. By the time everyone had landed, it was time for Thanksgiving Day dinner with chicken and rice but, fearful of a bombing attack from the Japanese, most of the aircraft gassed up and took off for Hengyang and Kueilin.

John Stewart, who flew one of the P-51s on this mission, gave this eye-witness account of the raid:

"This was my first mission in a P-51A. Of course we carried drop tanks under each wing, and were still on the drop tanks when we reached the Formosa coast at minimum altitude and the navigator in the lead B-25 called 'landfall!' As pre-planned, all of us climbed from cruising altitude of ten to twenty-five feet to about 500 feet or more. I dropped by tanks along with the others but failed to switch the external tanks and made my strafing run on a dead engine. All switches in a 51 cockpit can be hit in three seconds because I hit all of them and the engine caught just as I fired at a Betty bomber parked on the field and blew it up, killing the surprised crew chief who was sitting on top of the fuselage.

"The time from landfall, some three miles from the field, to completion of our run across the field, could not have taken more than three minutes. We were cruising at 180 to 200 miles mph when we hit the coast of Formosa and since it was about three miles to the field, plus two across, at the most, it didn't take long to do the job.

"Returning to Suichuan, the navigator in the lead B-25, must have dozed off because he failed to see the field behind a range of high hills to the west, and his pilot, Wells was on a heading for Canton, way the hell-and-gone into Jap-occupied territory. I called him to make a right turn, which he did, and all returned to base."[2]

XX

The Chinese-American Composite Wing

By the time that China and the United States joined forces in World War II, the Chinese Air Force had gone through a long struggle to stay alive against superior enemy aircraft and better trained pilots. And after the coastal cities came under blockade by the Japanese Navy, it was difficult to import new aircraft and parts. For a time the route lay through Hanoi in Indo-China via the French-built narrow guage railway to Kunming, but in 1940, that route was cut off and after that there was only one way to bring in equipment, namely, to fly it in from India over the Hump.

The background of the Chinese Air Force, as suggested in earlier chapters, was mixed. There were pilots and crewmen from provincial air forces and from national air forces. Some had trained under French or British instructors, some under Italian or American. They flew a hodgepodge of airplanes from all over.

Rivalries had arisen between Chinese commanders and between squadrons, and from 1927-1937, there was constant friction between air units in Nanking and units of the provincial government at Canton, birthplace of the Revolution. All these factors were reflected in one way or another by the uncoordinated command structure of the Chinese Air Force, still under the army.

The prospects of improving the situation in the middle of a war were dim. Chennault, who had watched several years of the internecine struggle, realized that the best thing the United States and China could do was to organize a combined air force of Chinese and American crews with all-American equipment, under his command. The outcome of this was the organization of the Chinese-American Composite Wing, known in Chinese as *Chung-mei Yun-ho.*

Before this idea was conceived, Chinese pilots, under the auspices of the Chinese Air Force, were being sent to Thunderbird and Luke fields in Arizona for training, some in basic, and some ready for advanced training. Out of this pool of cadets, some were brought back to India and assigned to a new training area at Malir Air Base located in the desert north of Karachi (today in Pakistan). Some of the trainees among the fighter pilots were also veterans of air battles over Chungking, but who had never flown a P-40.

Malir Air Base, although surrounded by the Sind Desert, lies not far from the coastline of the Arabian Sea. Between the desert and the sea lies a vast area of mud flats and low sand dunes, excellent for gunnery practice and skip bombing.

Some of the training planes at Malir were P-40Ks left over from north Africa, some from the Black Scorpion Squadron of the Tunisian campaign and several from the American Volunteer Group.

One of the officers shipped out from the states to activate the C.A.C.W. was Eugene L. Strickland.[1] In his diary he recalls the high accident rate at Malir, much of it the result of malfunctions in old engines with too many hours of bad

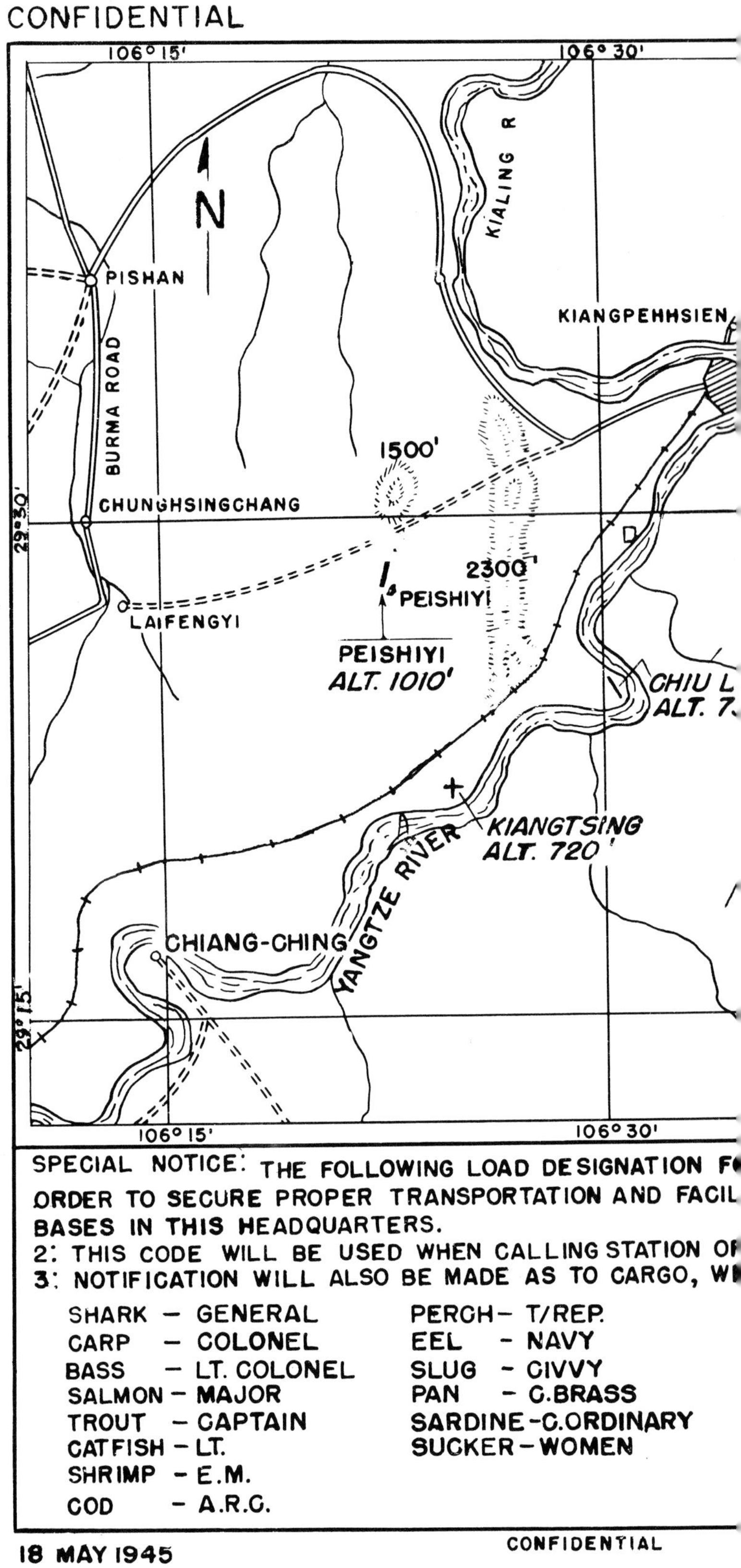

maintenance. For example: one officer of the 1st Fighter Squadron was injured when he bellied in on a P-40B because the engine cut out. The fuel strainer had been installed upside down!

Among the Chinese Air Force veterans at Malir was a Captain Wang. He pulled up the trouser on his right leg to show a scar which ran from knee to ankle. Said he: "Over Chungking I was flying a plane something like your AT6 (advanced trainer) when four Japanese Zeroes came down on me. I rolled over and the first one missed. I caught him in my sights and he went off smoking. But the next Zero came down on me and I had to jump." He looked at his leg again.

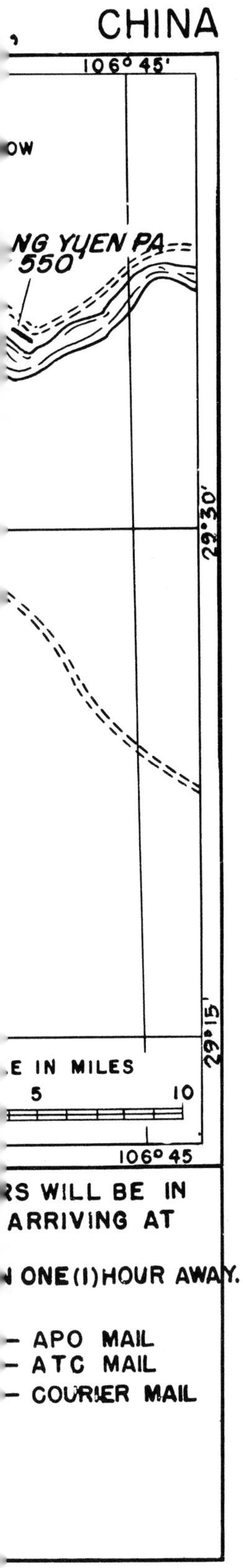

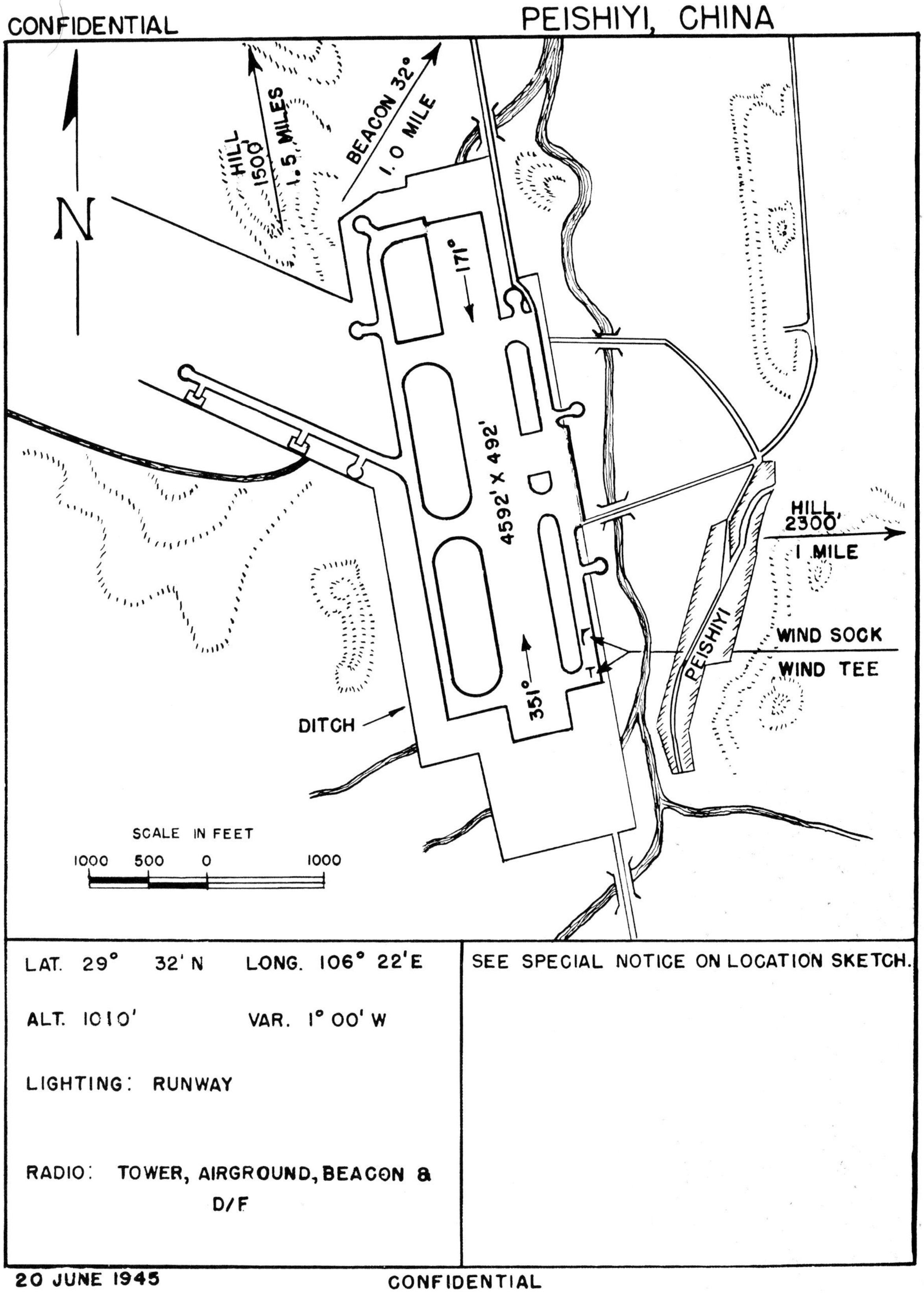

Overlays above taken from *Pilots' Handbook,* U.S. Army Air Corps 7 May 1945. Maps were used to guide pilots into field and provide navigational aids. Map at right was for commercial pilots approaching Peishiyi with passengers. Chungking was capital of China at this time and control tower had to prepare for proper reception of passengers according to rank. Incoming pilots used "fish" code found at bottom of map at right. Most of code words are self-explanatory, but a few may need clarification, e.g., "Shrimp" for "E.M." refers to enlisted men; "A.R.C." to American Red Cross, "Slug" to civilians, "Pan" to Chinese brass, and "Sardine" to field grade brass.

"My ankle is still stiff but in China we are so short-handed of pilots so they must keep me."[2]

In another section of the operations building a class in high explosives, attended by both Chinese and Americans, was in progress. The instructor, Kenneth E. Kay, was speaking to the Chinese through an interpreter, and later was having trouble trying to find the Chinese equivalents for "T.N.T." and "detonator." At one point the instructor forgot himself and talked on without stopping for his interpreter. When finally he stopped, the interpreter said, as an aside, *"T'a chiang-te t'ai to,"* (He said too much). No one smiled.

Strickland, who later commanded the 3rd Fighter Group,

says he arrived at Malir from the States on 29 July 1943, and spent the next three months at the OTU before going on with his outfit to China. Training at Malir consisted of little transition training, but a great deal of formation, dive bombing, skip bombing, aerial and ground gunnery. The airplanes were in the air from sun up to dark every day seven days a week.

By 29 October the first echelons under Strickland were on their way to China and by the end of the year, most of the personnel for the C.A.C.W. had taken off for China via the usual air route, i.e. Karachi to Agra to Gaya to Chabua, and from there over the Hump to Kunming, China.

On arriving in Chabua, Strickland learned that an advance echelon of sixteen men, American and Chinese non-coms, had been lost in a flight over the Hump while flying as passengers in a commercial transport.

The 28th Fighter Squadron eventually was stationed at Kueilin and often flew top cover for B-25s attacking targets in south China. After Kueilin had to be evacuated in October 1944, most of the C.A.C.W. squadrons, fighter and bombardment, moved to the north of the Yangtze River. The 7th and 8th fighter squadrons of the 3rd group headquarters, and the 28th squadron went to Enshih, and the 32nd to Hanchung.

An insight into the composition of a C.A.C.W. fighter squadron is provided in a poop sheet which survives from the 7th Squadron based at Liangshan. This squadron was commanded by William N. Reed, a former A.V.G. ace. Not long before the war ended, he was returning from a mission but found the field under enemy attack. Since he was already low on gas, he had only a short time to land, and it was already dark. The Japanese were running one of their night nuisance raids, one bomber at a time, coming in at intervals. Reed and his wingman were forced to bail out but Reed, in doing so, apparently hit his head on the empennage and was knocked unconscious. His squadron mates found him with his chute unopened. The entire squadron mourned his death.

The poop sheet distributed for the 7th Squadron is reproduced here:

December 10, 1943

Malir, India—This morning Lt. Col Summers, Group Vice Commander, talked to our pilots on fighter operations in China, from which he has just returned. He told us that our units are doing a good job, but our fighters lack air discipline which must be stressed. For the new Zero can outdive us now, and the only salvation is to stick in formation and fight. Our pilots are around Kueilin now, cooperating with the bombers and with the Chinese ground forces. It certainly made us all itch to get up there.

This afternoon we had nine ships in and got in twenty-seven indivudial missions of dive-bombing, ground gunnery and aerial gunnery. Major Reed went on a cross country to Larkham, Sukkur, Hyderabad and back, to see how it would be for navigation training.

The squadron flying personnel has been reorganized on orders from headquarters on the following basis:

HEADQUARTERS

Major William N. Reed	American CO
Capt. Hsu,. J.S.	Chinese CO
1st Lt. Yang, Y.K.	Chinese Vice CO
1st Lt. J.T. Hancock	American Operations Officer
Capt. Yieh, W.F.	Chinese Operations Officer
2nd Lt. Lin, S.C.	Chinese Ass't. Opr. Off.

Irving L. ("Twig") Branch, commanded 1st Bombardment Group, Chinese-American Composite Wing. He led four-plane formation of C.A.C.W. behind ten-plane formation of 11th Bomb Squadron in raid on Shinchiku 25 November 1943. He was awarded the Air Medal. He was test flying after WWII and died in a crash.

1st. Lt. Armit W. Lewis	American Senior Flight Ldr.
1st Lt. Wang, K.F.	Chinese Senior Flt. Ldr.

"A" FLIGHT

1st Lt. Donald J. Burch	American Flt. Ldr.
1st Lt. Tan, K.	Chinese Flt. Ldr.
2nd Lt. Ho, C.S.	Ass't Chinese Flt. Ldr.
1st Lt. Lee, C.K.	Pilot (Attached)
Sub Lt., Chang S.C.	Pilot
Sub Lt. Li, T.T.	Pilot

"B" FLIGHT

1st Lt. John T. Hancock	American Flt. Ldr.

Above: American major at left learning Chinese checkers from comrades in Chinese-American Composite Wing.

Below: Chinese and American airmen of Chinese-American Composite Wing, stand outside tower counting return of aircraft from mission. Photo probably taken near Chengtu, 1944.

1st Lt. Chang, L.M.	Chinese Flt. Ldr.
2nd Lt. Hu, H.K.	Chinese Ass't. Flt. Ldr.
2nd Lt. Li, H.	Pilot (Attached)
2nd Lt. Lin, S.C.	Pilot
Sub Lt. Hsu, S.Y.	Pilot

"C" FLIGHT

1st Lt. Armit W. Lewis	American Flt. Ldr.
1st Lt. Kung, S.H.	Chinese Flt. Ldr.
2nd Lt. Wang, M.T.	Chinese Ass't. Flt. Ldr.
Sub Lt. Yang, C.T.	Pilot
Sub Lt. Keh, H.S.	Pilot
Sub Lt. Chao, L.P.	Pilot

John A. Dunning at his desk in Chihkiang, commanding 5th Fighter Group, C.A.C.W. Banners behind U.S. flag are gifts from grateful Chinese to Dunning and his men.

Robert L. Van Ausdall, commanded 26th Fighter Squadron, 5th Fighter Group C.A.C.W. at Chihkiang, 1945 and became an ace.

The 5th Fighter Group, C.A.C.W., which went directly to Chihkiang from Malir, remained in Chihkiang for the duration. It probably saw more action than the 3rd Fighter Group because it became involved in the campaign launched by the Japanese to occupy the American air base at Chihkiang in 1945.

The fighters operating out of Chihkiang had been a constant source of trouble to Japanese ground forces, and boats in the Siang River, but at the time that the *Ichi-Go* campaign was launched, the Japanese ignored this threat on their right flank in an effort to get at the main air fields at Hengyang, Lingling, Kueilin, and Liuchow. Finally, in the early spring of 1945, they could no longer ignore the heavy drain on their supply lines caused by the 5th Fighter Group and they prepared to destroy this nest of hornets. After two months of fighting they failed to reach their objective. It was the second major defeat for Japanese ground forces in China after Changte in 1943.

The 5th Fighter Group was beefed up by the arrival of the 4th Bombardment Group (M) at Chihkiang. After the campaign ended, the Unit Citation went to both the 5th Fighter Group and the 4th Bombardment Squadron. The citation is signed by Chennault. It reads:[3]

> The 5th Fighter Group and the 4th Bombardment Squadron (M) both of the Chinese—American Composite Wing, are cited for outstanding performance of duty in action against the enemy in China from 10 April 1945 to 15 May 1945, opposing 22,000 Japanese ground troops driving from the Paoching sector towards the important American air base at Chihkiang. Those units waged a brilliant battle which was principally responsible for the first major allied land victory in China. In thwarting the enemy this small combination aerial force wrote a classic in the employment of tactical air power against ground forces. With four enemy prongs aimed at Chihkiang, the fighter group and bombardment squadron turned the full force of their power against spearheads and strong points, against lines of communication and intermediate strong areas. In these thirty-six days they killed more than one-fourth of the enemy troops, slashed their supply lines so thoroughly and destroyed such quantities of supplies that the enemy drive was completely paralyzed. Ill-equipped Chinese forces, heartened by aerial support, began their own offensive and rolled the enemy back to Paoching. Flying 920 missions in day and night attacks, the pilots braved vicious barrages of anti-aircraft and small arms fire to hit the enemy. The relentless attacks left 6024 enemy troops killed, 1491 cavalry and work horses killed, and ninety-seven gun positions destroyed and

damaged, 1639 small river boats knocked out, 304 vehicles, thirty-nine river vessels longer than 100 feet were sunk or damaged, while forty eight bridges or rail and road arteries were destroyed or damaged. Immense quantities of enemy supplies and equipment were lost in the 4006 buildings destroyed and damaged by the fighters and bombers. The boats and vehicles destroyed, in the majority, were laden with supplies. This carnage of destruction caused the cited units eighteen aircraft and two pilots lost. The extraordinary heroism, gallantry and determination and esprit de corps that evoked this brilliant victory are supplemented by the fact that the units are composed of personnel from two nations, Chinese and American. Differences of language, custom and temperament were overcome in the uniting aim of destroying the enemy. The achievement of the 5th Fighter Group and 4th Bombardment Squadron is worthy of the gallant traditions of both the American and Chinese military services.

(J.E. McCann, a member of the 3rd Bomb Squadron, 1st Bomb Group (M), recalls the following from his experiences with the C.A.C.W.)[4]

It was in July 1943 when I left Miami and started my long trip by air to South America, Africa and finally to Karachi, India, where we were sent out to Malir where no one seemed to know what was going on. This was in August and the 1st Bomb Squadron of the 1st Bomb Group was to be formed with a nucleus of U.S. Army Air Corps personnel and the rest would be Chinese crews. The officers were on the way from Luke Field and they arrived in late August, and the Chinese enlisted men began to arrive.

"We spent August, September and October training in the various fields. The pilots trained the Chinese pilots on B-25s, and one or two B-26s. The training was for low level skip bombing and strafing. The Chinese navigators, bombardiers, and line officers were all being trained by their American counterparts. All mechanics were trained by American crew chiefs, the armorers by American armorers, gunners and so on down the line.

"In November 1943 the 1st Bomb Squadron was ready to move up to Kunming on its way to Kueilin. I ended up in the base hospital at Malir with yellow fever. I was left behind for the 2nd Bomb Squadron when it began training.

"We lost our first American enlisted man on a B-25 that went down while teaching low-level gunnery. At that time I remember the order came down—no American personnel on any training flight unless the pilot is an American officer.

"We also had trouble making the Chinese gunners understand the automatic cut-off on the interrupter system when the turret guns came in line with our own tail fins. They shot the tail fins off a couple B-25s before they understood the interrupter did not stop the guns when firing, except at their own tail.

"As soon as I was released from the hospital I hitched a ride to Chabua and over the Hump to rejoin my squadron at Kueilin.

"In January 1945 a number of us were detached from the 1st Bomb and sent back to Karachi to train more Chinese for the 3rd Squadron.

"We began to get the new B-25Hs when we moved to Moran, India, in early 1944, for support missions, bridge

Colonel T. Alan Bennett with Major Yuan Ching-han, group commanders, C.A.C.W., probably at Peishiyi, January 1945.

James E. McCann, 3rd Bomb Squadron (M), Chinese-American Composite Wing, repairs top turret mount for 50 caliber machinegun. Heavy white gloves were used to make repair job easier, and cap is probably gift of Chinese Air Force crewman.

Above: Chinese noncoms listen to instructions by Americans in engine repair and overhaul. Chinese at left in black glasses is interpreter. In background a North American trainer.

Left: Headquarters building and tower for Chinese-American Composite Wing, at Chihkiang, June 1944, with Chinese and American flags waving in the wind, B-25s parked in the distance at right.

busting, rolling stock, and to help harrassing the Japs on the Burma Road. We flew from a British air field and were housed in tents. The monsoons kept us grounded a lot. We later moved up to Dibrugar, India. The runway was steel mat laid in the jungle and our missions were into Burma. We flew one to Lashio and Bahmo to knock out Jap fighter planes and a sugar refinery. For the balance we were bridge busting and all our missions were low level. It seemed like every bombing or strafing run we made was never higher than fifty feet above the tree tops. We used a homemade bomb sight and the pilot triggered the release of the bombs.

"One of the first missions in a B-25H proved the experts wrong. We were told the "H" would never fly on one engine. We climbed from 100 feet on one engine after putting out the fin, feathered the prop and all to 500 feet to scrape back through the pass into Moran. We threw out every thing we could to lighten the plane and landed safely.

"In the spring of 1944 we moved on to Kueilin where we continued our low level bombing. We started doubling up the bomb load by using wire snappings to hold a second bomb on the bomb hooked on the station. We also made sure that we did not salvo any bombs with the arming wire still in the fuse. All of our arming wires were wired to the bomb shackle so that everything we dropped was live. If the weather was socked in on the primary target we tried the secondary and if that was socked in, we brought our bomb load back and never salvoed any bombs, but landed with them in the bomb bay because they were a scarce commodity.

"We staged out of Kueilin and Liuchow to harrass shipping on the south China coast, and out of Chengtu and Sian to bomb the Yellow River bridge. We made three missions the last day before Kueilin fell to bomb and strafe Jap troops in sampans on the Li River. The last time we took off we strafed them as the gear was coming up and our own people were blowing the field behind us.

"We crash-landed on an emergency fighter strip near Chengkung. Our squadron was to be stationed at Peishiyi on the other side of the mountain from Chungking. We staged from there to various locations the rest of the year in 1944. We were at Liangshan, Laohoko, Chihkiang, Lingling and Hengyang. This was in late 1944 and early 1945.

"We were the crazy ones. We flew the missions no one else wanted. We flew single night flights out of Kueilin. We landed up or down the valley or you flew into a mountain. We learned to trust the Chinese navigator. The Japs just couldn't learn to drive a truck at night without lights.

With our double loads sometimes we couldn't get the bomb bay doors closed. We flew anyhow. A lot of times on the short fighter runways we would put up the gear and it was either fly or belly it in the rice paddy. Our pilots made 'em fly anywhere.

"We were the idiots in the B-25J, all silver (not camouflaged) that doubled our frag load from our normal capacity by hanging one more cluster on each station. Then we flew to Hankow to strafe and bomb the searchlights and ack-ack batteries and over to White Cloud at Canton and the Kowloon docks at Hongkong while the B-24s plastered the area with 500 pounders from altitude. We made seventeen bombing and strafing runs at low-level, knocking out the lights and ack-ack. We didn't get one hole in our plane. The volunteer crew was recommended for citations but the only one who got any was the pilot who got the Air Medal.

"The B-25s we used in China in 1943 were either the 'C' or 'D.' This airplane had pilot, co-pilot, navigator, bombardier and top turret gunner. The top turret was mounted in the waist compartment behind the bomb bay. Armament was twin 50 calibers, in the turret, one 50 caliber in the bombardier's compartment, or it could be a 30 caliber.

"Finally, we got the B-25H, a big improvment. The top turret, was moved up ahead of the bomb bay which eliminated the bombardier compartment. It also did away with the co-pilot and ended up with a pilot, navigator, top turret gunner, one or two waist gunners, and a tail gunner.

"The manufacturers also added four 50-caliber guns in the nose, firing forward and twin 50-caliber gun packages on either side of the fuselage, also firing forward, and all operated by the pilot. A 75 mm. cannon was installed along the left inner part of the fuselage under the pilot with the loading breech in the entrance way behind and below the pilot compartment. The 75 milimeter shells were racked on the left side of the aircraft in the front entrance area. These guns were controlled and fired by the pilot.

"The manufacturers also added a 50 caliber gun in each waist window and a new tail turret with two 50 caliber guns there. All in all, when everything was firing, there was a lot of shaking going on. All the 'H' models were painted OD (olive drab).

"Then along came the new B-25J. It had the bombardier's compartment with one 50 caliber mounted on the floor on the right hand side, fired by the pilot, along with four package guns, 50 caliber, two on each side.

"There was also a moveable 50 caliber gun set on a swivel mounted up to the left for the bombardier to use. It had a turret ahead of the bomb bay and the same armament in the waist and tail as the 'B.' No one at the factory bothered any more to paint the new aircraft. They came out to us shining like silver—all aluminum. We still added some 30 caliber guns in the camera port at the bottom of the plane, but they did not work too well on the 'J'. The 'J' took us back to the time when the pilot had a co-pilot.

William F. Carroll (left) and Gail Smith, 1st Bomb Squadron, 1st Bomb Group, C.A.C.W. in front of "Double Trouble," a B-25 ("strafing J"), at Hanchung, spring of 1945.

"The letters C.A.C.W. meant Chinese-American Composite Wing, but we were told that we were in the U.S. Army Air Corps and that we were on detached service to the Chinese Air Force. All our planes were marked with the Chinese insignia, and the tail fins were painted blue and white alternating stripes. Through the C.A.C.W., a handful of Americans were carrying out Claire Chennault's ideas. General Chennault was one of the greatest men I ever met. A handful of Americans trained and built a powerful offensive weapon from personnel of the Chinese Air Force. We taught them how to fly and how to skip bomb. We fought beside them and they beside us. We were members of a great team."

XXI

The 308th Bomb Group (H)

The support force of the 308th Bomb Group left San Francisco by ship on 14 February 1943 and arrived in Karachi on 15-16 March, en route to China. The airplanes were brought out by the flight crews and were already in Karachi when the ground crews arrived.

The group was divided into four squadrons, the 373rd, the 374th the 375th, and the 425th. The distinguishing tail mark of the 373rd Squadron was the red or light blue rudder; the distinguishing mark of the 374th Squadron was the yellow rudder; the 375th light blue rudder with vertical white stripe, and the 425th with black and yellow slanted stripes on the rudder.

Eugene H. Beebe brought the group to China and remained in command until about the beginning of November 1943 at which time he was replaced by William P. Fisher, who remained in command until 19 October 1944 when he was replaced by John G. Armstrong, who remained in command of three squadrons which moved to India, leaving the 373rd radar squadron at Luliang under William D. Hopson who technically became C.O. of the 308th Group, less three squadrons, on 30 July 1945.[2]

The 373rd was based for the most part at Yangkai, north of Kunming; the 374th ad 375th at Chengkung, and the 425th at Kunming air base. After the 58th Bomb Wing (VH) left the Chengtu area for Tinian in March 1945, three squadrons of the 308th took over former B-29 bases around Chengtu, namely, the 374th, 375th and 425th, while the 373rd moved over from Yangkai to Luliang.

On 8 August 1944 the 425th Bomb Squadron carried out a mission over Hankow which was a near-disaster. The fighter escort failed to rendezvous with the bombers over Hengyang, and rather than turn back, the squadron continued on to Hankow and made the mistake of attacking Wuchang Air Base first, which alerted the enemy fighters parked on

Monsoon Maiden, a B-24J from 425th Bomb Squadron, was first low altitude bomber (LAB) of 308th Bomb Group in China, and flew fifty-seven missions in sea sweeps off coast of China, accounting for eight naval ships and ten merchant ships sunk. Art work on nose of *Maiden* was done by Howard Colvin and photograph is preserved in collection of Robert L. Cowan who stands at extreme left, in leather jacket. Photo was taken at Chengkung, in September 1944 and includes flight, repair and engineering personnel of 425th B.S. The *Maiden* was converted to a tanker in latter part of 1944 to haul gas to advanced fighter bases in east China. Returning to Chengkung one night, the ship picked up some flak and the radio went out. The navigator lost his heading and, running out of gas, the crew bailed out, leaving *Monsoon Maiden* to the winds of war.

Hankow Air Base. As a result, they were waiting for the slow-moving Liberators.

The problem of coordination between fighter escort and bomber formations was never solved. For example, on 10 July 1943 B-24s made a raid on Haiphong. Here is what the 75th Fighter Squadron Journal has to say on this: "Major Goss led eight ships to Haiphong to intercept anything coming up to meet the '24s.' The B-24s were twenty-five minutes late and scattered from hell to breakfast. No opposition except light flak. The P-40s finally made contact with the B-24s and proceeded to the target which was difficult to get at as the weather favored the Japs . . . "[2]

After the Hankow raid of 8 August, the 425th was out of action for weeks and by that time, bad weather had moved in which kept most of the squadron grounded for the rest of the winter.

After the pasting the 373rd took over Haiphong on 15 September 1943, morale in the squadron was at low ebb, and the squadron commander, hearing that Eddie Rickenbacker was in India on an inspection trip, invited him to come to China to speak to the troops at Yangkai. Deno Renieri, who attended the meeting, said it was a pep talk, and afterwards the men enjoyed meeting the famous ace of World War I and survivor of the legendary bailout over the South Pacific in the first months of WWII.

After the weather improved in the spring of 1944, it was determined that the B-24s would be used mainly for sea sweeps to interdict, that is, to cut off or sink Japanese shiping along the south coast of China and in the Gulf of Tonkin.

An insight into the night sweeps by radar-equipt aircraft of the 308th Bomb Group is provided by William L. Dwyer who was a radar operator with the 374th Bomb Squadron in China. He explains:

Ground echelons of the 308th Bomb Group sailed to India by surface vessel and rejoined their squadrons there before flying over Hump to China. One who did not make it was PFC Clorada Willard, 425th B.S., who died on board army transport *West Point* in March 1943 and was buried at sea. Chaplain reads prayer as coffin is made ready for casting off for Davy Jones locker.

"Briefly, on the night searches, we called them sweeps or strikes—we carried ten 500-pounders and we were after Jap shipping through use of radar and what was called LAB or Low Altitude Bombardment. We had trained at 1000 feet at Langley but in combat we made our runs at 400 feet. A typical sweep would be for take off at dusk. This was done so that when we passed over Hongkong it was lit up, likewise Macao, and we used both for checkpoints. It was usually dark when we passed over and with the bottom of our aircraft painted black, we were not easily spotted, since the enemy could only see a brief dark silhouette. We continued out into the South China Sea and began to let down from over ten thousand feet to around three or four where we began our search.

"Incidentally, our B-24s had no belly turret gunner because we had installed the radar spinner in the bubble. We were a bit vulnerable without the belly gunner when we had altitude and were attacked by fighters, but the fact is, we were flying mostly at night, and low over the water, and had little need for the belly gunner.

"Shipping showed up as white blips on my scope. Indicentally, my station was over the bomb bay in an enclosure just above and forward of the waist. I, through the intercom, kept the pilot and bombardier posted as to our position, and the size of the convoy. I projected the same picture was tuning in on up to the bombardier's compartment in his scope which gave him the choice of selecting the blip, or ship, he wanted to make a run on. I instructed the pilot then as to the location, right, left mileage from our B-24. Meanwhile the pilot was letting down to 400 feet and we started our run, bomb bays open, bombs fused by the engineer. I guided the plane into a distance of three miles from target. The bombardier then took control from there on in and synchronized his bomb sight with the radar and, at the precise instant, a string of four bombs was triggered out. Usually, two hit and one straddled close enough to buckle the plates on the ship.

"After bombs away, the pilot began to climb to right or left, but usually the element of surprise left us undetected except on clear nights when the moon was high. At times like this we could be silhouetted against the light of the moon and we had to make a run with the moon behind us. We preferred rain or fog to the moon because we could see the target on radar but all the enemy had to go on was the sound of our engine noise.

"Occasionally, we were spotted and took some flak, but we were never seriously damaged. Incidentally, the navigator also had a scope, and the picture he got of the China coast line was invaluable for check points on our return to base.

(end of letter).

As a result of these long-range sea sweeps, the 308th Bomb Group amassed an impressive record of bombing operations and was awarded the Unit Citation. This is how it reads:[3]

> This group preyed relentlessly on the Japanese homeland and her conquests throughout southern Asia and adjacent insular territories. During most of this period, the 308th Bombardment Group (H) was the only organization among all the Allied forces in a position to conduct interdiction operations against the vital supply lines. Operating from bases in China, the group swept the East and

South China seas, the Straits of Formosa and Gulf of Tonkin through all kinds of weather, sinking and damaging nearly three-quarters of a million tons of vital Japanese shipping. They sank 107 merchant vessels and sank 12 enemy naval vessels, including 2 cruisers and seven destroyers. They probably sank 29 vessels and damaged 48, for a total of 427,252 tons of shipping sunk, and 102,765 tons probably sunk, and 187,045 tons damaged.

Pressed by the constant need for economy of operations from air-supplied China, the Group forsook the usual high altitude style of bombing to attain accuracy and minimum expenditure of bombs. Heroically and deliberately, the combat crews developed and employed low altitude tactics that brought their slow, heavy bombers down to within 400 feet of their target during attacks. This exposed their aircraft to murderous fire from the armed merchantmen and naval vessel they attacked. On a number of occasions, the crews carried out attacks at 400 feet over entire convoys of 8 to 12 armed merchant and naval ships. Throughout the cited period, the Group was forced to fly much of its own gasoline and bombs over the "Hump" (Himalaya Mountains) into China, and in the same period was forced to evacuate from three bases before the advance of enemy ground forces. For several months the Group launched its sea search missions from a base behind enemy lines in east China. The phenomenal achievements of the 308th Bombardment Group (H) in its interdiction of these vital enemy shipping lanes are the result of extraordinary heroisim, gallantry, determination, and esprit de corps demonstrated by the members of this organization. Their attainments are consonant with the finest traditions of the American military service.[4]

In April 1944, William D. Hopson, as project commander, left Langley Field with thirty B-24s for delivery to the 308th Bomb Group. These aircraft had a new radar bomb sight (Azon) which had a low altitude and night capability. In China the aircraft were divided between the four squadrons of the 308th, and, with the radar capability, they were used also as pathfinders.

In the spring of 1945 the radar aircraft were consolidated into one squadron, the 373rd, which had moved from Yankai to Luliang to continue strikes against enemy shipping. For several weeks there were take-offs from Liuliang every four hours, around the clock.

John Armstrong, Group C.O. at this time, took the other three squadrons from Chengkung to Chengtu where he thought there were more profitable targets for high altitude

The author hitch-hiked a ride across the Hump (his first) in October 1943 with a flight of three B-24s from the 375th Bomb Squadron who were returning from Chabua to base at Chengkung. This picture was shot through plexiglass nose while nose-gunner scanned the skies for enemy interceptors. The river in the canyon below at left is probably the Salween in northern Burma which flows south. The airplane seen here points east which means it is approaching the river. The next check point was Lake Tali in western Yunnan Province of China, a lake Hump pilots considered one of most beautiful in the world. From here most 308 pilots set a heading southeast for Chengkung.

nonradar bombing. The 374th and 425th faced each other across the flight line at Kuanghan, the 375th at Hsinching, closer to Chengtu. Only a few combat missions were flown to targets in north China for owing to the shortage of gasoline, there were extended periods of inactivity, and, moreover, the bombardiers were running out of worthwhile targets in north China.

Edward P. Zeidler, a pilot with the 374th, remembers this interlude at Chengtu:

"We flew the J's at 73,000 pounds which was about 6,000 over the maximum overload, but we had good runways, built for the B-29s, and when we got to the end of them we hauled back, inhaled, and usually got off. We were using 500-pound RDX bombs which were more potent than the regular 500-pounders. We also dropped napalm fire bombs, and bundles of tied propaganda leaflets over the cities of north China. I ran five bomb missions, forty-one combat Hump missions before VJ Day, and nine Hump missions after VJ Day, mostly hauling gas. My logged combat missions with time en route and date follows:

To Shihchiachuang	10:15 Hrs.	3-8-45
To Tsinan	12:40 Hrs.	3-10-45
To Chenghsien	10:00 Hrs.	3-24-45
To Shihchiachuang	10:10 Hrs.	3-31-45
To Taiyuan	10:40 Hrs.	4-2-45

After Hopson had delivered the new radar aircraft, William Fisher, then Group C.O., had Hopson assigned to Group as deputy, and he continued in this capacity until the end of July 1945 when he succeeded John Armstrong. One of the first missions he flew before VJ Day was against the Lloyd Triestino passenger liner, *Conte Verde,* tied along side the dock at Yangtzepoo on the Whangpoo River in Shanghai. The ship was caught in Shanghai on the day Pearl Harbor was attacked, and the crew, even though Italy was a partner of Japan, scuttled the ship. It was later raised by a Japanese salvage company, riding high on the water when Hopson saw a reconnaissance photo of the ship and determined to sink her. With his underbelly painted black, and with his radar bomb sights checked out, he took off from Luliang at night on a heading for Shanghai. The bombardier found the ship in the moonlight, zeroed in on the target, and *Conte Verde* went down.

The 373rd Bomb Squadron was now advised by the Joint Chiefs of Staff that bases were opening up on Okinawa to attack Japanese shipping, but three aircraft were soon lost in sea sweeps against enemy shipping and to kamikaze air attacks.

The three squadrons which had moved to Chengtu earlier, were now in Rupsi, India, in the Bramaputra Valley, hauling gas across the Hump. The Hump gas operation continued until well after the war, since Chinese troops were being transported to north China and Manchuria from west China and it was easier to get the gas to these transports from India than from the port cities of China on the east. The three squadrons began flying home in October 1945.

The 308th Bomb Group probably lost as many airplanes to navigational error, the weather, overloading and engine malfunctions as it did to the enemy. An example of what could happen is described by Leonard Gelhaus, a co-pilot with the 425th Squadron who, on the flight described here, was in the right seat of the "Monsoon Maiden." He writes:

"On 19 November 1944, we took off from Kunming on a

"Cocky Bobby" a B-24L, of 425th Bomb Squadron, Robert Boyce, pilot, William Roby, co-pilot. Photo probably taken near Chengtu winter of 1945.

trip to Suichuan to transport some 500 lb. RDX bombs, gasoline and .50 caliber ammo to this advance base. Our crew consisted of Robert Revard, pilot, John Turner, engineer, Ray Daniels, radio, Tony Lopez, navigator, and myself, co-pilot. Our ETA at Suichuan was around 2100 hours. When we arrived, the base was under attack by some Japanese bombers. We were re-directed to Kanchow, a nearby fighter strip. When we "dragged" Kanchow we found the strip very short, especially with a full load of bombs etc. We were able to get "The Maiden" down but had to pull the parachutes of Ray and John out the waist ports anchored to the machine gun mounts to give us braking power.

"We were treated as heroes by the P-51 Squadron who had been out of gas, ammo and bombs for over a week. The next day they flew fifteen sorties, using our load.

"On the trip back, Lopez suddenly announced that we were over Kunming! Bob and I figured we were at least 200 or 250 miles away. We dropped down to take a look and started to pick up flak. Lopez had navigated us to Hengyang which was now held by the Japs. We got the hell out of there before they sent up some Zekes after us.

"Some time later, Ray reported he was having difficulty with his radio and could not raise Kunming. John went back to the command deck to check out the radio equipment and found that we had taken some flak hits which not only wiped out the radio but also hit our gas lines. You will recall that the cross-feed went through the command deck. By this time it was dark as pitch and we were at 17,000 feet.

"Pretty soon we could smell the gas and were scared that we would blow up. We shut down everything we could and were particularly cautious about our parachute "snaps" lest they hit something and cause a spark. We opened the bomb bays for ventilation. We figured we were over Kunming by the amount of air traffic but there was a heavy undercast. Time for decisions. Without a radio, any kind of instrument let-down was impossible. We didn't know about Luliang and Yunnani being open and with the gas leak didn't know how much gas we had left. We turned on the cross-feed and decided to bail out. The only dissenter was Lopez. Bob put on the automatic pilot and told Lopez if he wanted to fly it, go ahead. Ray went out first, John second. Lopez was supposed to be third but he kept hesitating. Bob told me to go ahead so out I went. Bob told me later that Lopez pleaded with him not to make him bail out. Well, he finally jumped, Bob followed. As Bob jumped, one engine stopped, apparently out of gas. The plane went into a spiral and centrifugal force kept Bob on the cat-walk. He finally managed to shove himself out. He ended up at the outer edge of the spiral and "The Maiden" kept coming at him, one time even spilled his chute. He saw her hit the ground.

"We all got together after a couple of days of wandering

Left: B-24s from 308th Bomb Group in raid on Kowloon docks, 16 October 1944. This was a combined mission which included fighters and medium bombers and is considered the most successful ever made by the group. Several ships were hit. Black smoke suggests target area. Near side of picture is "Hongkong side," with downtown area running along foreshore of bay. Japanese had ships sunk at lower right as barrier to shipping.

"Miss Beryl," 373rd Bomb Squadron, having engine change. Markings show crew already has sunk seven enemy vessels and one probable in sweeps off coast of China, on more than seventy missions. Marking behind nose shows this to be a B-24 J, serial number 44-40832. Squadron was first based at Yangkai and later moved to Luliang for low altitude bombing with radar.

around China and lived happily ever after."[6]

Only one man in the 14th Air Force won the Congressional Medal of Honor, Horace S. Carswell, Jr., 28, of Fort Worth, Texas. The medal was presented posthumously to his family.

On the night of 16 October 1944, Carswell and his crew sank a Japanese destroyer off the south China coast, and less than two weeks later, the same crew went back on another sea sweep, looking for a target. A convoy of enemy ships was sighted. Despite the odds, Carswell made a low-level attack on one of the ships without knowing whether it went down. Nor did he realize, in the darkness, how many ships there were in the convoy, probably as many as twelve. Anti-aircraft guns seemed to explode in all directions, and two engines on Carswell's airplane, dubbed the "Tail Heavy" were knocked out before he could climb for cover. He and his co-pilot nursed the aircraft back to the mainland of China but could not gain altitude, and when a third engine sputtered, he ordered the eight members of the crew to bail out.

Remaining on board were Carswell, his co-pilot and the bombardier. The latter could not jump because his parachute was riddled with flak. Carswell, refusing to jump himself, thought he could "smooth out." Carlton S. Schnepf, the last man of the eight to bail out, remembers seeing the bombardier brooding over his ruined parachute which lay on the deck behind the cockpit.

Carswell finally managed to get his airplane back to 4000 feet by bringing it into a stall, then diving a little to regain speed, and he did this several times. Finally, the aircraft began to wobble from side to side, losing altitude slowly at which point a mountain rose up in the night to meet the oncoming plane. It exploded on impact. There were no survivors.

Carswell Air Force base at Fort Worth is named for this intrepid flier.

Eugene H. Beebe was first commander of 308th Bomb Group and led group to China in March 1943. He is shown here just after being decorated with Distinguished Flying Cross by Maj.-Gen. Claire L. Chennault, commanding the 14th Air Force.

William D. Hopson, last C.O. of 308th Bomb Group. This photo taken at Rupsi, India, July 1945.

Jay E. Levan (center) was pilot on 308th bomber which sank Japanese light cruiser off coast of China in August 1944. At right is his bombardier, John D. Shytle, Jr., who is being congratulated by William P. Fisher, who took over command of 308th when Eugene Beebe was rotated home.

"Axis Nitemare," first ship of 308th Bomb Squadron to leave Salina, Kansas, December 1942, on orders to proceed to China, arriving Chengkung 13 March 1943.

Troops of 373rd Bomb Squadron arriving at processing center at their new base at Yangkai, spring of 1943.

Armorers and crew loading bombs in bomb bay on B-24 for upcoming mission.

Katy-Did

Temptation

"TAIL HEAVY"
FIRE EXTINGUISHER

"STORK CLUB"

DOUG
SHOOTIN' STAR

"Bonnie Belle"

BELLE
STARR

Pistol
Packin'
Mama

RUM RUNNER

ESCALATOR-II
CHARLEY

"LAKANOOKI"
'GEORGE'

"Thelma C."
Lonesome Polecat
"Jeannette"

"FLIGHT DECK"
"TOMMY"
SUSIE'S SISTER

Jay Le Van
"NITEMARE"

RELUCTANT
DRAGON

FIRE
EXTINGUISHER
ENNESSEE
BELLE"
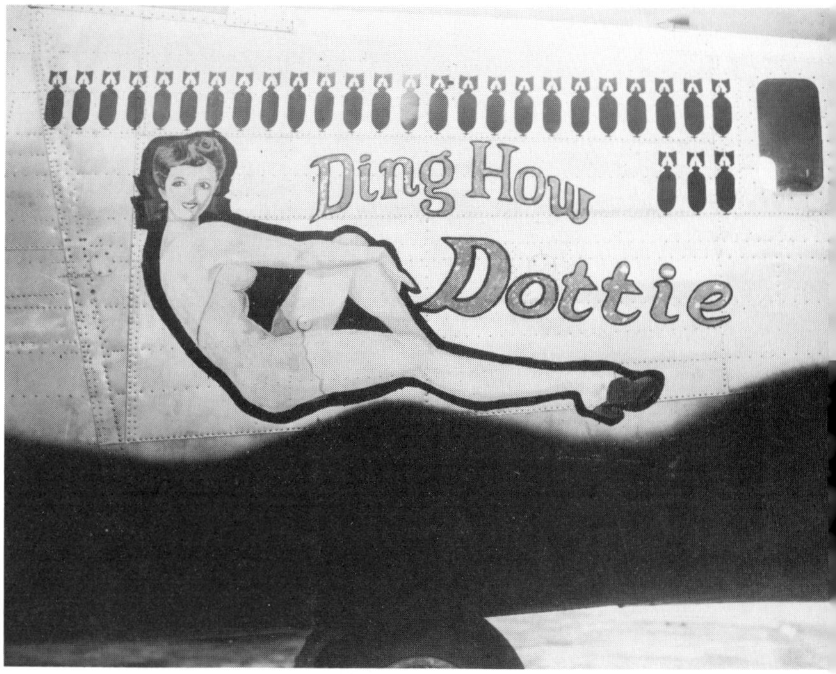
Ding How
Dottie

BUZZ-Z
BUGGY
"Dead-Eye Dick"

LILI
MARLENE
BARRACK
GATE

TAYLOR
MAID

MISS LACE

Above, Loose formation of B-24 "Liberators" 375th Bomb Squadron, probably over Tungting Lake area, 1944.

Below: Direct hit on factory, probably at Kowloon.

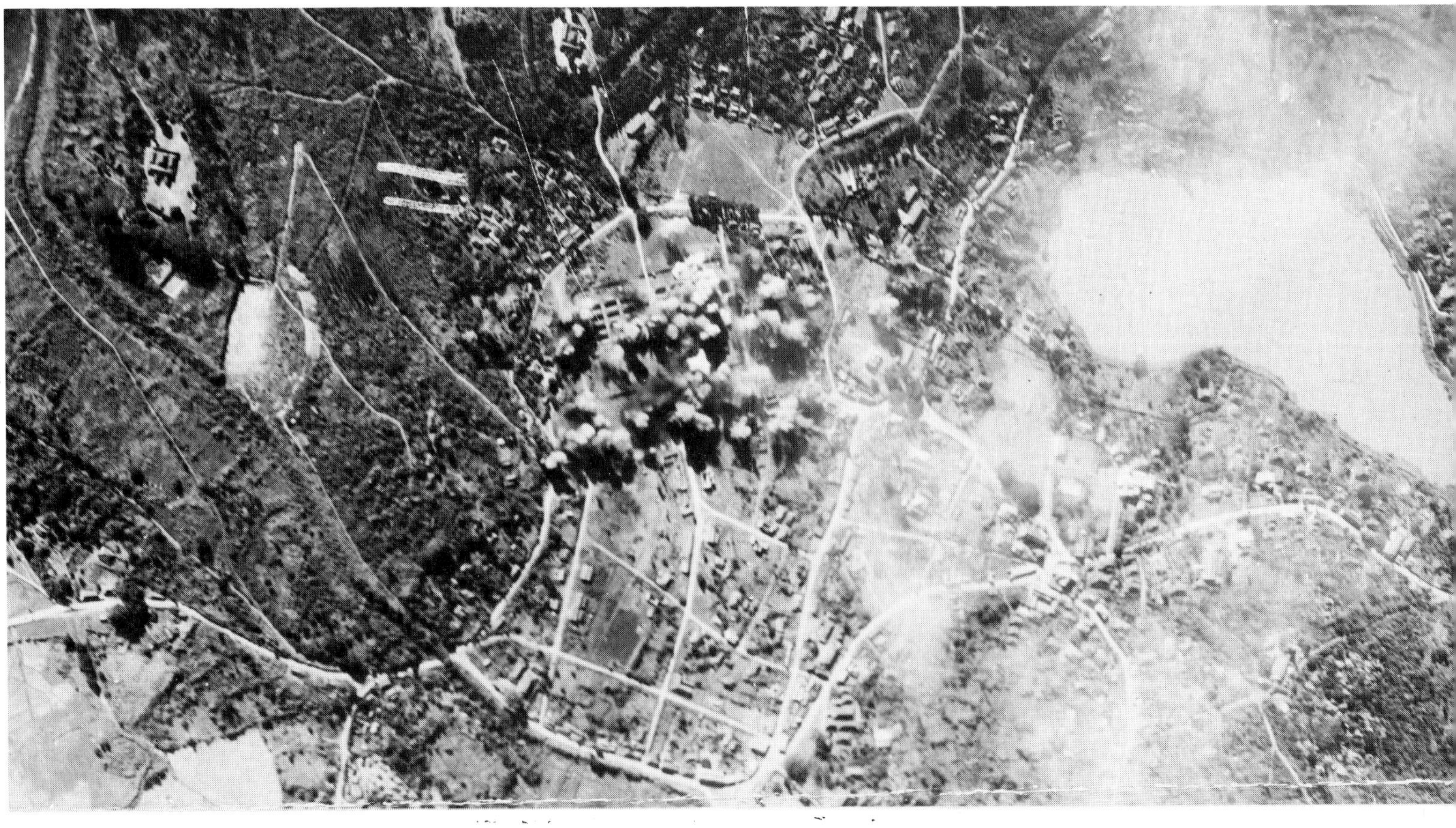

The 58th Bomb Wing (VH)

Line service in progress on Boeing at Kuanghan, China, advance base of 444th Bomb Group, 58th Wing. This is K-304, serial number 224732 which was government contract number awarded in 1942. Diamond on tail section was logo of 444th, but ball below undertain, probably a sign that it was flown by "lead crew" where best bombardier was riding. Airplane stands unprotected. There were no revetments for these aircraft in China or India. Two crewmen in flight suits seated in foreground seem dwarfed by huge plane. Behind them are bombs waiting to be loaded. Control tower stands in center distance, and main runway on other side of tower and to right.

A personal visit to the cockpit of "Fifi," a B-29 (very heavy) bomber which may be seen at national air shows almost every year, reveals a large area in the nose, almost like a room with no doors or panels, but plexiglass in every direction except the rear. The pilots sit so far back in the "room" they have to crane their necks to see the runway.

"Fifi" was built, with several thousand others, at Wichita, Kansas, in the early 1940s, flight tested and turned over to the Army Air Corps. Today, the only B-29 still airborne, she spends most of her time in a hangar of the Confederate Air Force, at Harlingen, Texas.

In 1943, four groups of the 58th Bomb Wing, equipped with these new airplanes, were formed in Kansas. One, called the 40th, was located at Pratt, another, the 444th, at Great Bend, another, the 462nd, at Walker, and the last, the 468th, at "Smoky Hill" (Salina). In September of that year all four groups assembled at Salina where pilots and crews were put through an intensive course of training in a plane already referred to as the "Superfortress" as opposed to B-17 "Fortress" which, up to the advent of the B-29, was the heaviest bomber in the air.

The B-29 training program was continued throughout the winter of 1943-44 in order to have pilots and crews ready for take-off in the spring of 1944. Their destination: the China-Burma-India (CBI) theater of operations. But despite the intensive training program, most of the crews were still learning to fly the airplane when the first flights began taking off in early April.

Kenneth B. Wolfe of the materiel command, went out to India to inspect the fields and arrange for housing and mess facilities. The first B-29s to leave Kansas were elements of the 40th Group led by Laverne ("Blondie") Saunders who got airborne early in April. Thomas W. Coffey, in his book written on General Henry Arnold, has this to say of this stage of development:

> Two planes, the first one flown by Blondie Saunders, had preceded the Group. Saunders arrived in Chakulia, India, on April 7, and was followed within the next ten days by several other planes. But toward the end of April five of the newly arrived bombers crashed near Karachi on the Arabian Sea, owing to overheated engines. Two planes were demolished and five men were killed. K.B. Wolfe, who was on the scene, grounded all the surviving planes and told Arnold in an urgent message that some method would have to be found to cool the engines . . . [1]

Robert Wix, co-pilot on the *Bugs Bunny,* recalls that owing to all the repairs and engine changes made en route, it took his crew fifty days to reach India, and after arrival, *Bugs Bunny* had a red stripe on the fuselage, meaning it could fly but not very well.[2]

Wolfe remained in India until July 1944 when he was ordered back to Kansas to prepare for the movement of the 73rd Bomb Wing to the Marianas.

Blondie Saunders commanded the 58th Wing in India-China from the time he arrived until the arrival in late August of Curtis E. LeMay who had been transferred to the CBI out of the 8th Air Force in England where he led the 3rd Air Division and survived the Regensburg-Schweinfurt disaster over Germany. Saunders had already been assigned to

Formation of B-29 Boeings of 58th Bomb wing on way to target from bases in China, probably 1944.

form a new wing in Kansas, but stayed on in the CBI to help LeMay get acquainted. It almost cost him his life. On 18 September he was returning from a farewell dinner with fellow officers at another base, flying a B-25, and on the return flight, both engines cut out on take-off and he crashed into the jungle two or three miles from the end of the runway. LeMay ordered an air search but no airplane was found. He then guessed that Saunders had crashed on take-off and made a personal search in the night accompanied by one of his aides holding a flashlight and compass, and found the missing aircraft. Saunders was alive with multiple injuries to bone and body with one leg pinned under an engine. Saunders' first comment was: "What took you so long?" He had to be invalided home, refusing at first to allow anyone to remove his battered leg, but finally had to bow to fate.

Meanwhile, after the four groups of the Wing finally reached India, they settled down on four different bases, the 40th Group at Chakulia (near Calcutta) with Hsingching as its forward base in China; the 444th Group at Charra, India, later Dudhkundi, with its forward base in China at Kuanghan; the 462nd at Paradoba with forward base at Shuangliu, and 468th Group at Kharagpur with its forward base at Pengshan. Wing headquarters in India was established at Higley, an old prison complex near Calcutta which once housed a prisoner jailed on charges of civil disobedience — Mohatma Gandhi.

The code name for the four advance bases in China were A-1 for Hsingching, A-3 for Kuanghan, A-5 for Shuangliu and A-7 for Pengshan, and the code name for the entire 58th Bomb operation in China was MATTERHORN.

The forward or advance echelon bases in China were all located within a distance of thirty miles or so from Chengtu, an old and large city which lies on a vast plain hugging the mountains of Tibet to the west. These bases were rushed to completion at tremendous cost with a work force of more than a quarter of a million Chinese who trudged back and forth with baskets balanced on bamboo poles while thousands of women and children broke up large rocks into small rocks with hammers for top cover, the whole rolled and packed by huge concrete rollers pulled by scores of men, most of them wearing straw sandals to protect their feet.

The secretary of the 58th Bomb Wing Association, Denny D. Pidhayne, believes that when all the airplanes reached India in the spring and summer of 1944, there were not less than 160 and not more than 180. As they struggled in from the long flights out of Kansas, four squadrons were assigned to each group, with about ten airplanes to a squadron.[3]

The original group commanders, often referred to as Hap Arnolds' "Boys", were hand-picked by Arnold himself; namely, Lewis R. Parker, 40th Group; Alva L. Harvey, 444th Group; Richard Carmichael, 462nd Group; and Howard P. Engler, 468th Group. Parker and Harvey were put on detached service to the 8th Air Force in England to gain experience in bombardment and each made five combat missions over Germany. On the fifth, Parker was shot down and spent the war as a POW. He never reached India. He was replaced by William R. ("Butch") Blanchard, who was followed by Henry R. Sullivan and finally by William K. Skear.

Harvey, C.O. of the 444th, was succeeded in April 1945 at Tinian by James G. Selser, while Harvey became deputy, 58th Bomb Wing.

Carmichael went down in the second Yawata raid 20 August, but survived. This was a daylight mission. The first mission over Yawata had been run at night and was relatively ineffective. Carmichael was hit by flak and fire broke out in the front compartment. Fire extinguishers were no

Control tower and operations building at Hsinching, China, forward base for 40th Bomb Group. It was also headquarters for 58th Wing. Sign board at lower left, waiting to be hung up, reads: "40th Bomb Group Tactical Operations." Jeep (left) and weapons carrier were most widely used vehicles by crews in China.

Chinese workers rolling gas drums unto storage platform at Shuangliu (A-5) near Chengtu, China, with B-29 packed at upper left. This was how aviation gas was transported to China.

Right: Curtis LeMay talking with C.C. Meyers, pilot, and two crew members, probably at Hsinching. Aircraft recognition poster of Japanese fighter hangs on wall.

help because no one had thought to replenish them after a previous fire.

Robert S. Clinkscales, C.O. of the 94th Squadron, was on the same raid, leading his unit over the target when he was rammed by a kamikaze, probably flying a Kawasaki Ki-45, a two seat fighter (U.S. code name "Nick"). The suicide pilot hit Clinkscales' plane between the No. 1 and 2 engines on the port wing, which sent him into a spin, killing the crew, including Clinkscales' black cocker spaniel, "Sally" who was also on the mission.

The kamikaze plane, on impact with Clinkscales, bounced upward, slamming into the No. 2 engine of Ornell H. Stuaffer's ship. Irving Newman, navigator on Stuaffer's crew, said he was standing between the pilot and co-pilot, observing the attack, ready with the compass heading for home when he saw this apparition, a kamikaze, slam into the No. 2 engine and mortally wounding Stuaffer.[4]

The action, he relates, took place at 25,000 feet. The plane went into a flat spin, allowing free movement of the crew. Co-pilot James Wine came back to brush off the escape hatch, and found a brief case caught in the lock mechanism of the hatch. This made it necessary to hold the hatch open manually. He signalled for the crew to bail out while he returned to his seat to "talk out" the men in the rear and to help keep the plane airborne as long as possible.

Benjamin Blom, bombardier, and Newman held the hatch open for two others to jump and then Blom signalled for New,man to jump, indicating he would follow down. Instead he changed his mind and went down with the plane. There were no survivors from Clinkscales crew and only three from Stauffer's.

Of the four B-29s that went down over target that day, there were eighteen survivors. Carmichael, Harold Mann, Ernest Pickett and Newman were separated from the others. The Japanese charged them with responsibility for atrocities committed on the Yawata raid. Pickett was suffering from severe burns, and was taken to a hospital where he spent several months. The remaining three spent six and a half months in solitary confinement at Kampetai headquarters in Tokyo while undergoing days and nights of interrogation. Finally, their status was changed from "war criminal" to "special POW."

Although emaciated from lack of proper food, all four survived. Newman recalls that when a person gets as hungry as he and the others were in prison, a man can hear anyone cracking a peanut at 100 yards![4]

After Carmichael went down, he was replaced in Group by Alfred P. Kalberer who carried on to the end of the war. Engler was transferred to XXth Bomber Command in August 1944 and was replaced by Theodore S. Faulkner who was KIA over Singapore in November 1944, and his replacement was James V. Edmundson who ended the war with the Group and went on to fly in two more wars, Korea and Vietnam.[5]

On 6 June 1944 a "shakedown" mission was run by the "Superforts" against the railway shops and sheds in Bangkok, Thailand. Although some damage was done, it was agreed that the mission was, as one commander said, "loused up." While plans went ahead for attacks on other targets in Malaysia, the main effort in early June was directed against the Yawata iron works in western Japan. It would mark the first entry of United States aircraft into the Japan Air Space since the Doolittle raid two years earlier. The Doolittle raiders flew from east to west; the B-29s would fly from west to east.

More than two weeks was spent hauling gas and bombs to the forward bases in China. Taking part in the Yawata raid were elements of all four groups. Sixty-eight aircraft reached the target, one was shot down and seven were lost to various malfunctions. It was an audacious mission, but it did not shorten the war much because the weather precluded visual bombing, and radar in the early B-29s was quite rudimentary. But, all four groups were awarded the Presidential Unit Citation, not for results, but for courage and endurance. With time over target, some of the aircraft flew non-stop 3000 miles that night and into the next day.

Nearly a dozen newspaper and radio correspondents accompanied different flights of this mission, and among them was F. Tillman Durdin for the *New York Times*. His own feature story of the raid, written under great tension inside the airplane on the way back, is a model of skillful reporting.

The *New York Times* featured the Durdin story on top of page one. Another story by a wire service, also on page one, gives details of the raid, that is, damage assessment, most of which was based on speculation since no recon mission had yet reached the target area. Bombing a target at night, without weather information and without precision radar, was found to be a gamble and this was a gamble that lost.

After Yawata, targets were attacked at Sasebo, the big naval base in Japan, and the Anshan iron works in south Manchuria. The fifth mission was made in another direction, namely, Sumatra. Admiral Lord Louis Montbattan of Burma, commanding Southeast Asia, wanted the Pladjoie Oil Refinery at Palembang bombed, and the Moesi River mined above the refinery. This was an entirely new concept for the high altitude airplane, to come down on the deck and lay mines. After considerable discussion, the Americans agreed to the experiment. It was another "first." The mines were laid and the refinery partially destroyed, but not without losses. An artist has recaptured the drama of this mission in a painting which shows tracers, flak and searchlights threatening the intruding aircraft, titled "Shootout at Palembang." But the experience gained on this mission led, a year later, to the far-reaching decision by LeMay to use the B-29s to mine Japanese home waters.

Palembang lies on the southeast end of the island of Sumatra, and to reach it, the B-29s took off from Ceylon (Sri Lanka). Nearly the entire flight was made over the Indian Ocean and the round trip covered at least 3950 miles. This is not a significant statistic in the jet age but for propeller-driven aircraft, it was an outstanding feat. However, it was never tried again.

One of the most successful strikes made by the 58th Bomb Wing was on a second visit to Anshan, Manchuria on 8 September 1944. The new wing commander, Curtis LeMay, flew with the mission although he was under orders not to fly more than this one. But he wanted to know, among other things, how good the Japanese fighter pilots were against unescorted aircraft, holding to a personal maxim "what we don't know, we'll find out."[6]

By the morning of 8 September 115 airplanes had assembled at their advance bases around Chengtu. Out of this number 108 became airborne. Ninety-five actually reached Anshan. The weather was CAVU and follow-up photos by the F-13 squadron showed excellent results. The cost of the raid was one crack-up in India, two forced landings in China, one partly damaged and one listed as missing although the crew walked out.

The third and last raid on Anshan was made on 26 September 1944. Harry Changnon was not on it but was on one of the fields when the bombers returned. In his diary he has this to say.

> . . . . Later we went out to watch the B-29s coming back from the third Anshan mission. They reported that the planes were not bothered by flak and only a few fighters, but there was some air-to-air bombing. No planes were lost which was good news. The bad news was that it was a terrible mission for results . . . . they had to bomb through an overcast without seeing the target. One formation dropped their bombs miles too soon . . . .[7]

From what LeMay saw of the Japanese fighters on the first mission to Anshan, he did not think they would be "as tough as the Germans." This statement was made in the light of experience he gained with the 3rd Air Division but he had seen practically nothing of the Japanese fighters and, as far as the pilots in Manchuria were concerned, they were not, as he believed, the cream of the crop. On another mission to Anshan, the B-29s were actually attacked by a single Nakajima 97 with fixed gear (code name "Nate"), a relic of the 1930s.

When LeMay replaced Saunders, he early on flew to Hsinching to inspect the fields around Chengtu and to get acquainted with the troops. Saunders had been chided by Arnold for not getting enough airplanes over target, but some of the blame for this lay with the Bengal Air Depot where engine overhaul was being done, mostly by Indian mechanics. After Saunders left, LeMay closed the Bengal Air Depot and engines were shipped back to the States for overhaul.

The new commander came to China anxious to get a

B-29 CREW STATIONS

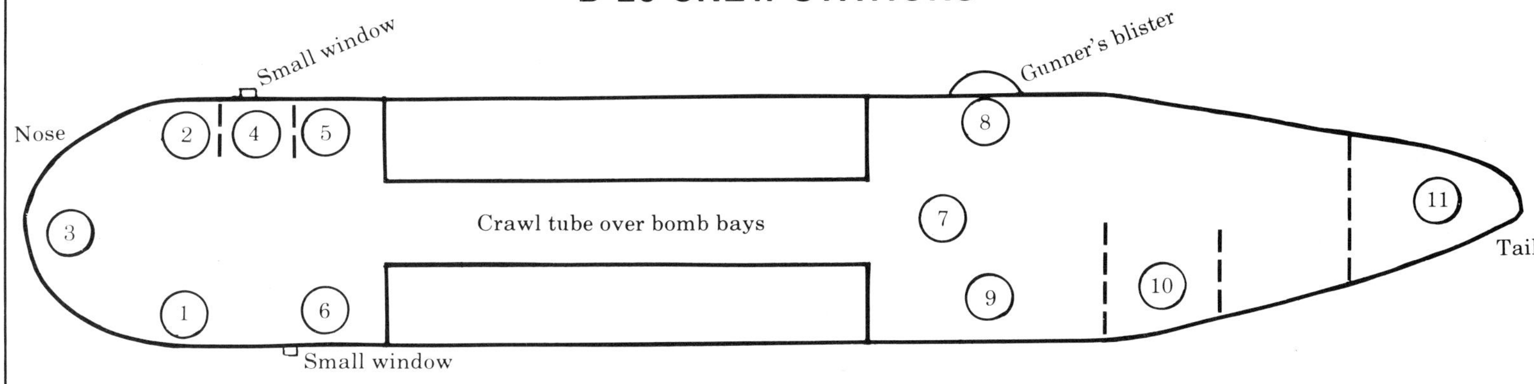

1—Pilot (aircraft commander)
2—Co-pilot
3—bombardier (who also functioned as nose gunner)
4—flight engineer (he faced aft looking at an instrument panel)
5—radio operator
6—navigator
7—Central Fire control gunner (his seat was elevated so that his head and gunsight were in a plastic bubble on top of the plane)
8—right gunner
9—left gunner
10—radar operator (his compartment was sealed off so no light entered it)
11—tail gunner

S. Stanley Andrews, 25th Squadron, 40th Group, in right seat of B-29 #298, at Hsinching, 1944.

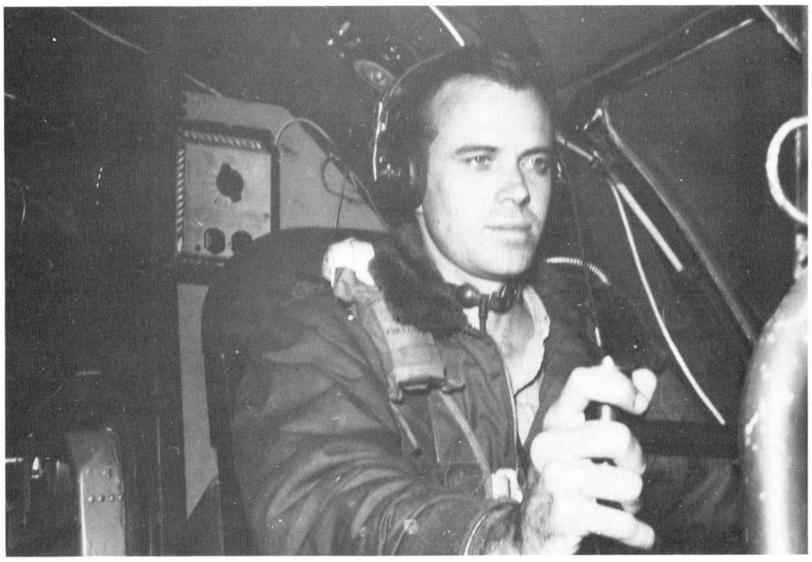

Ira V. Matthews, 40th Bomb Group, in left seat of No. 49-24579 after returning from raid on Omura, Japan, 21 November 1944, to base at Hsingching (A-1). Matthews was flying the "Eddie Allen," named for a Boeing test pilot who was killed flying one of the first B-29s off the production line, February 1943

Robert S. Clinkscales, C.O. of 94th Bomb Squadron, in cockpit with his "Sally," a black cocker spaniel. Both were KIA over Yawata on 20 August 1944.

James V. Edmundson, squadron leader with 58th Bomb Wing, later group commander. Wristwatch hooked to shirt pocket was common sight for servicemen in hot climates to avoid rust to watch and also, for working pilots or crew chiefs, to avoid breaking watch on arm without protection of sleeve.

LaVerne ("Blondie") Saunders who brought 58th Bomb Wing from Kansas to CBI in April-May 1944.

Left: Original crew of "O'Reilly's Daughter," taken at "Smokey Hill" (Salina) Kansas before leaving for China with 468th Bomb group. Front, kneeling, l. to r. Eddie E. Winkler, co-pilot; Casimer F. Stelmach, flight engineer; James V. Edmundson, pilot; Donald G. O'Brien, navigator; and Franklin J. McDonald, bombardier. Back, left: Walter N. Alspaugh, radio operator; James P. Meehan, gunner; Harry G. Brissenden, gunner; William A. Beckham, radar operator; Ernest J. Brundage, top gunner; and Rollin B. Heffernan, tail gunner. Most of this crew died in action or in line of duty flying with other airplanes and crews and only Brundage and Edmundson survive. Below: Alva L. Harvey, commanding 444th Bomb Group, 58th Bomb Wing, in China, 1944-45.

fresh start and correct the faults of his predecessor which is Standard Operating Procedure. But "Young Iron Pants," as the troops called him, got little farther than his predecessor. He could not make gasoline in China; he had to go back to India to get it, spending altogether too much time flying back and forth across the Hump to build up enough backlog to fly one airplane on one mission. (The B-24s took nine and ten trips for the same.)

The Hump slowed down LeMay the same way it slowed down every pilot who flew an airplane into China. To bomb Japan from China sounded like the logical thing to do when Chiang Kai-shek met with Roosevelt and Churchill in Cairo in 1943 and agreed to build bases in China to accommodate the big bomber than on the production line. The Chengtu plain was selected as the best place from which to bomb targets in Japan because it was the closest on the east side of the Hump and yet somewhat isolated from Japanese bombers. It was not the best place but, at the time, it was the only place.

In his biography, LeMay agrees with an article in *Air Force* magazine which refers to the 58th Bomb Wing in China as "a comparatively minor effort" where it was possible to run only "about four missions a month . . . ."[8]

The mission of the 58th Wing was primarily to attack Japan, not targets in China, but at the request of Chennault, LeMay made two strikes against Hankow and Nanking. Great stores of Japanase military supplies were hidden in warehouses along the Yangtze River at Hankow and Wuchang, and Chennault felt it would be a target worthy of the B-29s. Earlier attempts by Chennault's own 308th Bomb Group (H) to attack these targets had not come off well. One was a disaster.

On 18 December 1944, LeMay's planes made a raid on Wuchang-Hankow with 511 tons of bombs, mostly incendiaries. Eighty-four ships reached the target area and ten failed to return to base immediately owing to engine trouble and other causes, but not because of fighter interception. Japanese fighters had been destroyed or chased away by fighter and medium bomb squadrons of the 14th Air Force who went in first to neutralize the two airfields at Hankow and Wuchang. As for the B-29s, they arrived over target in haphazard formations and, as a result, did little damage to enemy stores. It was a badly-run mission.

In his book, LeMay expresses admiration for Stillwell and Mao Tse-tung and mentions Chennault only on page 351 and then in a derogatory manner:

> . . . .I remember running these missions for Chennault out of the Chengtu (sic). He was always pestering me to do things for him. I ran that mission against Nanking, and the one against Hankow. Incendiary attacks, warehouses and supply installations at Nanking, and docks and storage facilities at Hankow. Everything was fouled up there, beginning with Chennault's last minute request for a change in our Time over Target . . . . people dropped [their demolitions] in the wrong sequence, smoke obscured the primary areas, so on . . . .

The smoke also obscured LeMay's vision. Actually, Roosevelt had promised Chiang Kai-shek additional air support and the B-29 raids on Hankow and on Nanking fulfilled, in small measure, this promise. Chennault was therefore not "pestering" LeMay and it is safe to say that nearly everyone in the China Theater wondered why the B-29s could not be used against suitable targets in China as well as Japan. But LeMay had his orders from Arnold, who had his orders from Marshall: bomb Japan! To back him up, Arnold gave LeMay priority over all gas allocations coming over the Hump at the expense of every other squadron in the China Theater.

The shortage of gasoline the last year of the war is made abundantly clear in a stack of "buck slips" kept by William D'Ettore (A-1) covering all incoming and outgoing messages for 14th Air Force headquarters from September 1944 to the end of the war in 1945.[9]

Above: William Rooney, S-2, back to door, listens to Edwin Glass in debriefing session after Yawata raid, 15 June 1944. Glass and most of crew seen here were later lost in storm over Hump.

Above, right: Edwin Glass (standing left) explains plans for first raid on Yawata Iron Works with Charles Olavson (center) and John J. Madeline (right).

Above: William McNair, S-2, end of table, interrogates crew after Yawata Raid. Four men in back all general officers, some from Washington, who will report back to General Arnold.

Squadron mess for 40th Bomb Group, Hsinching.

Crew of the "Bugs Bunny," 45th Squadron, 58th Bomb Wing. Front row, l. to r.: Walter R. Gustavson, top blister gunner; LeRoy E. Bane, right gunner; Herbert Murrath, left gunner; Durdan S. Athey, Jr., tail gunner, and Henry Frees, Jr., radio man; back row, l. to r.: Robert Wix, pilot; Jack Riggs, co-pilot; Everett Miller, navigator; Alfred F. Eiken, bombardier; John K. Hilliard, flight engineer. Gustavson, Eiken, Riggs and Frees died in another airplane which crashed into Mt. Fuji in bad weather, on mission to drop food to POWs after VJ Day.

Black and white reproduction of oil painting commissioned by 58th Bomb Wing Association, called "General Arnold—Admiral King Strategy," depicts B-29s (left) coming out of China in their first raid against Japanese homeland (Yawata) 15 June 1944, on same day as island of Saipan (far right) was invaded by U.S. Marines, with Navy TBF "Avengers" (lower right). flying top cover for invasion forces, but momentarily flying over island of Tinian. Original painting hangs in U.S. Naval Academy at Annapolis, Maryland.

This B-29 was from the 462nd Bomb Group, 58th Wing. In the CBI, each group had its own marking, and when the wing moved to Tinian, new markings were issued and the 58th Wing was assigned a big triangle on the empennage, each Bomb Group bearing a different letter. The one shown here is a triangle with letter 'U,' which was assigned to the 462nd Bomb Group. Photo above was taken soon after move was made to Tinian from China because the rudder bears red coloring. Photo was taken from "blister" in another 29 and since there is another plane in picture, it is fairly certain that both are en route to Japan for a daylight formation mission. The 462nd called themselves the "Hellbirds," and the group commander at the time was Alfred F. Kalberer.

But Arnold was no longer interested in China, only in the primary target—Japan—which was understandable but also debateable. He seemed to forget that without the 14th Air Force in China there would be no B-29 bases.

The fire bombing of Hankow, as suggested by LeMay, was not successful because "people" were dropping their bombs in the wrong sequence. It was the first attempt in the China Theater to fire-bomb a target, but this was not LeMay's idea either. T. Alan Bennett, a former group commander in China, recalls that he was present at a staff meeting in Kunming attended by both Chennault and LeMay.[10] During the discussion, Chennault urged LeMay to try incendaries on Hankow as well as on the inflammable, wooden-built cities of Japan.

Confirmation of the Chennault thesis appears in a long memorandum he wrote on the origin of the First American Volunteer Group under date of 19 October 1941, long before LeMay came to China.[11] Chennault was in Toungoo at the time struggling to create an aerial combat team of volunteers, later known as the "Flying Tigers," and he wanted his pilots and crew chiefs to understand, fully, the nature of their mission in China. On page 11 of this 34-page memorandum he lists four main objectives, and the fourth reads:

> 4. If the political situation warranted to subject the Japanese islands to long range bombardment from Chinese airports, using incendiary bombs to burn the highly inflammable Japanese industrial areas.

In his biography, LeMay fails to mention the staff meeting shared by Bennett. But the passage of time has mellowed LeMay's antipathy for Chennault. On 15 December 1972 both LeMay and Chennault were enshrined in the Aviation Hall of Fame at Dayton, Ohio, and in his acceptance speech, LeMay expressed guarded admiration for Chennault, now deceased.[12]

Defending the air bases around Chengtu against night bombing was nearly impossible. The four fields were defended primarily by the 311th Fighter Group consisting of the 528th "Dragonflies," the 529th "Gold Tigers," and the 530th "Yellow Scorpions." All three squadrons had seen action in support of Merrill's Marauders and other units in Burma before moving to Chengtu. Fighter and medium bomber squadrons of the Chinese American-Composite Wing stationed at Enshih, Liangshan, Ankang, Hanchung, and Sian also served as a defensive net around the big birds. But no one could stop the enemy night raiders. In all the Japanese ran about ten missions against the B-29 bases with a total of forty-three bombers, or on an average of four-plus per mission. Since they were operating mostly at night, they also found it difficult to locate the fields and their bombs usually landed far off target. Yet, the raids kept men awake and edgy the next day.

Official records of the 58th Bomb Wing show a raid on

John Radford (back center) and crew of "20th Century Unlimited" October 1944, probably at Hsinching, advance base for 40th Bomb Group.

one or more of the Chengtu fields on 8 September, when only minor damage was done. The most serious raid occurred on the night of 26-27 September when five B-29s were damaged, two seriously.

On one occasion, at least, Japanese bombers avoided all interceptors and bombed one or more of the fields around Chengtu in daylight. Harry Changnon remembers one of these daylight attacks over Hsinching:

> .... A 2-ball alert sounded, so we all took off in various directions, some to slit trenches, rice paddies, etc. ... Two fires were going strong right away and we could see that one of them was a C-46 and the other a C-87. There were a few men wounded by shrapnel as there were a few anti-personnel bombs mixed in their bomb loads ... We knew that several more planes had been hit by shrapnel and flying pieces of metal, as well as a few bullets that the Japs fired. We all went over to Hostel #2 for a nervous dinner ... then sat around drinking coffee and eating pudding until bedtime. They used engine warmers to help keep the mess hall warm for us ... In the morning, we ate a quick breakfast before going down the line to check on our ship. It was full of holes .... We found that in addition to our C-109 [a tanker] the Japs had knocked out a C-46, two C-87s, two fighters, and a C-54. Luckily, none of the B-29s had any major damage (only) a few holes to patch, but nothing like old 846. We started counting the holes in the main fuselage and quickly got to 611 holes from back of the engine. That didn't include the double tail assemblies which looked like sieves ... The engine didn't appear too badly hurt, although we could see they would all have to be changed .... "[13]

Between 6 June 1944 and 29 March 1945, squadrons of the 58th Bomb Wing in the CBI flew forty-nine missions ranging from Singapore to Manchuria to Japan to Formosa. They did what they could against targets too far away with gasoline too hard to get. Blondie Saunders summed up the situation most in a letter addressed to former comrads of the 40th Group in convention in 1980.

> You had a fine Group under trying conditions .... trying to bomb Japan from lousy bases in China (and) trying to get enough stuff over the Hump, we put everything in one end and got too little out the other ....[14]

In January 1944, LeMay flew to Guam to meet Lauris Norstad, deputy chief of staff to Arnold. Norstad told LeMay that the 58th was being moved to the Marianas in the western Pacific, and for him to return to China and make the necessary arrangements to phase out the wing and move to the Marianas. LeMay then flew back across the Pacific and China to India. He promoted Roger Ramey to brigadier-general and made him the new commander of the wing with the understanding that he would bring the wing to the Marianas as soon as the necessary staff and paper work had been completed. The new home of the wing would be Tinian, south of Saipan. On a clear day the two islands are visible to the other.

LeMay took off from India for the last time on 19 January 1945, topped his tanks at Luliang, and raced back across the Pacific to Guam where he spent the rest of the war as commander of the XXth Air Force.

Listed below are the missions flown in the C.B.I. by the 58th Bomb Wing, and date of each mission.

Bangkok, Thailand, 6 June 1944; Yawata, Japan, 15 June 1944; Sasebo, Japan, 7 July 1944; Anshan, Manchuria, 29 July 1944; Palembang, Sumatra, 10 August 1944; Nagasaki, Japan, 10 August 1944; Yawata, 20 August 1944; Anshan, 8 September 1944; Anshan, 26 September 1944; Okayama, Formosa, 14 October 1944; Okayama, 16 October 1944; Heito, Formosa, 16 October 1944; Einansho, Formosa, 17 October 1944; Omura, Japan, 25 October 1944; Rangoon, Burma, 3 November 1944; Singapore, Malaya, 5 November 1944; Omura, Kyushu, 11 November 1944; Omura, 21 November 1944; Bangkok, 27 November 1944; Mukden,

Harold T. Oblender, navigator, computes ETA for Anshan, Manchuria in one of raids run by 40th Bomb Group in the "20th Century Limited."

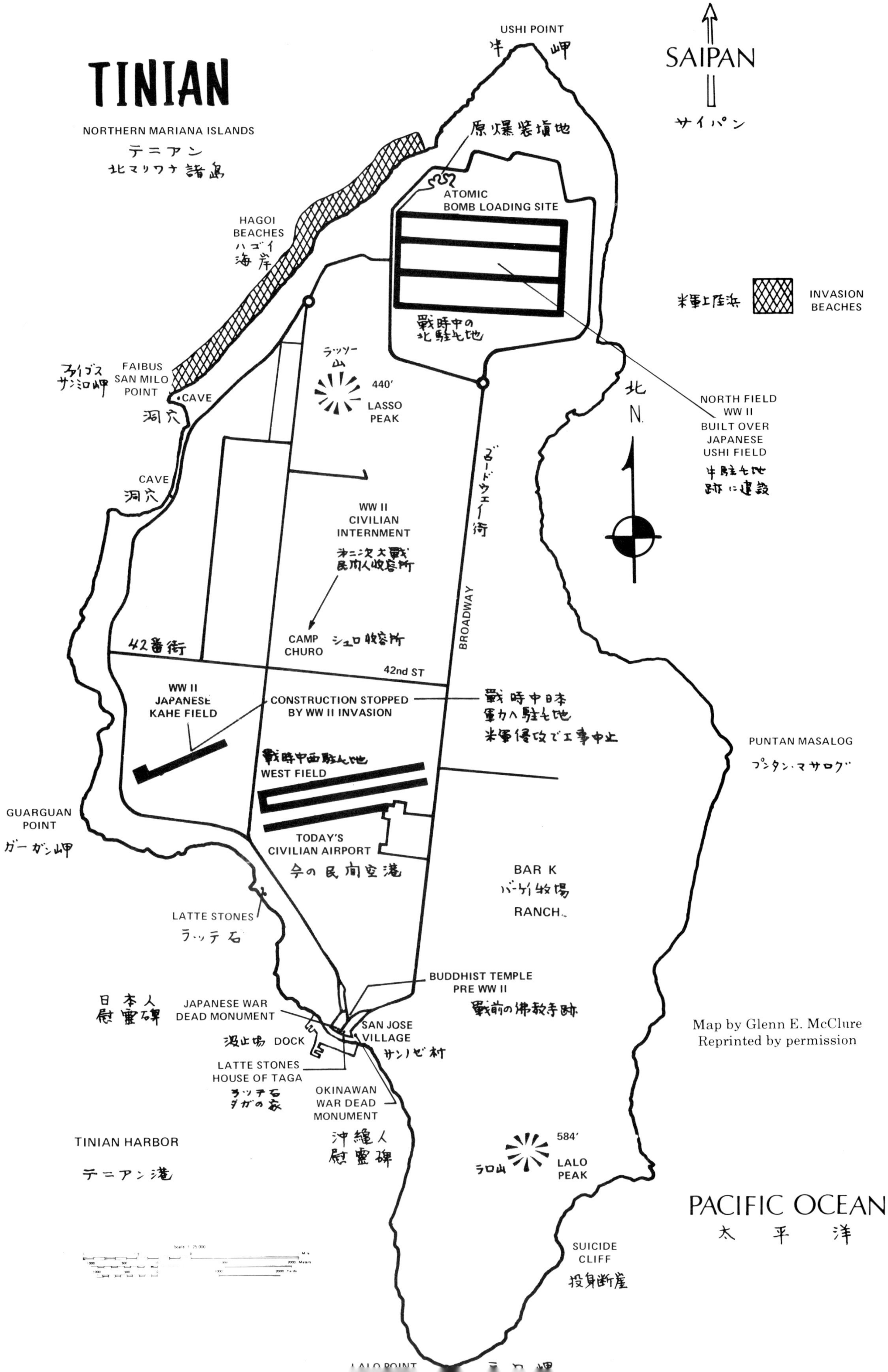

Map by Glenn E. McClure
Reprinted by permission

Manchuria, 7 December 1944; Bangkok, 14 December 1944; Hankow, China, 18 December 1944; Omura, 19 December 1944; Mukden, 21 December 1944; Bangkok, 2 January 1945; Tachieri, Formosa, 6 January 1945; Heito, 9 January 1945; Singapore, 11 January 1945; Kagi, Japan, 14 January 1945; Shinchiku, Formosa, 17 January 1945; Saigon, Indo-China, 25 January 1945; Singapore, 25 January 1945; Saigon, 27 January 1945; Singapore, 1 February 1945; Saigon, 7 February 1945; Bangkok, 7, February 1945; Rangoon, 11 February 1945; Kuala Lumpur, Malaya, 19 February 1945; Singapore, 24 February 1945; Singapore, 27-28 February 1945; Shanghai, 4-5 March 1945; Singapore, 2 March 1945; Singapore, 12 March 1945; Kuala Lumpur, 10 March 1945; Rangoon, 17 March 1945; Rangoon, 22 March 1945; Shanghai, 28 March 1945; Saigon, 28 March

West Field on Tinian with two main runways at right (see map opposite page).

1945; Singapore, 28 March 1945; Singapore, 29 March 1945.

Meanwhile, the assault on Saipan by U.S. Marines began on the same day as the B-29s ran the first raid on the Yawata Iron Works, 15 June 1944, but it was not until 12 October before the waste of war could be cleared up on Saipan sufficiently to allow the first flight of XXI Bombardment Command to land on Saipan, although fighting was still in progress on the north end of the island around Marpi Point.

When the planes of the 58th Bomb Wing began taking off from the Chengtu fields for the Marianas, they flew directly into Tinian which became their permanent station and also future station of the *Enola Gay,* bearer of the atomic bomb to Hiroshima.

But the bombing of Japan was already under way before the 58th Wing got to Tinian. The first raid, launched 24 November 1944, was led by Emmett ("Rosie") O'Donnel who got 110 aircraft airborne, bombed Tokyo and lost two of his ships.

After LeMay reached Guam on 19 January 1945, the former 58th Bomb Wing, or XXth Bombardment Command, was combined with the XXIst Bombardment Command, already in the Marianas, to create a new air force, the XXth. On 8 and 9 March, LeMay sent his bombers over Tokyo, dropping mostly M-69 incendiary clusters which created the greatest man-made fire in the history of the world. More than 80,000 people were killed, a million made homeless, and sixteen square miles of Tokyo were leveled.

Not long after this, LeMay learned from his crews and photo reconnaissance people that the Japanese anti-aircraft guns were quite effective at high altitude, but aircraft approaching the target at 10,000 feet or lower, were meeting little flak. It appeared that the Japanese had overproduced heavy anti-aircraft guns at the expense of medium range guns.

All this was hard for LeMay to believe, but he came to the conclusion that if he sent his bombers over target at 10,000 feet or less, he would escape most of the flak. He also removed all armament except the tail gun and used the tail gunner chiefly as an observer, while depending on the speed of the airplane to evade Japanese fighters. In one of the greatest gambles of the entire war, LeMay outguessed the Japanese anti-aircraft gunners, and B-29 losses fell off sharply. Today, Robert Wix, who flew thirty-five missions in the CBI and from Tinian, believes, with many other veterans of the XXth Air Force, that LeMay's tactics changed the course of the air war in the Pacific.

And now the air war would come full circle. As the Japanese Air Force had bombed Chungking for three years—without mercy, the XXth Air Force would bomb Tokyo for three months, without mercy.

On 13 August 1945, two days before the Emperor of Japan accepted terms of surrender, Curtis E. LeMay made the cover of *Time* magazine. There could be little doubt about it; he was No. 1 in V.L.R. (very long range).

Omura, located in western Japan, was hit four times from China, as the above target list shows. On the third attack, launched 21 November 1944, there were tragic consequences for three of the crew. Had they been shot down in Japan, they might have survived, but they were shot down near Hankow on the way back to base in west China.

This is roughly what happened: On 16 December, members of the crew, all from the 770th Bomb Squadron, 462nd Group, were flying about 150 miles to the north of Hankow when the bombardier, in the nose, reported a Japanese fighter dead ahead doing acrobatics. Suddenly, the fighter plane turned in and opened fire on the B-29 bomber, hitting the No. 2 engine which caught fire, cutting the control cables and knocking out the power system on all the guns. A second engine began acting up and at 13,000 feet, the pilot sounded the alarm bell for the men in the rear to bail out. Frank Sommer, radar man, William McHahan, right gunner, and Benjamin Lisowski, left gunner, went out in that order, and a short time later, at 1000 feet altitude, Anthony Tomaczak, tail gunner, jumped, all landing at different points along the Hankow-Peking railway zone where they

Admiral Lord Louis Mountbatten of Burma (left), supreme commander Allied forces in Southeast Asia, comes to inspect B-29 base somewhere in India. Officer in overseas cap is American pilot who brought visitor in Beech 18 (Army C-45). Admiral wears short-sleeved bushjacket favored by British officers in India. In distance at left (four stripes on fin) stands aircraft No. 224908.

First arrivals of 58th Bomb Wing on Tinian from China in March 1945 had outdoor cooking and tents—catch-as-catch can. Some men wear steel helmets, some not.

Fuji, a sacred mountain to the Japanese people, became a deadly check point for approaching bombers of the 58th Bomb Wing after it moved to Tinian and began to attack targets in Tokyo, a target never attempted from China bases. The 58th Wing joined other wings stationed at Guam, Tinian and Saipan to form XX Air Force under Curtis E. LeMay.

were rescued by guerrillas of the Communist 8th Route Army.

Before the officers in the cockpit could bail out, Henry Wheaton, central firing control, reported that his parachute was shredded and he could not jump. The pilot then decided to belly in, but as the big airplane came down, full flaps, it was met with heavy ground fire from Japanese guard posts along the railway zone. The pilot, bombardier and navigator were all killed and co-pilot Vernon Schaefer was wounded in the neck by shrapnel and lost consciousness. Lester W. White, James Forbes and Henry Wheaton, gunners and radar men, also rode the plane down and survived the crash. All were captured. Schaefer was pulled from the airplane and regained consciousness long enough to recognize the three other living members of the crew at a Japanese guard post, but they were soon parted and Schaefer, despite his severe wound, was given no first aid. He was thrown into a dirty cell and held there for a couple of weeks. Luckily the wound began to heal of itself.

Unknown to Schaefer, the other three men, Wheaton, Forbes and White, were taken to Hankow where on 16 December 1944, they were paraded through the streets of the city in a propaganda ploy by the Japanese army to influence Chiang Kai-shek to urge the Americans to ease up on the bombings of Chinese cities in Japanese occupied areas (mostly Manchuria), and, indirectly, to relax the bombings on the Japanese homeland by the B-29s. The three men were paraded all day, spit upon and beaten with sticks by Japanese posing as Chinese. In the evening, completely exhausted from the constant walking, lack of food and water, and the torture, the three men were nearly dead when they were taken in hand by a detail from the secret police, strangled with rope cords, and thrown into a crematorium.

Chinese custodians at the crematorium were told by the Japanese to throw the ashes into a pond nearby because the ashes were "bad things." The Chinese expression used here was probably *huai tung tze,* one of the most common expletives in the Chinese language, meaning something or someone morally perverse or rotten. But instead of following orders, the Chinese buried the ashes and after the war, American officers were guided to the site and recovered the ashes.

Also, after the war, eighteen Japanese army officers and enlisted men, gendarmes and civilian employees of the Japanese consulate in Hankow were bound over for trial in Ward Road jail in Shanghai, charged with war crimes. All the accused, including Major-General Masataka Kaburagi, chief of staff of the 34th Army who had connived in the propaganda parade and death of the three Americans, pleaded not guilty. After a trial lasting seventeen days in February 1946, all but one were found guilty. Four, including Kaburagi, were given the death sentence, and the others were sentenced from two years to life.[15]

Vernon Schaefer survived. He was forwarded to Hankow later, paraded through the streets in a down-river city, held in Shanghai and eventually sent to Japan where he spent the last months of the war in a POW camp.[16]

Two Gooney-birds bringing supplies across the Hump.

XXIII

THE JAPANESE AIR WAR IN CHINA

Japanese Navy had at least three carriers operating off Chungming Island in mouth of Yangtze River against targets in China. Below, His Imperial Japanese Majesty's ship *Kaga,* the first to reach China, rebuilt from a battleship in 1920s and first to have bridge on flight deck. Other two carriers were *Soryo* and *Ruyjo.* These ships carried Nakajima A2Ms and A4N biplane fighters as well as Yokosuka B4Ys and Aichi Dia bombers. Hankow was out of range for fighters, but Mitsubishi 63Ms from Formosa could reach it. The three ships were part of 1st Carrier Squadron under Rear Admiral Shiro Takasu which left Sasebo on 12 August 1937 for China. In China squadron become part of 3rd Fleet commanded by Vice Admiral Kiyoshi Hasegawa, whose flagship, *Idzumo,* became target of Chinese bombers on 14 August 1937. *Kaga* and *Soryu* both went down at Midway in June 1942, and *Ryujo* went down in Solomons in August 1942.

Immediately after VJ Day (14 August 1945) and after Japanese forces in China had acknowledged their Emperor's wishes for them to lay down their arms, a massive effort was made by Japanese commands everywhere to destroy all documents and official papers in their possession. Although defeated, the Japanese were unwilling to admit any culpability which official documents might confirm.

As a result of this premeditated effort to destroy the past, post-war historians were faced with a dilemma. With no documentation, how could the events of the past be reconstructed with any degree of accuracy? They could not, but there was an alternative, namely, to interrogate living witnesses. As a result, American army and navy historians, based in Tokyo with Far East Command headquarters, were assigned the task of recording these memories. Unfortunately, the evidence suggests that none of the interrogators had ever been in China, or they might have been more critical of the responses they got from the Japanese witnesses.

It was not an easy assignment for the Japanese either, but most of them who were willing to cooperate seemed to realize that truth in history is just as important as national pride. Today their oral accounts of Japanese campaigns in the Pacific, the Philippines, Southeast Asia and China form a national treasure of literature and history. Textural materials, maps, and orders of battle have been painstakenly reconstructed and are now available in a multivolume series of books called the *United States Strategic Bombing Survey.*[1]

Despite these accolades to the Japanese who participated in this oral history program, it seems clear that they found it difficult to admit defeat in any given campaign, at least in China where their land forces were defeated four times in full-scale campaigns or in campaigns of limited objective, namely, twice before Changsha, once at Changte and once at Chihkiang. Never is there any admission of defeat, only of "withdrawals" or "stalemate."

Two themes seemed to dominate official Japanese thinking on China in the 1930s, one, that China had grown proud, and two, China needed to be taught a lesson, i.e. punished, for growing proud.

A rather typical editorial on this appeared in *Shanghai Mainichi*, a Japanese daily, which charged that China did not have the slightest wish to see Sino-Japanese relations readjusted, and warned that "if China does not come to her senses at this time, we shall meet with a great commotion in the near future."[2]

The "great commotion" began the very next day at Lukouchiao, sometimes called "Marco Polo Bridge" in Peking, when war broke out between Chinese and Japanese forces. But the idea that China needed to be chastised is echoed again and again in the Japanese press and in statements by public officials.

In October 1937 the Japanese *United Press,* another Japanese daily published in Shanghai, said it was Japan's intention "to chastise China for obstructing Japan's efforts to establish lasting peace in the Orient."[3]

General Iwane Matsui, commander in chief of Japanese forces in central China, reaffirmed this general line of thinking when he told newsmen in Shanghai that the Japanese

Mitsubishi A5M2, Type 96 carrier fighters with open cockpit. Arresting hooks visible under fuselage on plane in rear. (U.S. code name, "Claude.") This picture taken on Chinese field near Shanghai, 1937. Later version of "Claude" replaced this one, but all were obsolete by 1941.

aims were "the protection of Japanese lives and property and punishment of the Chinese government and army." He also felt sorry, he said, "for the innocent persons who were unfortunately living in the area of hostilities, and it goes without saying that the Japanese Army has no enmity toward the Chinese people in general . . . "[4]

This news conference in Shanghai was followed a few weeks later by the Japanese "Rape of Nanking" in which some 40,000 Chinese prisoners of war were executed, many in cruel and inhumane manner, and thousands of Chinese civilians were tortured, murdered, and women raped. It was the most savage assault on the human condition ever witnessed in contemporary Chinese history. (After WWII, Matsui was tried before the International Tribunal of Justice in Tokyo, and was one of seven top officials executed for war crimes.)

Few Americans had an opportunity to listen to the torrent of excuses which poured over Radio Tokyo immediately after the surrender and before the entry of General MacArthur into Tokyo in late August 1945. American personnel in China listened to these broadcasts with amazement, and to the excuses that were being made to avoid responsibility for events of the past seven years which had left an entire nation ravaged, and millions of dead, suffering or homeless people.

Here, then, is a much abbreviated account of the Japanese air war in China as found in the Strategic Bombing Survey with no change in the wording of the text per se.[5] The author has made a few literary "assists" which appear in brackets to help clarify some point in the text. The spellings follow the text, except "Nanching" is changed to Nanking, and "Hankou" to Hankow. Reference to the "enemy" here means Japan's enemy, China, until 1942, when the word can mean either China or the United States:

On 7 July [1937] as soon as news of the Lukouchio Incident was received, the Kwantung Army dispatched part of the air force . . . to the Tientsin airfield and from there supported ground operations . . .

Immediately after the outbreak of fighting in Shanghai, the Provisional Air Group transferred one air squadron to the Shanghai Expeditionary Army. Although at the end of September enemy planes near Yangchu were completely wiped out . . .

On 21 November, the Air Group carried out a surprise attack against Loyang airfield. Several enemy planes were damaged on the ground and one plane was shot down. On the 12th the Air Group carried out a surprise attack against Changan [Sian] and destroyed an unknown number of planes on the ground.

With the landing of the Shanghai Expeditionary Army at the mouth of the Whangpoo River on 23 August 1937, the Navy Air Force attacked enemy bases at Shanghai, Nanchang, Hangchow, Nanking and Chujung from carriers and at bases in Formosa and Saishu Tao [island]. This force destroyed the greater part of the advance units of the Chi-

Photo by author taken along boundary of International Settlement of Shanghai and Chinese municipality of Chapei, August 1937, shows Nakajima E8NI, 95 Reconnaissance seaplane (U.S. code name "Dave").

Japanese carrier fighter operating off H.I.M.S. Kaga *1929-42.*

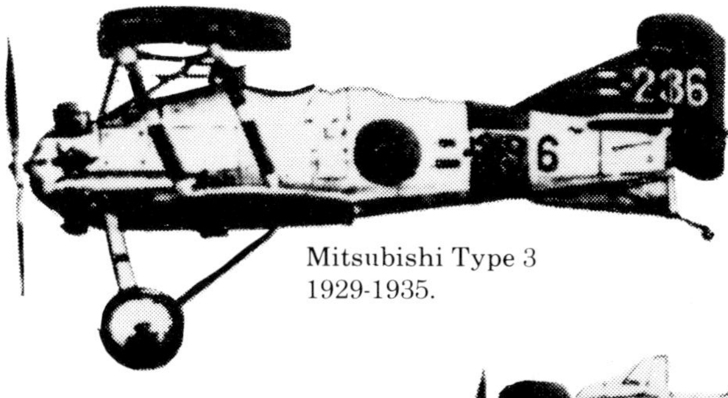

Mitsubishi Type 3
1929-1935.

Nakajima Type 90
1935-1937.

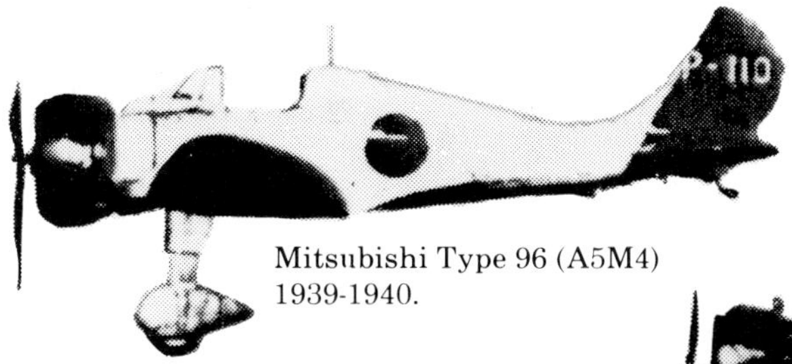

Mitsubishi Type 96 (A5M4)
1939-1940.

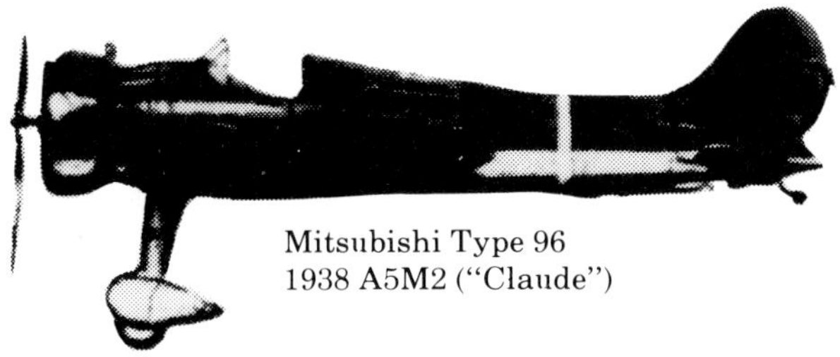

Mitsubishi Type 96
1938 A5M2 ("Claude")

Mitsubishi Type O, A6M2
1941-1942 ("Claude")

nese Air Force and secured our supremacy in the Shanghai area . . . It was estimated that the enemy air strength in this area was between 70 and 80 planes. The Japanese Navy had 72 planes . . . both land and carrier based . . . During September the Navy Air Force shot down 32 planes in the vicinity of Nanking. On 2 October, 9 enemy planes were destroyed on the ground at Hankow and, on the same day, 20 enemy planes were shot down in air combat. On 24 October three planes were shot down. Later, most of the Chinese planes withdrew from the area.

It was estimated that enemy planes in the early part of 1938 numbered between 350 and 450. Half these planes were of Soviet manufacture, and in addition, between 70 and 80 new planes were being imported from Russia each month . . . [6]

The strength of the Japanese Army Air Force in China from January to April 1938 was 24 squadrons. Fourteen squadrons (four reconnaissance, three fighter, three light bomber and three heavy bomber squadrons) . . . were stationed in north China while ten squadrons (five reconnaissance, three fighter, one light bomber and one heavy bomber squadron) were placed in central China. In January [1938] the Giba Air Brigade shot down 12 enemy planes over Loyang airfield . . . The Provisional Air Group . . . attacked the [Sian] airfield on the 11th and 14th with considerable success with five enemy planes shot down on the 11th.

From the beginning of 1938 the 3rd Air Brigade had been unsuccessful in repulsing enemy air attacks on the air fields at Nanking, Wuhu and Hangchow due to the fact that the Russian-made planes used by the Chinese were superior to the Type 95 then being used by the Brigade. In mid-March, however, the Brigade was reinforced by a few Type 97 planes [Nakijima B5N carrier attack bomber] . . . In January two enemy planes were shot down, in March three and in April six.

Operations to annihilate enemy strength [along the Peking-Hankow railroad] were carried out [from April to July] . . . During these engagements a total of 34 enemy planes were destroyed, four on the ground . . .

After moving to the airfield near Hankow in early November the main force of the Air Group raided Hengyang and destroyed 16 enemy planes on the ground . . .

. . . From the middle to the end of November [1938] the main force of the Air brigade, using Paotow and Ani [Ansi] as its bases, bombed Lanchow continuously . . .

About the beginning of November [1938] enemy air strength was estimated to be about 49 squadrons with 340 planes, half of which were fighter planes. Their bases were at Lanchow, Nancheng [Hanchung], Chungking, Chihkiang and Liuchow with Changan, Ankang, Liangshan, Enshih, Henyang and Chian as front line bases.

In November the Navy Air Force raided Liangshan, Chihkiang, Kueilin, Chungking and Chengtu, and in December, Liuchow [and] shot down or destroyed on the ground 70 planes . . .

By the end of October 1938 Japanese forces had occupied Canton and Wuchang-Hankow areas . . . On 12, 20, and 22 February [1939] approximately 30 planes cooperated in attacks on Lanchow. One hundred enemy planes, both in the air and on the ground, were reported damaged during these attacks. At this time it appeared as though the morale of the enemy air force was high and air battles were often carried on for a long period of time. As both the 12th and 98th Air Regiments suffered heavy casualties the attacks were suspended. [Most of the attacks around Lanchow were against Soviet Russian pilots].

. . . The Navy then suggested a combined Army-Navy operation . . . against Lanchow late in December. The strength of the bombing force for this operation was approximately 100 planes, consisting of about 36 Army Type 97 (K-21 "Sally") of heavy bombers of the 60th Air Regiment, 36 land medium type attack planes . . . and 28 land medium type attack planes . . . The combined Army-Navy force . . . shot down 20 enemy planes and inflicted severe damage on . . . Lanchow, but the enemy air force retreated and

Left: Nakajima Ki-44s, Army Type 2, single seat fighters, photographed in China, location uncertain. (U.S. code name "Tojo").

Right: Nakajima KI-84 Army Type fighter (U.S. code name "Frank"), 1944-45. No. 302 may be position assigned to it with Japanese *Sentai* (Group).

Japanese Aichi AB-3, single seat trainer with 130 hp engine. Picture shows plane being lifted over side by derrick from a Japanese tender, probably H.I.J.M.S. *Notoro* off Woosung, 1932.

frustrated the Japanese Air Force efforts to destroy it.

During 1939 the Navy Air Force . . . made thirteen raids against Chungking, two against Chengtu and Liangshan . . . one against Chengtu and two against Lanchow in November and three against Lanchow in December.

In February 1940, intelligence reports were received stating that enemy aircraft planned to attack Nanking and Shanghai in an effort to disrupt the establishment of the new Japanese-sponsored central government [under Puppet Premier Wang Ching-wei]

On 6 June the army [air] units attacked Chungking. Later in the month they launched seven more attacks against Chungking . . .

The Navy Air Force attacked Chungking eight times, [1943] was at 1103 hours. The 90th Air Regiment of 16 light bombers, and the 33rd Air Regiment of 18 Type 1 fighters, attacked the runway of Kueulin airfield . . . The second attack was launched at 1600 hours. The 90th Air Regiment of nine light bombers, and the 33rd Air Regiment of 9 Type I fighters, attacked the airfield and railway station at Hengyang . . . At 1625 hours eight light bombers and seven fighter planes attacked the runway at Kueilin . . .

On February [1943] our units succeeded in engaging three enemy P-43 fighters, one of which was shot down [over Liangshan].

Intelligence reports state that on the afternoon of 31 March [1943] the U.S. had advanced about 40 fighters and seven B-25s to Kueilin and Lingling and that they were

This is probably a Nakajima Ki-43 (U.S. code name "Oscar"), first known in China-India-Burma theater as "Jim," but to the Japanese airmen as "Hayabusa." This was Army Type I fighter and not a Zero. Only Navy planes could be referred to as "Zeros."

Chengtu twice and Liangshan twice in May [1940], Chungking eleven times and Suining once in June, Chungking four times . . . in July and Chungking eight times in August . . .

In the early hours of the morning of 8 December [1941] the Army Air unit attacked Kaitak airfield [Hongkong] and destroyed all but two of the enemy's planes [i.e. British planes] . . . On the morning of the 12th, the main force attacked Stonecutter Island (and) part of the unit hit Kueilin airfield and destroyed two enemy planes that were attempting to attack the Japanese forces from the rear . . . The 1st Air Brigade gave very effective support to the land forces, especially during the capture of Changsha and later during the *difficult withdrawal through strong enemy resistance* [italics mine].

It was estimated that at the beginning of 1943, the United States Air Force had 44 fighters and 20 bombers and the Chinese [Air Force] 148 fighters and 43 bombers (with) their main force at Yunnan . . . The Chinese Air Force appeared to be concentrated on the Szechwan plain and undergoing further training . . . The [U.S.] air strength was gradually increased, making it daily more difficult to effectively bomb enemy air bases.

The first [Japanese] attack on Kueilin airfield on 7 January demanding weather reports for 1 April for the coast of Fukien Province with Fuchow as the center, showing clearly that they were planning not only attacks on the China front but also against the Homeland [i.e. Japan].

On 31 March, the 3rd Air Division ordered a squadron of the 33rd Air Regiment, using six Type 1 fighters. [Nakijima Ki-43 "Oscar"] to attack Kueilin. A few enemy planes were sighted but they retreated before they could be engaged.

On 1 April twelve fighters (and) six Type I fighters and two Type 2 (Ki-44 "Tojo") from the 33rd Air Regiment advanced to Hengyang and Lingling and at about 0900 engaged more than twenty P-40s over Lingling. Four enemy planes were shot down. The 1st Air Brigade lost four planes.

At 1444 hours on 24 April [1943] 23 fighters and nine light bombers of the 1st Air Brigade flew over Lingling air base . . . No enemy planes were encountered . . .

From 1410 on 4 May, . . . six B-24s, under the protection of eleven [U.S.] fighters, made repeated attacks . . . on Hainan Island . . . This was the first time B-24s had been used since they had attacked the Kailin coal mines in north China the previous winter.

On 6 May approximately ten B-25s with strong fighter protection bombed Canton city and airfield, inflicting slight

Japanese plane which crash landed on shore of Gulf of Tonkin and was recovered by Chinese salvage crew and brought to Liuchow where it was rebuilt by Gerhard Newman. It appears to be a Mitsubishi A6M2 Navy Type O carrier fighter (U.S. code name "Zeke"). Zero airplane commemorated 2600th anniversary of ascension of Emperor Jimmu to Japanese throne. Plane had range of 1118 miles, double that of best U.S. fighter at beginning of WWII. More than 10,400 were built by Mitsubishi from 1942 to 1945. Above: same airplane being test flighted.

This Nakajimi Ki-27b, Army Type 97, was recovered by Chinese forces in central China and flown or hauled to Shuangliu air base, near Chengtu, where picture was taken. Original American code name for Type 97 was "Abdul" but changed to "Nate" in south Pacific. On this photo Chinese Air Force has already painted own logo under wing. American Air Force personnel stand at left and one is in cockpit checking out controls.

damage in both areas.

At 1525 hours on the 10th [of June] sixteen light bombers and twenty-seven Type 1 fighters attacked the air base at Hengyang and bombed five enemy planes on the ground. This force also engaged about ten P-40s in the air, shooting down and destroying five of them.

As it was considered absolutely essential that U.S. air bases in Yunnan be destroyed . . . On 15 May [1943] the 5th Air Division attacked the Yunnan [Kunming?] air base and destroyed eight large planes and severely damaged five others. It also shot down eight fighter planes and one transport plane . . . Between 23 July and 22 August the 3rd Air Division attacked Hengyang aerodrome nine times, Lingling three times, Kueilin twice (fighters only), Chienou nine times and Paochi and Chihkiang each twice. The results achieved were

P-40 . . . . . 49 planes (seven unconfirmed) shot down.
P-38 . . . . One plane shot down.
B-24 . . . . Two planes shot down.

Enemy fighters at Kueilin, Lingling, Hengyang and vicinity were estimated to be between 60 and 70 in the middle of August. Headquarters reconnaissance planes reported that there were seventeen B-24s, ten B-25s, six twin-engine planes [P-38s?] and thirty-four small type planes on Yunnan airfields, and thirteen large planes and three small planes at Chanyi [northeast of Kunming] . . . between 23 August and 8 September planes shot down in east and west China were

P-40 . . . . 11 planes (three unconfirmed)
P-43 . . . . 5 planes (three unconfirmed)
B-24 . . . . 6 planes
Unknown type . . . . 1 plane

In the period 9 September to 7 October [1943] the 3rd Air Division advanced its bases to south China (and) during this time, Lt. Gen. Moritaka Nakazone, together with his staff officers in charge of operations and intelligence, flew from Kagi, Formosa, to the command post at Canton. At 1745 hours on 9 September the plane in which General Nakazone was flying was shot down over Hsiaochou, 10 kilometers southeast of Canton, and the General was killed . . .

During its summer operations [in 1943] the 3rd Air Division shot down or destroyed on the ground approximately

Above: A Kawasaki Ki-48 Army Type 99 MKI, a twin-engine light bomber used by Japanese in China. This airplane was developed after 1937 when Soviet air mission to China surprised Japanese with the Tupolene SB-2 which was superior to any bomber Japanese had in operation. This photo, probably taken in early 1940s, show camouflage along length of fuselage, probably to confuse enemy fighters from above. U.S. code name for this bomber was "Lilly."

Below: Mitsubishi Ki-21s, Army Type 97 heavy bombers on raid over Chungking c. 1940. Same airplane was used in extensive bombing raids on Pingliang and Lanchow to halt Soviet war supplies and airplanes from reaching China in 1938-39 period. First U.S. code name for bomber was "Jean" after General MacArthur's wife, but the general took a dim view of this honor and the code name was changed to "Sally." Later, after Japanese made some modifications to airplane, U.S. code name was "Gwen," and still later was "Sally III."

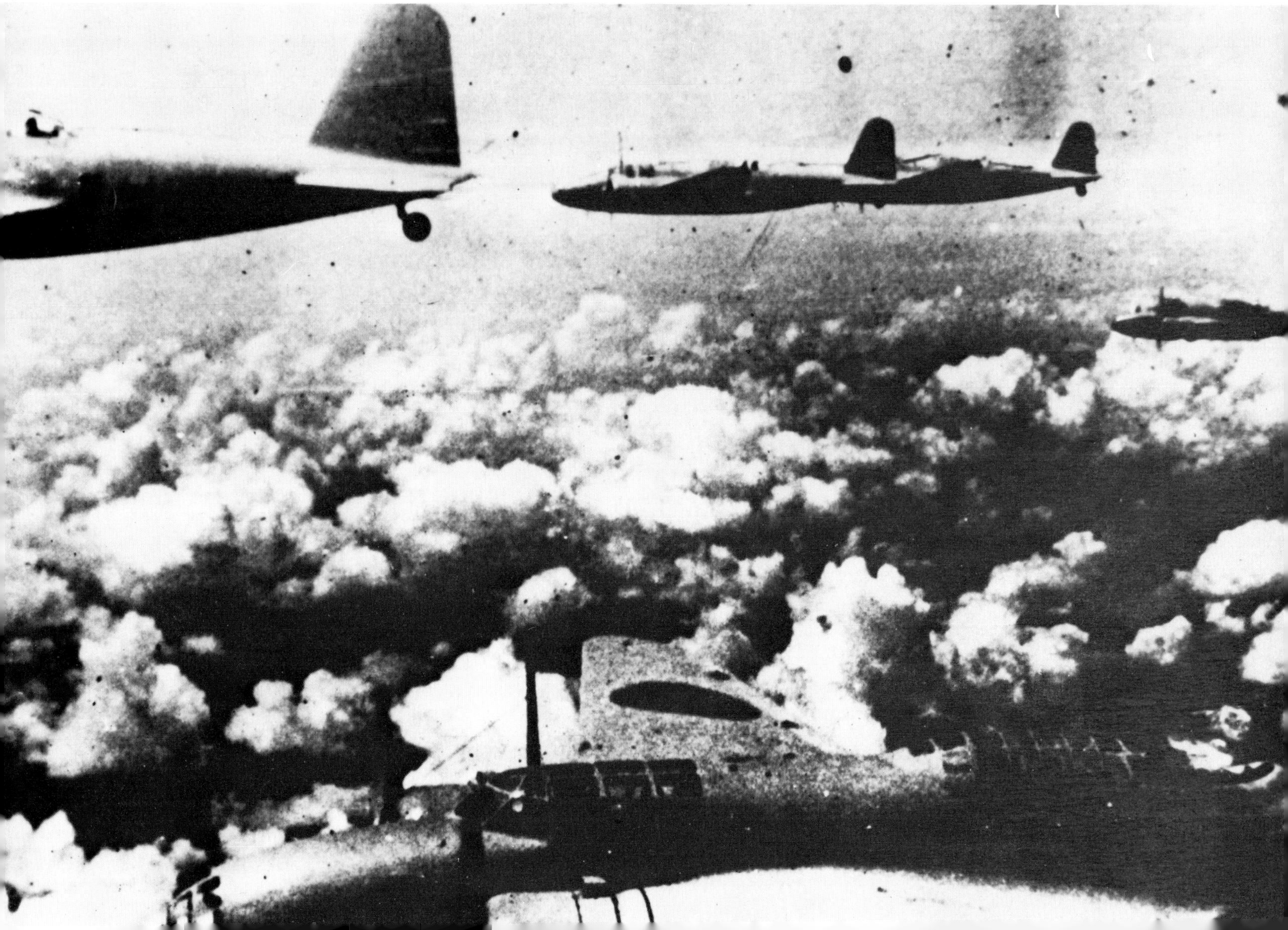

Cloth banner for good luck (right) carried by Japanese pilot shot down in China, and at left, his *hachimaki,* cloth headband. But not all Japanese pilots wore *hachimaki* or carried good luck banners any more than all American pilots wore St. Christopher medals.

103 enemy fighters and twenty-one bombers. Japanese losses were twenty-five fighters and fifteen bombers missing, seven fighters and one bomber destroyed . . .

Before dawn on 10 December [1943] fifteen P-40s, one P-54(?) and three B-25s were shot down or burned on Hengyang air field. At dusk on the same day, eight P-40s, one P-38 and two B-25s were destroyed on Lingling airfield. In addition, the next day forty-nine fighters and bombers of the 3rd Air Division attacked and destroyed eleven of the fifteen enemy planes on the ground in Lingling . . .

This was the first occasion on which the small Japanese air force in China had severely damaged the powerful enemy air force. It had skillfully taken advantage of the darkness and had operated throughout all areas in China.

In January 1944 Imperial General Headquarters approved the plan for *Ichi-Go* [code name One-Five] operation calling for . . . the destruction of Kueilin and Liuchow airfields in order to alleviate the threat of Japan.

It is estimated that by early March 1944 the U.S. Air Force in China had increased its strength to 340 planes of which approximately 160 were fighters and ninety were bombers . . .

The fighting strength of the Fifth Air Army at this time was less than one half its T/O strength although an average of fifty planes (mostly Type I fighters and light bombers) were being received each month from Japan . . .

On 6 May [1944] approximately forty-five enemy fighters and bombers carried out a daylight attack against the Wu-Han area, inflicting heavy damage.

With the opening of the *Ichi-Go* operation [27 May 1944] the number of pilots and planes gradually decreased and, in spite of every effort it was found impossible to replace them . . .

The Fifth Air Army sustained heavy losses during this operation and received little replenishment. It was compelled, therefore, to avoid air combat with the superior enemy air force whenever possible . . .

During the early part of May [1944] it was estimated that the total number of [U.S.] airplanes at the front was approximately 520, composed of 340 U.S. Air Force planes, 109 U.S.-Chinese Air Force [C.A.C.W.] planes and 80 Chungking Air Army [Chinese Air Force] planes . . . It was estimated that they had increased to 600 planes by mid-June and 750 by mid-August . . . It was characteristic of the entire *Ichi-Go* operation that land troops depend on just one line of communication to penetrate a group of enemy air bases and, in consequence, lines of communications were always exposed to enemy attacks. The enemy air force furiously attacked not only the front lines but also the rear . . . including ground troops, naval craft, automobiles, vehicles and trains (which) were forced to move at night.

The 22nd Air Regiment, equipped with Type 4 Fighters [Nakijima Ki-84 "Frank"] . . . proved of great help to the Air Army [but] with the withdrawal of reinforcements, Japanese Air power in China proved vastly inferior to that of the enemy.

The number of serviceable planes of the various air units around mid-November [1944] were 152 (which included Type 2 fighters, Type 1 fighters, Type 4 fighters, light bombers, biplane fighters . . . As veteran pilots were continually lost they were replaced by inexperienced men . . . . the supply of planes and parts from Japan proved inadequate, making it necessary to put back into service even obsolete planes . . . used as trainers . . . Loss of planes due to mechanical failure became increasingly frequent . . .

Great publicity had been given to the stepped-up production of the U.S. B-29 long-distance bombers and it was estimated that the time was not far distant when the enemy . . . would commence large-scale bombing attacks on Japan and Manchuria from bases in China . . . With the appearance of the B-29s, the need for bombing enemy air bases in the vicinity of Chengtu became vital . . . On the night of 8 September, eight heavy bombers and ten twin engine ligth bombers of the 8th Air Brigade bombed . . . the airfields around Chengtu for the first time. They burned seven B-29s and destroyed eleven others on the ground . . . Two of the heavy bombers [Japanese] were reported missing in action. These attacks were repeated . . . as long as Chengtu was used as a B-29 base.

It was estimated that the number of enemy front line planes based in China in July 1945 was somewhere in the vicinity of 1,100 planes. (Footnote: the Japanese army estimated, in April, the number of enemy planes at between 1500 and 2000.)

On land, attacks were centered (by the Americans and Chinese) on north and central China where the P-51s and

Tower at Nanking was built by Chinese before war began, but after occupation by Japanese in December 1937, this supertower was added to the roof for better ground control. Name "Nanking" on roof was painted either by Chinese or Americans after VJ Day and jeeps in front are probably for American personnel.

Hangar at Nanking airdrome, built by Chinese, used by Japanese and bombed by Americans. Roof is gone, doors are hanging off center, and walls are pock marked with machine gun bullets.

Japanese aircraft above and upper left abandoned on Nanking air base after surrender in August 1945. These are Nakajima Ki-84 Army Type 4 Fighters *Hayate* (Gale), U.S. code name "Frank." Chinese Air Force has already painted own logo on fuselage, Wind sock seems to be improvisation erected by Chinese or Americans after they occupied field in August 1945.

P-40E captured by Japanese forces and removed to Yakata air base where it was found after war.

P-47s fighters concentrated on attacking rolling stock and other vital transportation targets. They seriously damaged the railway system and made it almost impossible to operate it during the day.

The most difficult problem in regard to the preparation of lines of communication was the lack of aviation fuel. At the beginning of 1945 fuel supplies stopped . . . Experiments in regard to the manufacture of substitute fuel were conducted but the rate of production fell far short of the rate of demand . . . The fighter units were instructed not to engage the enemy except under extremely favorable circumstances . . . (end of Japanese record).

These are the highlights of the air war in China, roughly 1937 to 1945, as viewed by Japanese air force witnesses. Aside from exaggerations of victory claims made against Chinese and American airmen, the most serious charge that can be made against the post-war historians is their failure to account for their own losses more accurately. In the entire account there is an admission that the 3rd Air Division lost 25 fighters and 15 bombers, and seven fighters and one bomber destroyed on the ground in the summer of 1943. In this same period the 3rd Air Division allegedly shot down or destroyed on the ground "approximately" 103 enemy fighters and 21 bombers. Since the Chinese Air Force was no longer a fighting organization in 1943, the losses would have to be mostly American losses which are highly exaggerated.

The Japanese also admit "missing" two bombers over Chengtu in attacks on B-29 fields, and they admit "heavy casualties" over Lanchow, the latter against Soviet Russian airmen. These casualties were so great that the Japanese had to withdraw from the fight. This is the only admission made of defeat in the air. Finally, the post-war historians admit that four planes from the 1st Air Brigade were lost over Lingling and Hengyang in April 1943.

The exaggerated "kill" ratio claimed by the Japanese can be attributed largely to two factors; one, poorly trained debriefing officers—often none at all — and to a system that did not require a pilot to have a witness of any of his "kills." It seems obvious that if the ratio of kills to losses had been sustained, the Americans and Chinese airmen would have been annihilated early in the contest. But, as the chapter on Japanese air operations in China closes, the reader is left to wonder what became of the big Japanese bomber formations which once raided Chungking, or the fighter squadrons that were shooting down American P-40s right and left over Lingling? Suddenly, the Japanese historians begin to excuse themselves; they don't have any airplanes left to fight with; they are flying obsolete aircraft; they are out of gas!

The Americans lost many airplanes in China and on the Hump, but half of them, perhaps, were lost not to enemy action. They were lost to pilots pushing their luck too far on empty gas tanks, pilots losing their compass bearings, pilots losing their wits. But at no time or place, in the time frame cited above, did the Americans ever lose the number of aircraft claimed in the narrative.

The most important air battle ever waged against the Americans by the Japanese in China fell on a three day period from 23 to 26 July 1943 when the Japanese made a supreme effort to knock the Americans out of the sky. The victory claims in this fight by the Japanese are noted above, although the air battles also raged over Lingling, the next field to Kueilin, and over Hengyang, on the same days. But after the Japanese withdrew, Colonel Vincent, commanding, sent a commendation to all personnel, forward echelon, 14th Air Force, under date of 28 July, and this is how the commendation reads:

This Kawasaki Ki-61 Army Type Fighter, called "Hien" (swallow) by Japanese, was parked in a hangar at Yakata air base when WWII ended. It was built under license from German engineers. This one appears to have no armament. Ki-61 became a deadly foe in last year of war but there were not enough of them. (U.S. code name "Tony.")

> During the period July 23 to 26 inclusive, the Japanese directed eight bombing attacks against our bases in this area, employing an estimated 200 planes. During this critical period I have noted with pride and gratitude the conscientious devotion to duty and unselfish cooperation displayed by both officers and men. It was only through this superior performance of duty that we were able not only to defend our bases from serious damage and destroy fifty (50) Japanese airplanes (and probably thirty-four (34) more, but also to put our planes into the air and thus carry the fighting to the enemy.[7]

Colonel Vincent failed to mention that at one point, against standing orders, he took to the air himself and, with four other earlier victories, made ace. Whether or not his statistics on the number of Japanese aircraft destroyed is exaggerated, the fact is that the Japanese were seldom out looking for a fight in daylight again. Bombing missions against U.S. bases were run mostly at night from now on, and staggered to keep American personnel in slit trenches, but seldom hitting any worthwhile target except for a stray airplane parked on the line, not in a revetment. Most of the night raiders used "banana" bombs, i.e. anti-personnel clusters meant for parked aircraft, not to damage buildings.

The claims of destruction to B-29 bombers on fields around Chengtu are also inaccurate. Although a number of B-29s were damaged, at no time was any B-29 completely destroyed. Most of the raids on these fields were also made at night except on one occasion when a formation of light bombers sneaked through the warning net in daylight and caught a number of P-51s parked on one of the fields. But that was it.

The Japanese historians fail completely to mention the devastating air strike made by Chinese and Soviet airmen against the Hankow air field in 1939, and they fail to mention the equally devastating strike made by a combined force of B-25s, P-51s and P-38s on Shinchiku air base on Thanksgiving Day 25 November 1943.

The most serious error in the Japanese account is the failure to mention the defeat their airmen suffered at the hands of the 1st American Volunteer Group ("Flying Tigers") over Burma and West China. In the six months of operations, the Flying Tigers destroyed 294 aircraft in the air and on the ground against a loss of twelve of their own, and not all to combat.

The value of this chapter in the original version of the *Strategic Bombing Survey* lies not in the number of claimed air victories, but in the numerous tables of organization charts and the order of battle maps which, on the surface, appear to be excellent.

Colonel John ("Pappy") Herbst, who commanded the 74th Fighter Squadron in 1944, became the leading American ace in the China theater. In a critique of Japanese airmen, he believed the average American pilot easily outclassed the average Japanese pilot. Herbst found that in most enemy formations of four or more planes, the leader was good, the others quite inferior. However, he also found that a good Japanese pilot, flying an "Oscar," one on one, was practically impossible to shoot down.

Emperor Hirohito asked his countrymen to lay down their arms on 15 August 1945 and accept the terms of surrender offered by the United States and its allies in World War II. Several regional surrenders of Japanese armies in the field were held in China, and in the picture above, a camouflaged transport brought Japanese officials to a ceremony held at Chihkiang air base, late August 1945. Plane is a Mitsubishi Ki-57-11, 15-place Army transport (U.S. code name "Topsy"), a modified Ki-21 heavy bomber. In foreground American air corps personnel get first close-up look at a Japanese plane. Below, left: Japanese officers (at right) conferring with Chinese and American officers at Chinkiang over details of army withdrawals. Lower right: Japanese military and civilian leaders were ordered by General MacArthur to fly to Ie Shima, a small island off Okinawa, flying two Mitsubishi bombers ("Bettys"), painted white with green crosses on fuselage. The code word for their approach to Ie Shima was "Baatan." At Ie Shima, they boarded a C-54 for the flight to Manila, 750 miles south, where the first major surrender ceremony was held in August 1945, before the formal surrender on the battleship *Missouri* in Tokyo Bay 2 September 1945.

The document below is a photographic reproduction of a news release issued by the Office of War Information in Chungking dated 15 August (1945). Asian time is one day ahead of the United States. The Emperor's message was heard and recorded in Chungking and the translation into English was probably made by Japanese language officers attached to the OWI, but the text is imperfect and there are several misspellings.

From U.S. OFFICE OF WAR INFORMATION August 15, 1945.
Chungking

DOMEI NEWS SERVICE

THE SPECIAL ISSUE OF THE JAPANESE EMPEROR'S MESSAGE OF ACCEPTING POTSDAM PROCLAMATION.

Tokyo 15/8: His Imperial Majesty the Emperor was graciously pleased to personally read the Imperial rescript accepting Potsdam declaration in a nation-wide radio broadcast at noon today. It is the first time in Japan(s history that the Emperor spoke to nation over radio.

The full text of the Imperial rescript is as follows: "To our good and loyal subjects; after pondering deeply on the general trends of the world and the actual conditions obtaining in our Empire, we have decided to effect a settlement of the present situation by restoring to an extra-ordinary measure. We have ordered to our government to communicate to the governments of the United States, Great Britain, China , and Soviet Union that our Empire accepts the provisions of their joint declaration. To strive for the common prosperity and happiness of all nations as well as the security and well-being of our subjects is the solemn obligation which has been handed down by our Imperial ancestors, and which we lay close to heart.

Indeed, we declared war on America and Britain out of our sincere desire to ensure Japan's self-preservation and the stabilization of East Asia, it is being far from our thought either to infringe upon the soverei~~g~~nty of other nations or to embark upon the territorial aggrandizement. But now the war has lasted for nearly four years. Despite the best that has been done by every ~~xxxxx~~ one--the gallant fighting of military and naval forces, the diligence and assiduity of our servants of the state and the devoted service of our one hundred million people, the war situation has developed not necessarily to Japan's advantage, while the general trends of the world have all turned against her interest.

Moreover, the enemy has begun to employ a new and most cruel bomb, the power of which to do damage is indeed incalculable, taking the toll of many innocent lives. Should we continue to fight, it would not only result in an ultimate collapse and obliteration of the Japanese xnation, but also it would lead to the total extinction of human civilization. Such being the case, how are we to save the millions of our subjects; or to atone

ourselves beforethe hallowed spirits of our imperial ancestors. This is the reason why we have ordered the acceptance of the provisions of the joint declaration of the powers.

We cannot but express the deepest sense of regret to our Allied Nations of East Asia, who have consistently cooperated with the Empire toward the emancipation of East Asia. The thought of those officers and men as well as others who have fallen in the fields of battle, those who died at their posts of duty, or those who met with untimely death and all their bereaved families, pains our heart night and day. The welfare of the wounded and the war-sufferers, and of those who have lost their home and livelihood, are the objects of our profound solicitude. The hardships and sufferings to which our nation is to be subjected hereafter will be certainly great.

(continued on next page)

We are keenly aware of the inmost feelings of all ye, our subjects. However, it is according to the dictate of the time and fate that we have resolved to pave the way for a grand peace for all the generations to come by enduring the unendurable and suffering what is insufferable. Having been able to safeguard and maintain the structure of the imperial state, we are always with ye, our good and loyal subjects, relying upon your sincerity and integrity. Beware most strictly of any outbursts of emotion which may engender needless complications, of any fraternal contention and strife which may create confusion, xxxxx lead ye astray and~~cause~~ and cause ye to lose the confidence oft the world.

Let the entire nation continue as one family from generation to generation, ever firm in its faith ofthe imperishableness oft its divine land, and mindful ofits heavy burden ofresponsibilities, andthe long road before it. Unite your total strength to be devoted to the construction forthe future. Cultivate the ways of recetitude; foster nobility of spirit; and work with resolution so as ye may enhance the innate glory oftheimperial state and keep pace with thep rogress of theworld." (end imperial rescript)

Left: H.M.S. *Eagle,* a British carrier, probably off Chungming Island, c. 1937. *Eagle* went down in Mediterranean on 11 August 1942 after making delivery of fighters which were flown off for Malta.

Below: A Mitsubishi Ki-21 army type 97 heavy bomber, (U.S. code name "Sally"). This one was found abandoned on field at Hanoi 29 August 1945. It was one of best Japanese planes for long-range operations, and was used over Chungking repeatedly in WWII.

XXIV

Ready For Wheels Down

(Miscellaneous Pictures and Documents)

A Dewoitine D-5-10, this one was assigned to the 41st Squadron of the Chinese Air Force at Kunming in 1938. The Chinese had twenty-four on order with the French manufacturer, but Henry L. Mazurie, a mechanic who was sent to Kunming from Hanoi to help service the aircraft, believes only six reached Kunming. He was never able to learn what happened to the others before the Japanese invaded Indo-China in the summer of 1940. The 5-10 shown here had a Hispano-Suize engine of 12 cylinders and 860 HP. Radio masts point up and down behind cockpit. Spats have been removed probably for easier take-off in muddy fields. Chinese blue and white stripes appear on stabilizer and, under port wing, logo of the CAF.

A.L. ("Pat") Patterson, first man to skywrite in the China Air Space, ready for take-off from Lunghua air base, Shanghai, 29 October 1936. Special occasion was celebration of fiftieth birthday of Chiang Kai-shek. Patterson improvised smoke-emission system from tail of this Porterfield aircraft and wrote Chinese ideograph for *shou,* meaning "long life." Feat was performed over west side of Shanghai then known as Jessfield Park. Patterson was sales agent for American aircraft. At his left stands Major "American Lee," from San Francisco, who went to China in 1930s to help develop aviation. He was KIA in Japanese bombing of Kiangwan air base in mid-August 1937.

Photo taken in early 1944 outside radio liaison station (DS8) established by John Birch near city of Changsha in 1943. Birch sits in front row flanked by Chinese army officers and, at far right, Malcolm Rosholt. Enlisted men in back are radio operators and code clerks, l. to r. Jerome Lassiter, William Hayes, Edmund Cain and Arthur Eickenberry. Rosholt was sent to Changsha to relieve Birch who was being assigned to new field of operations in north China. On 25 August 1945, a few days after war ended, Birch in a dispute with Chinese Communist guerrilas, was shot and killed. A rescue mission of Chinese, Americans and Japanese recovered his body and he was buried with military honors near Hsuchou, northern Kiangsu Province. The John Birch Society was later named after him.

Air Routes
INDIA-CHINA DIVISION ★ AIR TRANSPORT COMMAND
AFGHANISTAN
TIBET
NEPAL
BHUTAN
INDIA
BURMA
CHINA
FRENCH INDO-CHINA
THAILAND
MALAY STATES
SUMATRA
BORNEO
PHILIPPINES
FORMOSA
ARABIAN SEA
BAY OF BENGAL
SOUTH CHINA SEA
KARACHI
JIWANI
JODHPUR
NEW DELHI
AGRA
LAHORE
BOMBAY
NAGPUR
BANGALORE
COLOMBO
CHINA BAY
GAYA
CALCUTTA
BARRACKPORE
TEZGAON
CHITTAGONG
COX'S BAZAR
AKYAB
RANGOON
TOUNGOO
MANDALAY
BHAMO
MYITKYINA
LALMANIR HAT
MISAMARI
TEZPUR
LHASA
MOHANBARI
SOOKERATING
CHABUA
JORHAT
DERGAON
FT. HERTZ
YUNNANYI
LIKIANG
KUNMING
CHENGKUNG
LULIANG
YANGKAI
CHANYI
WEINING
IPIN
LOSHAN
CHENGTU
CHUNGKING
NANCHENG
HSIAN
KWEILIN
LIUCHOW
WUCHOW
CANTON
HONGKONG
AMOY
TAINAN
CHANGSHA
NANKING
SHANGHAI
HANOI
SAIGON
BANGKOK
SINGAPORE
BRUNEI
MANILA
SAMAR
LEYTE
DAVAO

Call letters and navigational aids jotted down by Pilot Robert J. Kennedy (27th Troop Carrier Squadron). 1944-45.

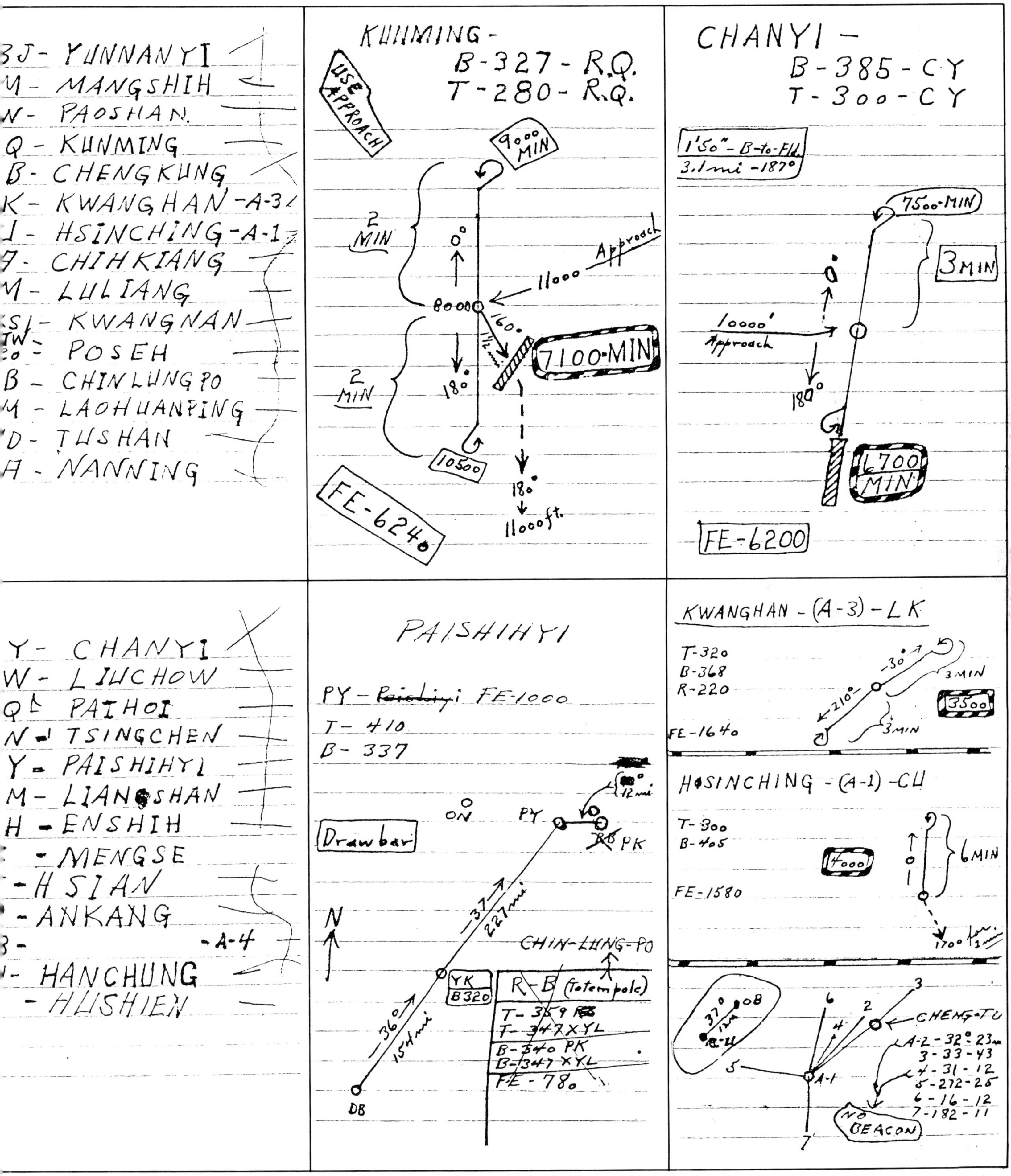

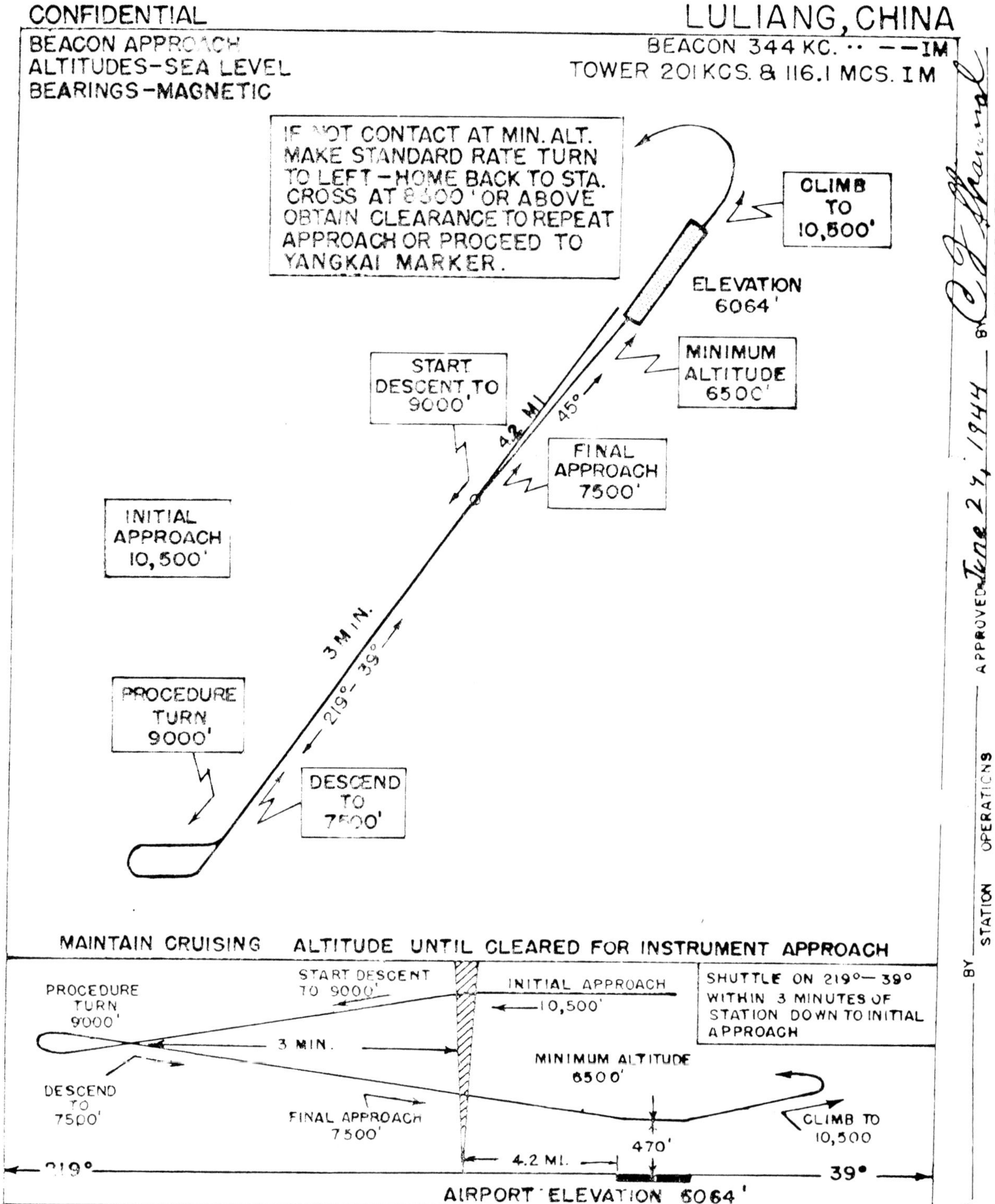

TIME-DISTANCE STATION TO FIELD 4.2 MILES

110 MPH	120 MPH	130 MPH	140 MPH	150 MPH	160 MPH
2 MIN. 17 SEC.	2 MIN. 6 SEC.	1 MIN. 57 SEC.	1 MIN. 48 SEC.	1 MIN. 41 SEC.	1 MIN. 34 SEC.

27

ICD ATC NAVIGATION & BRIEFING

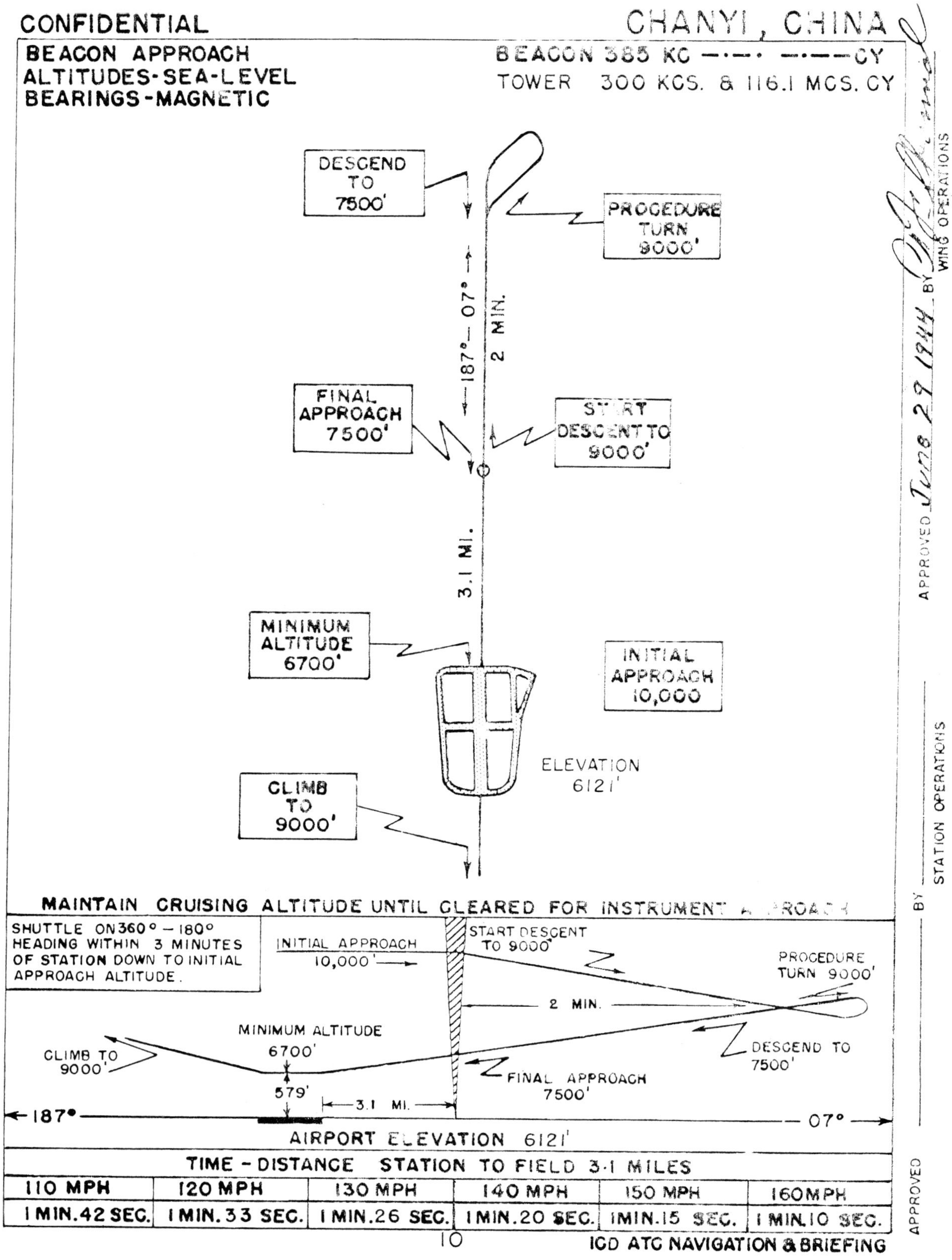

The charts above and opposite page are reproduced, actual size, from a manual to aid pilots in World War II called "Instrument Letdown & Departure Procedures, India-Burma-China," which was issued by command of Maj-Gen. [George] Stratemeyer, commanding the 10th Air Force in India, and Brig-Gen. William H. Turner, commanding India-China-Burma division of Air Transport Command, the latter a military organization which carried troops, equipment and supplies to theaters of combat around the world.

WAR DEPARTMENT
AAF FORM NO. 5
APPROVED DEC 7 1942

INDIVIDUAL FLIGHT RECORD

(1) SERIAL NO. O-828[illegible] (2) NAME ZEIDLER, EDWARD P. (Last, First, Middle) (3) RANK 2ND LT. (4) AGE

(5) PERS. CLASS (6) BRANCH AIR CORPS, (7) STATION APO 210, BOX 3

(8) ORGANIZATION ASSIGNED — Air Force: 14TH; Command: 308TH BOMB GP; Wing: ; Group: 374TH BOMB SQ (H); Squadron: ; Detachment: (Attached for flying)

(9) ORGANIZATION ATTACHED — Air Force; Command; Wing; Group; Squadron; Detachment

(10) PRESENT RATING & DATE PILOT (11) ORIGINAL RATE & DATE

(12) TRANSFERRED FROM (13) FLIGHT RESTRICTIONS

(15) TRANSFERRED TO (14) TRANSFER DATE

(16) Do not write in this space

Pers Class	Rank	Rtg.	A. F.	Command	Wing	Group No.	Group Type	Squadron No.	Squadron Type	Station	Mo.	Yr.

(17) MONTH FEBRUARY 19 45

DAY (18)	Aircraft type, model & series (19)	No. landings (20)	Flying inst. (Incl. in 1st pil. time.) S (21)	Commd. Pilot C CA (22)	Co. Pilot CP (23)	Qualified Pilot Dual QD (24)	First Pilot Day P (25)	First Pilot Night P N or NI (26)	Rated Pers. Non-Pilot (27)	(28)	(29)	Non-Rated Other Arms & Services (30)	Non-Rated Other Crew & Passgr (31)	Instrument I (32)	Night N (33)	Instrument Trainer (34)	Pilot Non-Mil. Aircraft Over 400H.P. (35)	Pilot Non-Mil. Aircraft Under 400H.P. (36)
1	B-24M	1				T-	2:55											
2	B-24M	1				T-	4:55							3:30				
10	B-24M	1				T-	1:00											
14	B-24M	1				T-	11:05							10:05				
15	B-24M	1				T-	6:20							2:00				
16	B-24M	1				T-	6:20							3:00				
17	B-24M	1				T-	4:45											
18	B-24M	1				T-	4:30											
20	B-24M	1				T-	6:30											
21	B-24M	1				T-	6:45											
22	B-24M	1				T-	7:25											
25	B-24M	1				T-	3:00											
Column Totals							65:30							18:30				

*** CERTIFIED CORRECT ***

[signature]
ROBERT M. SHEEKS,
CAPT., AIR CORPS,
ASST. OPER. OFFICER.

	(42) Total Student Pilot Time	(43) Total First Pilot Time	(44) Total Pilot Time
(37) This Month		65:30	65:30
(38) Previous Months this F.Y.		165:00	271:25
(39) This Fiscal Year		230:30	336:55
(40) Previous Fiscal Years			213:15
(41) To Date	213:15	230:30	550:10

Do not write in this space

Aircraft 19	NI 20	Card No. 1: 21	22	23	24	25	26	Card No. 2: 27	28	29	30	31	Card No. 3: 32	33	34	35	36

INDIVIDUAL BOMBARDIER'S INTERROGATION REPORT

Group 40th Squadron 25th Date Dec. 21, 1944

Bombardier LT. J. J. O'Keefe Field Order No. 23

Interrogator Maj. [illegible] Airplane No. 233

1. Target Bombed: Primary ✓ Secondary — Other —
(If other give location and description)

2. Aiming Point: MANCHURIAN A/C MFG PLANT MUKDEN

3. Number, type and size bombs carried: 8 M-64 GP; 8 M-76 INC

4. Fuse setting: Nose (.01) (INST) Tail (.025) (NON-DELAY)

5. Number released visually (16) ALL Number released blind NONE

6. Number jettisoned NONE Number brought home NONE

7. Sighting Operation: Principal ____ Secondary ____ on leader ✓

8. Intervalometer setting 15 feet. Heading (mag.) 55°

9. Calibrated Indicated Air Speed 192 mph Alt.(true) 21,700

10. Number of runs 1. Reason for extra runs: —

11. Length of run: 500 seconds. Time of Bombs Away: 0213 Z.

12. A.B.C. used: NO (yes - No) A.C.L.D. used: NONE (Yes - No)

13. Malfunctions of bombing equipment and cause, if known: NONE

14. Were all arming wires returned: YES (Yes - No) (If answer is "no" explain).

15. Comments and recommendations: NONE

16. Sketch your formation and indicate your position in formation.

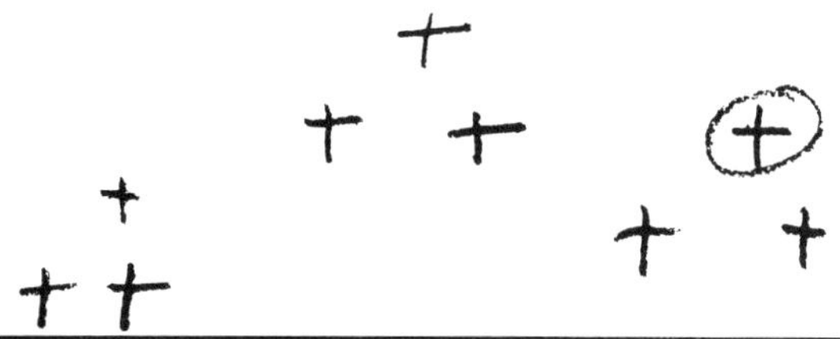

Army Air Corps had careful accounting system of all missions flown by air crew in China Theater. The flight record (opposite page), called a "Form 5," shows that Pilot Zeidler spent all of his time in February 1945 flying gas across the Hump, the "T" designation meaning "transport," not combat.

Below: In the last weeks of World War II, copies of this leaflet were dropped over cities of China coast by aircraft from 5th Air Force in Philippines. Illustration explains that Manila (meaning the headquarters of MacArthur) is as strong as a shining saber though "U.S." on hilt of saber suggests where real power originates. Point of saber is aimed at Tokyo. On back of leaflet, Chinese people are told Americans are going to get Japanese out of China and every other place they occupy because Americans understand what it means for people to suffer under Japanese occupation. Americans admit there will be danger to Chinese from air attack and for them to seek cover whenever they hear airplanes approaching overhead.

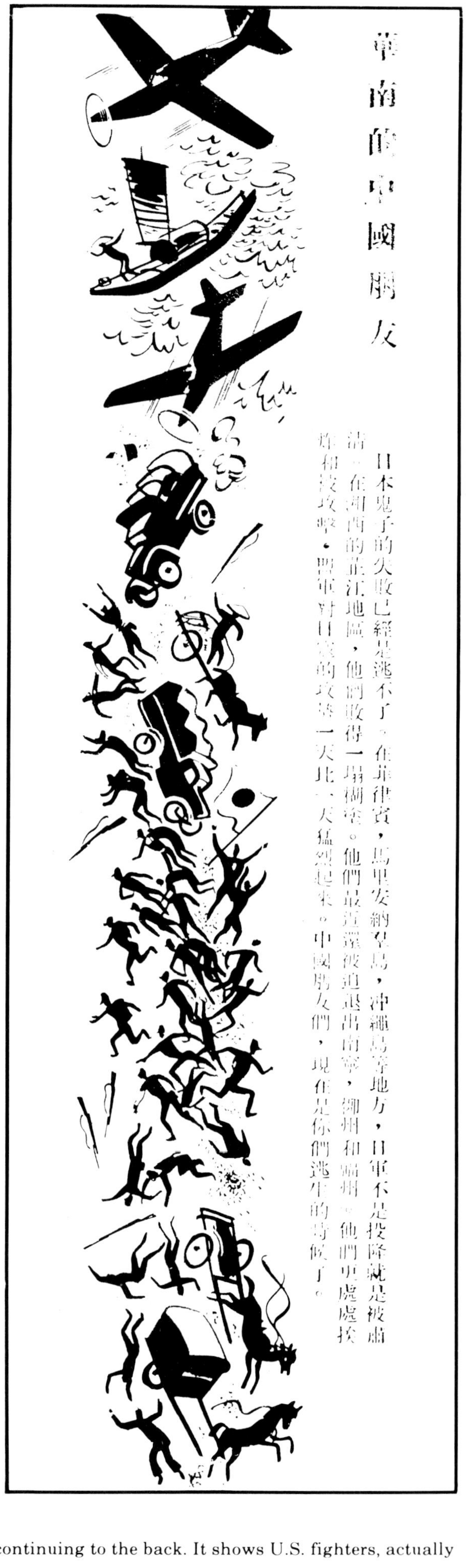

Right: propaganda leaflet, reproduced actual size, is illustrated on one side with message continuing to the back. It shows U.S. fighters, actually P-51s, strafing enemy troops and transport on side translated from Chinese, reads:

People of south China: The Japs are doomed. In the Philippines, the Marianas and, on Okinawa they have surrendered or been wiped out. At Chihkiang they were bloodily routed. They have been forced out of Nanning, Liuchow and Foochow. They are being bombed and attacked in a thousand different places. Every day the fury is mounting. Now is the time for you to save yourselves.

Here a pilot has crashed-landed. He approaches a Chinese farmer, pointing to the rescue patch sown inside his flight jacket. The patch asks that he should be rescued and sent back to his unit, but since most Chinese farmers could not read much, the patch was often of little value. Most Chinese farmers did not need to be told; they knew what to do. Nevertheless, the Office of War Information in Kunming, a branch of the United States Government, made every effort to improve the chances for the rescue of a downed pilot. Millions of leaflets were dropped over Japanese-occupied areas of China. The leaflet shown above is one of them. On the opposite side is the message which, translated from Chinese, has this to say:

TO [OUR] CHINESE FRIENDS: The air offensive started by the United States has become more intensified every day. Therefore it is possible that some of the American fliers will fall in your area. The Americans have come to help you defeat the Japanese, and today the Japanese air force has already become less rampant. The Americans deserve your help and they must have your help.

Should any American flier land in your area, the first thing to do is to lead him away from the landing spot and hide him. Then, put him in Chinese clothes and send him back to Free China [i.e. to that part of China not occupied by the Japanese]. On the way there will be friends to assist you.

The war situation has already changed, the U.S. is counter attacking from the Philippines. Japanese and Chinese traitors will soon receive their retribution. They have sown terror and they will reap terror. "In sowing melons, one reaps melons, and in sowing beans one reaps beans." This is an old saying you are all familiar with.

Good is rewarded by good and evil by evil. This is absolutely the truth. PLEASE REMEMBER THIS ENSIGN.

Nên tránh xa những đường sắt, cầu và tàu bè. Quân đội Nhật đang từ Trung-hoa dút lùi. Rất nhiều những quân đội này đã từ biên thùy chạy sang Việt-nam, va sẽ lại theo thói quen cướp cóc hay đè nén các người. Máy bay của Đồng-minh sẽ theo đuổi đằng săn những tụi giặc cướp này và dội bom xuông đầu chúng bất cứ chúng nó đi tới một chỗ nào. Giặc Nhật sẽ không có một con đường nào đề thầu chôn.

Vì giặc Nhật đã lợi dụng những đường sắt, cầu và tầu bè ở đất nước các người, những nơi này là mục-đích của chúng ta. Hãy tránh xa giặc Nhật. Không nên đáp các truyền tàu hỏa hay những thuyền bè. Không chữa những cầu và đường sắt, cũng không dúp cho Nhật đề sửa chữa lại những chỗ đó, khi mà bị bom của chúng ta đánh hỏng. Máy bay của chúng ta sẽ lại đến, và nếu các người ở gần nơi mục đích đó, các người rất có thể bị chết lây.

在華美軍 AFA-107

Leaflet above, front and obverse side, is dated 12 July 1945, and was composed at APO 627 (Kunming) probably under auspices of Office of War Information. The message is written in Annamese and the purpose is to warn the population in Indo-China to stay away from railroads, bridges and ferries which may be bombed. Also, it warns the people not to ride on trains, or repair bridges, or tracks or help the Japanese to repair them after they have been bombed. "The (U.S.) planes will come again and if you are near these targets you are apt to be killed."

大陸報

The Chi

No. 5846 Registered at the C. P. O. for Transmission in China.

SHANGHAI, WE

AMERICAN KILLED IN AI

Unyielding Chinese Turn J

Definite Check Given Attempts To Advance Line

Defending Soldiers Beat Back All Efforts To Take Miaohangchen; Kiangwan Held Firmly By Chinese Troops; Japanese Entrench Selves

The big drive of Japanese Army forces on Chinese troops around Shanghai yesterday was brought to a practical standstill.

The right flank of the invading army late last night still wrestled with determined Chinese troops outside of Miaohangchen, their objective ever since the opening of the drive early Saturday morning, while the long line of Nipponese soldiers from near Wentsaopang down to Kiangwan and to Chapei fought to prevent Chinese forces from piercing their positions, and feverishly entrenched themselves in an effort to hold positions established on all fronts of battle.

Kiangwan Village remained in Chinese possession. Japanese field headquarters remained at Lienloh Temple, east of the Shanghai-Woosung Railway, and Japanese still fought to stem Chinese sallies and cope with Chinese bombardments in Chapei.

Severe Fighting Continues

Up to midnight, severe fighting between Japanese and Chinese troops, which commenced at 8 o'clock on Monday evening, still north of Kiangwan village opened with artillery on Chinese troops firmly entrenched in strategic posts on roofs of buildings and nestled among the bamboo groves. Bombardment covered an advance that was short-lived, and after fighting at range so close that smoke from Chinese guns obscured their vision, Japanese troops finally scratched the village from the calendar of the day's activities. Japanese retired to a healthier distance of two kilometers

JAPANESE BOMB

AIR FIELDS ARE OBJECTS OF ATTACKS

Local Hangars Damaged Least Because Of Poor Aim

While one plane squadron plowed the terrain and pierced the hangers of Hungjao Airdrome, West Shanghai, with bombs, a second group of Japanese air raiders attacked the Chinese military aviation field at Soochow yesterday morning, wrecking the plane shelters and machine shop and tearing up the surface of the port.

A combined fleet of navy and military bombing planes escorted by other fighting craft went to Soochow in the morning and the assault caused sufficient damage to completely cut off any future threat from that source, according to admiral Shimada, local Japanese naval

Japanese Of Local Publicity

Foreign Corresponde General Tsai Visit His

Taking action following p tion of gruesome details of undeclared war against Chi vealed in the local press, t panese Consulate-General yes sent notice to all the foreig papers here that passes to jo ists permitting trips to the ines and to the army headq had been indefinitely suspen

The notice read: "Due to and confusion, the validity press passes or the courtes to the Japanese Army Headq and the front lines issued Japanese Consulate-Gener Army Headquarters is sus until further notice."

Despite the "danger and sion," however, no notices pension of the passes issu foreign correspondents had

a Press 大陸報

, FEBRUARY 24, 1932 14 PAGES 10 CENTS

BATTLE WITH JAPANESE

nese Offensive Into Failure

OCHOW AIRDROME

nd Passes Men After Atrocities

pted From Order; Journalists To e Today

d by any of them up to early orning.

arp contrast to the Japanese e, General Tsai Ting-kai, nder of the 19th Route Army, an invitation yesterday to all orrespondents and represen- of the local newspapers to e Chinese defense lines to- Conveyance to the front has arranged and tiffin will be at field headquarters.

known that the Japanese been exceedingly displeased e publicity given in the local o such facts as reverses in big push," the piles of dead e civilians within their lines, eir constant firing of the of poor peasants, as well as eneral destruction of proper-

U.S. AGAIN TO PROCLAIM OPEN DOOR

Stimson Understood To Be Planning Restatement Of Policy

WASHINGTON, Feb. 23.—(Reuters) — Political correspondents assert that the secretary of state, Mr. H. L. Stimson, is contemplating "clearing decks by a re-statement of America's policy in the Far East.

It is understood that the keynote of this statement will be John Hay's program in 1910 relating to the open door.

Delegates Hurry To Geneva

LONDON, Feb. 23.—(Reuters)—The hasty departure for Geneva of Sir John Simon, the Secretary of State for Foreign Affairs, 24 hours before the time given for his departure by Reuter last evening, was due to the receipt of information

Dies For China

Lieutenant "Bob" Short, (above) American aviator and intrepid stunt flier, Monday met death in the cloak of a Soldier of Fortune. "I would be glad to die in an airplane," Lieut. Short is said to have remarked at one time.

Intervention By French In East Thought Remote

Police Directed Toward Protecting Interests, And Citizens

PARIS, Feb. 22.—(Reuters).—

"BOB" SHORT SHOT DOWN OVER SOOCHOW

Nippon Lieutenant Also Slain In Desperate Duel Of Skies

SIX INVADING PLAINS INVOLVED

U.S. Aviator Last Year Planned To Span Pacific Ocean

Lieutenant Robert Short, youthful American airman, today had written his name large in the scroll of honor of Chinese history as a foreign friend who dared and died to defend the land against the Japanese invasion.

Short was shot down over Soochow in a battle involving six Japanese airplanes Monday afternoon.

A year ago Short told a China Press reporter he would be happy to die in an airplane because he loved the skies he had so long roamed. He got his wish.

Probably a Japanese Navy Type 2 fighter seaplane (U.S. code name "Rufe"). Picture may have been taken off Chungming Island in Yangtze River during hostilities around Shanghai in 1937.

Charles S. ("Chillie") Vaughn (in white coat at right) was pilot on this Sikorsky 43 which flew for China National Aviation Corporation between Shanghai and Canton in mid-1930s. Chinese mechanics are refueling from 50-gallon drum. Company had two S-43s in China, each with capacity of forty-five passengers. Photo by author taken at Lunghua.

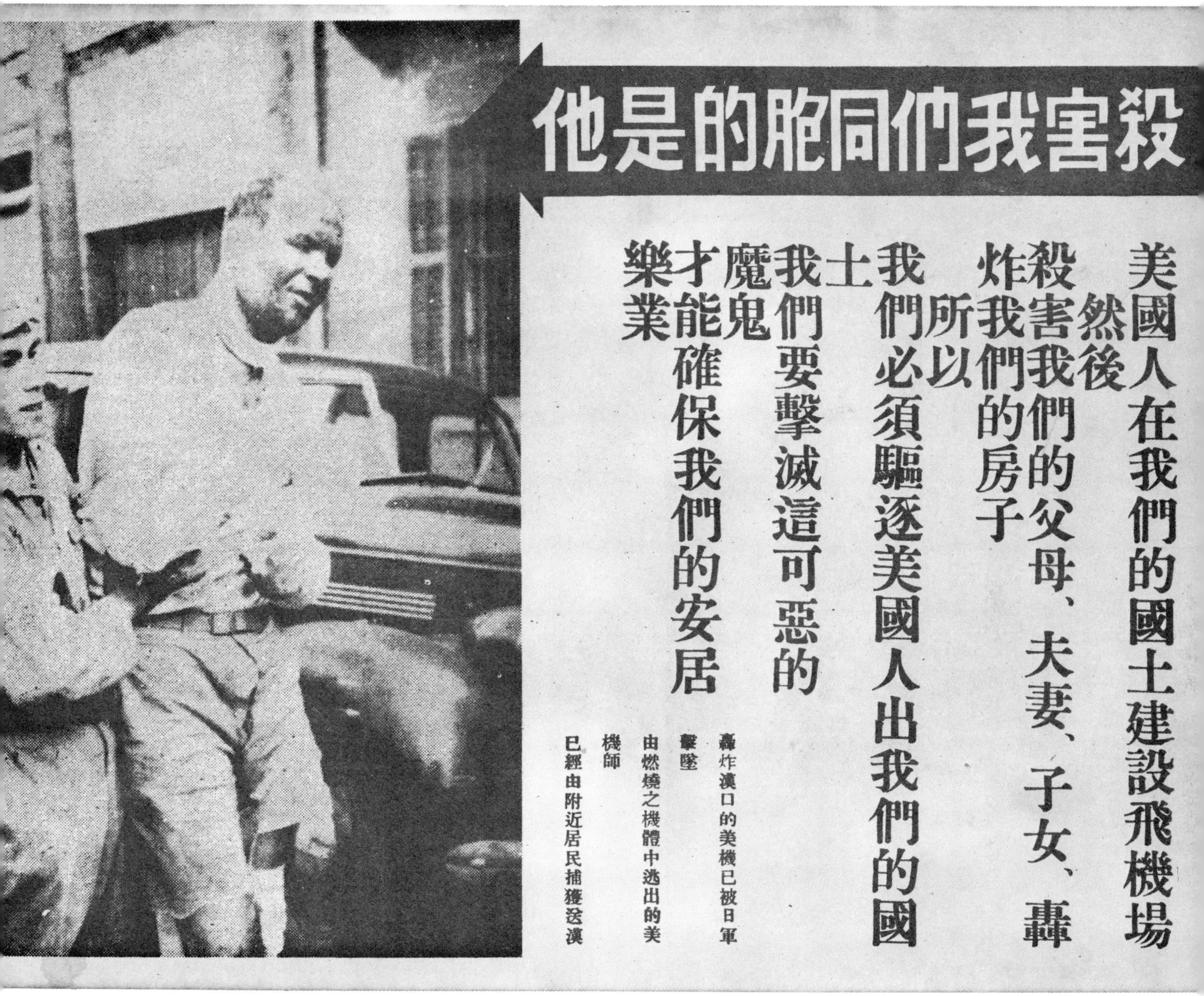

In 1943 this propaganda leaflet was dropped by Japanese aircraft over Chinese cities along the Yangtze River then under Japanese control. The leaflet is written as though it originated in a Chinese headquarters, although there is no "chop" or official source given. The foreigner at the left, in obvious pain, is Major Harry Pike, CO of the 16th Fighter Squadron (51st Fighter Group) who is being supported by a Japanese soldier. Pike was flying top cover for a bombing mission over Hankow when he was hit by flak. Fire broke out in the cockpit and before he could bail out, he was burned in the face and hands. He parachuted in near the field and was captured by a Japanese patrol. The large writing at top right in the leaflet reads: "The one who is killing our compatriots is him." Below this: "Americans are building air bases on our territory. They are killing our parents, husbands, wives and children, bombing our homes. Therefore we must drive them out of our country. We want to attack and destroy the hateful devils (*ma kuei*). Then we shall be able to make secure the peace and happiness and enterprise of our people. This man was shot down while bombing Hankow. He was burned in the airplane and has been taken into custody." This leaflet was picked up by Chinese agents working behind Japanese lines and sent to DS8, the U.S. liaison station near Changsha. From here it was forwarded to 68th Wing Headquarters in Kueilin. From the appearance of Major Pike, American airmen in the squadrons believed that the major had been tortured and paraded through the streets of Hankow. However, after the war, Malcolm Rosholt learned from Pike's parents, who then lived in Portage, Wisconsin, that Japanese army doctors had given their son expert medical treatment for burns which saved his face and hands from permanent injury. He spent the rest of the war as a POW in Japan and returned to the U.S. in 1945.

BIBLIOGRAPHY

Adams, Jean, and Kumball, Margaret, *Heroines of the Sky,* New York, 1942.

Anderton, David A., *The History of the U.S. Air Force,* New York, 1981.

Chennault, Claire Lee, *Way of a Fighter,* New York, 1949.

Chichester, Francis C., *The Lonely Sea and the Sky,* New York, 1964.

Coffey, Thomas M., *Hap,* The Story of the U.S. Air Force and the Man Who Built It: Gen. Henry H. "Hap" Arnold, New York, 1982.

Francillon, Rene J., *Japanese Aircraft of the Pacific War,* London, 1970.

Greenlaw, Olga S., *The Lady and the Tigers,* New York, 1943.

Jane, Fred T., *All the World's Air-Craft,* London, 1914 to 1928.

Tolley, Kemp, *Yangtze Patrol,* The U.S. Navy in China, Annapolis, MD 1971.

Leary, William M., *jr., The Dragon's Wings,* Athens, GA. 1976

—*Perilous Missions,* University, AL 1984.

LeMay, Curtis E. with McKinley Kantor, *Mission With LeMay,* New York, 1965.

Leonard, Royal, *I Flew for China,* New York, 1942.

Thomas, Lowell and Jablonski, Edward, *Doolitttle, A Biography,* New York, 1976.

McClure, Glenn E., *Fire and Fall Back, Universal City, TX 1975.*

—*Saipan—Then & Now,* Universal City, TX 1972.

—*Tinian—Then and Now,* Universal City, TX 1977.

McGregor, Carter J., *The Kagu-Tsuchi Bomb Group,* Wichita Falls, TX 1981.

Masatake, Okumiya and Horikoshi, Jiro with Martin Caiden, *Zero!* New York, 1956.

Maurer Maurer, ed., *Combat Squadrons of the Air Force in World War II,* Washington D.C. 1969

Morrison, Wilbur H., *Point of No Return,* The Story of the Twentieth Air Force, New York, 1979.

Pawley, William D., *Americans, Valiant and Glorious . . .* New York, 1945.

Pistole, Larry M., *Pictorial History of the Flying Tigers,* Orange, VA 1981.

Rosholt, Malcolm, *Dog Sugar Eight,* (novel), Rosholt, WI 1977.

—*Days of the Ching Pao,* Rosholt, WI 1978.

—*Claire L. Chennault, A tribute,* Rosholt, WI 1984.

Sakai, Saburo with Martin Caiden and Fred Saito, *Samurai!* New York, 1973.

Seagrave, Gordon S., *Burma Surgeon,* New York, 1944.

Smith, Sebie Biggs, *"Reminiscences,"* Oral History Research Office, Columbia University, 1982.

Tong, Hollington K., *Chiang Kai-shek,* Taipei, Taiwan 1953.

Tuckman, Barbara W., *Stilwell and the American Experience in China 1911 to 1945,* New York, 1970.

United States Strategic Bombing Survey, (108 volumes), Washington, D.C., 1946.

Wagner, Ray, *American Combat Planes* (3rd Edition), New York, 1982.

ARTICLES

Burton, Wilbur, "Mandate From Heaven," *Asia* Magazine, August 1935, pp. 458-465.

Chang Kia-ngau, "Development of Civil Aviation in China," *The China Quarterly,* August 1939, pp. 601-613.

Crow, Carl, "Bombs over Chungking," *Liberty* Magazine, 16 December 1939, pp. 49-50.

Field, John, "Tokyo Raid," *Life* Magazine, 3 May 1943, pp. 90-100.

Forman, Harrison, "Following the West," *Aviation,* May 1934, pp. 148-150.

—"Aircraft Manufacturing in China," *Aviation,* June 1934, pp. 172-175.

—"China Spreads Her Wings," *Aviation,* June 1936, pp. 24-26.

Fritsche, Carl H., "The Day We Rode the Storm," (a flight across the Bruma Hump) *Columbus Dispatch* Magazine, 1 December 1974, pp. 15-18.

Georges, Michel, *"L'aviation chinoise de 1909 a 1949," Aviation Magazine International,* Paris, 1974.

Gibbon, Elwyn H., "Commuting to War," *Colliers,* 12 November 1938, pp. 44-50.

Hiramoto Michitaka, "Our Little Visits to Nanking, *Asia* Magazine, January 1938, pp. 7-12.

Jouett, John H., "War Planes Over China," *Asia* Magazine, December 1937, pp. 827-832.

Kent, Walter C., "Wings For China," *Atlantic* Monthly, May 1937, pp. 644-648.

Koester, Hans, "Four Thousand Hours Over China," *"National Geographic Magazine,* May 1938, pp. 571-614.

Leary, William M., *jr.,* "Aircraft and Anti-Communists: CAT in Action," *China Quarterly* 52 (1972) pp. 654-669.

Leiser, Edward, "Memoirs of Pilot Elwyn H. Gibbon, the Mad Irishman," JOURNAL, American Aviation Historical Society, Spring 1978, pp. 3-18.

Newcomb, Richard F., "The Divine Wind," *American Legion* Magazine, September, 1982, pp. 20-21; 58-61.

Pentecost, Walter, and Sloan, James J., "Advance of the Flying Tigers," JOURNAL, American Aviation Historical Society, Summer 1970, pp. 137-144.

Rice, Lillian Shively, "Somewhere in Shanghai, *"Atlantic Monthly,* November 1937, pp. 648-650.

Wagner, Ray, "The Chinese Air Force 1931-1940," JOURNAL, American Aviation Historical Society, Fall 1974, pp. 162-171.

Wright, Nancy Allison, "Yankee on the Yangtze," *Aviation Quarterly,* Second Quarter, 1980, No. 2, pp. 131-151.

NOTES

Chapter I.

1. *South China Morning Post* (Hongkong), (hereafter as SCMP), 22 December 1911, microfilm.
2. *All the World's Air-Craft,* 1913, p. 69.
3. SCMP 29 May 1911, microfilm.
4. SCMP 5 June 1910, microfilm.
5. SCMP 19 July 1910, microfilm.
6. Major De Senn Chang, "Aviation in China," appears in Jane's edition of 1922, p. 30b. De Senn Chang studied aviation science in the United States after World War I and remained active in aviation all his life. He served as station manager for Civil Air Transport in Canton, 1948-50.
7. SCMP 8 May 1911, microfilm.
8. SCMP 28 March 1911, microfilm.
9. SCMP 26 August 1912, microfilm.
10. SCMP 28 January 1911, microfilm.
11. *Ibid.*
12. SCMP 7 February 1911, microfilm.
13. SCMP 18 March 1911, microfilm.
14. *Manila Times* (Philippine Islands) 12 April 1914, p. 2.
15. SCMP 10 July 1915, microfilm.
16. SCMP 16 October 1915, microfilm.
17. Photo in collection of Glenn H. Curtiss Museum of Local History, Hammondsport, NY.
18. SCMP 17 May 1921.
19. Telephone conversation with Mrs. Renee Robertson from Hongkong, January 1984.
20. SCMP 14 May 1914, microfilm.
21. SCMP 28 April 1914, microfilm.
22. Telephone conversation with Leigh Wade from Washington, D.C., May 1983; also Wade to the author 22 November 1983.
23. *San Francisco Examiner,* 26 April 1971, p. 26.

Chapter II.

1. From original file of SCMP (Hongkong) 12 January 1925 in library of Supreme Court, Hongkong.
2. *Hongkong Daily Press,* quoted in SCMP 3 August 1926.
3. *Central China Post* (Hankow) quoted in SCMP 23 August 1926.
4. Interview in San Francisco, March 1983.
5. Lincoln C. Reynolds, "Aviation in China," May 1935 in State Department File 839, 796, National Archives; "Aviation in China, The First of the Handley Page Machines," in SCMP 22 September 1919; "China Air Service" in SCMP 25 October 1919.
6. *Chiang Kai-shek,* p. 152.
7. *China Weekly Review* (Shanghai), 17 June 1936.
8. *Ibid,* 16 February 1936.

Chapter III.

1. SCMP 10 March 1919, microfilm.
2. Rodney Gilbert, "Aviation in China," State Department Report No. 5896, dated 6 July 1926.
3. Richard J. Hellinger (U.S. Military Academy) to the author 29 November 1983.
4. Janes, 1928 edition, p. 99.
5. "China's Aeronautical Future," a service report to the State Department No. 2735, dated 29 January 1921, in National Archives.
6. "Development of an Air Force in Manchuria," submitted to Defense Department (G-2) by Captain Parker C. Tenny, from Peking 30 September 1931, National Archives No. 8115.
7. The author was in a party of student tourists who attended the funeral rites which were still in progress in Mukden on 4 July 1928, a month after Chang Tso-lin was killed.
8. Jane's 1925 edition, p. 37a.
9. Edwin P. Howard actually wrote most of the report which Captain Tenny used in his own report to G-2.
10. Also quoted in Tenny report.

Chapter IV.

1. Reed Knight to the author 10 November 1982.
2. Copy of contract in Reed Knight Collection, American Heritage Center, University of Wyoming, Laramie, WY.
3. *Ibid.*
4. Olga Greenlaw, author of *The Lady and the Tigers,* died in California in 1983.
5. John Schweitzer to the author 26 October 1982.
6. In "War Planes Over China," *Asia* Magazine, December 1937.
7. Date of death taken from diary of Rolfe Watson, courtesy of Mrs. (Ivaloo) Watson.
8. Sebie Biggs Smith, "Reminiscences," p. 33.
9. Interview with W.A. ("Al") Spencer, Mesa, AZ, April 1983.
10. Reed Knight to the author, 10 November 1982.

Chapter V.

1. Loiwing is not identified on modern English maps of China, including the *CIA Administrative Atlas of Communist China* (1968), but it is found on a map published of Southeast Asia by the National Geographic Society, 1946.
2. Ivaloo Watson to the author 22 May 1983.
3. A photograph of the Seagrave hospital at Namkham was sent to the author by Mrs. Watson. It is the same which appears in *Burma Surgeon.*
4. Telephone conversation with Mrs. Edna Pawley from Miami Beach, March 1984.

Chapter VI.

1. A.L. ("Pat") Patterson to the author 21 September 1982.
2. Cited in Lincoln Reynold's "Aviation in China," SD file 839, 796.
3. *Shanghai Evening Post & Mercury* 4 July 1935.

Chapter VII.

1. *The Dragon's Wings,* p. 21.
2. *National Geographic Magazine* May 1938.
3. *China Weekly Review* 4 September 1937.
4. *Ibid.* 3 July 1937.
5. *The Far Eastern Review* (Shanghai) June 1934, p. 246.
6. CAT *Bulletin,* Vol. 5, No. 2, February 1952.
7. *Way of a Fighter,* p. 358.
8. Interview with Dorothy Leuba, Phoenix, AZ March 1982.
9. *Perilous Missions,* p. 102.
10. Interview with Robert Lee, Kansas City, MO August 1983.

Chapter VIII.

1. Copies of Carl Kjus letters in author's collection.
2. A reference to an important political conference of Chinese

military and civilian leaders.

3. Charles and Anne Morrow Lindbergh landed on Lotus Lake in Nanking 19 September 1931.
4. Kjus' information is correct. Harold M. Bixby, vice-president of Pacific American Airways, was in China to explore future air links between California and Far East for Pan American "Clippers."
5. Most of the Sian Incident is described in a book written by Mme. Chiang Kai-shek called, in English, *Sian: A Coup d'Etat,* but in Chinese as *Sian: pan yueh chi, Sian, hui i lu* (Sian, half a month: Sian, the way back). It was published by Kelly & Walsh in Shanghai, 1937, and printed on Chinese bamboo fibre paper with red silk covers.

Chapter IX.

1. *Domei news agency,* 14 October 1937.
2. Interview with H.S. ("Newsreel") Wong at Shanghai, 17 August 1937.
3. *China Press* (Shanghai), 8 October 1937.
4. *Japanese Aircraft of the Pacific War,* pp. 344-345.
5. *China Press* 16 August 1937.

Chapter X.

1. Copy of translation held by Ray Wagner, San Diego, CA, on loan to the author.

Chapter XI.

1. *Osaka Mainichi* 21 August 1937.
2. *New York Times,* 23 November 1937.
3. *Way of a Fighter,* pp. 63-64.
4. *"Memoirs of Pilot Elwyn H. Gibbon,"* p. 10.
5. *United States Strategic Bombing Survey,* Vol. 9-1, Air Operations in the China Area (July 1937-August 1945) Monograph No. 76.
6. *Red Star* (Moscow) No. 157 (13292) 7 July 1967, p. 3, in collection of Ray Wagner on loan to the author.
7. *Way of a Fighter,* pp. 66-67.
8. *Ibid.*
9. *Samurai!* pp. 28-29.
10. The three Soviet advisors were seen by the author at a Christmas theater party hosted by General Hsueh Yueh but no contact was made.

Chapter XII.

1. Translation in collection of Ray Wagner on loan to author.
2. *Chicago Tribune,* 29 September 1937.
3. *Domei* from Tokyo 16 November 1937.
4. *The Lady and the Tigers,* pp. 20-21.
5. *Way of a Fighter,* p. 68
6. Original document on loan from Ray Wagner.
7. Interview with Hsieh Mang at San Francisco, March 1983.
8. Interview with William D. Hopson in Washington D.C. May 1983; also letter to the author from Guadalajara, Mexico.

Chapter XIII.

1. *New York Times* 15 January 1939.
2. *Liberty* Magazine 16 September 1939.
3. Interview with Hollington K. Tong, Tokyo, Japan, May 1954.

Chapter XIV.

1. *Domei* from Tokyo 4 August 1937.
2. Sebie Biggs Smith, "Reminiscenses," p. 119.
3. Julius Barr to Major King Kia Sze in "Memoirs of Pilot Elwyn H. Gibbon," p. 4.
4. *Ibid,* p. 10.
5. Sebie Biggs Smith "Reminiscences," p. 122.
6. *"Memoirs of Pilot Elwyn Gibbon,"* p. 12.
7. Sebie Biggs Smith, "Reminiscences," p. 117.
8. In Gibbon File, San Diego Aero-Space Museum.
9. Diary of Rolfe Watson, 1936-1940; copy in author's collection.
10. *Way of a Fighter,* pp. 88-89.
11. Telephone conversation with Robert A. Fausel, Winter Park, FL, February 1984.

Chapter XV.

1. Interview with P.Y. Shu at Palo Alto, CA, 14 March 1983.
2. Interview with John M. Williams, Washington, D.C. May 1982.
3. "The Role of Defensive Pursuit" was first published in *Coast Artillery Journal,* December 1933. A copy is also in the archives of the Air Force Academy at Colorado Springs and it was included in a commemorative issue of the *Pictorial Magazine* of the 14th Air Force Association published in 1983 and edited by Malcolm Rosholt.
4. *Time* Magazine 6 December 1943.
5. *American Volunteer Group,* a paperback (N.D.), was published by the General Headquarters of the Chinese Air Force, P.O. Box 7083, Taipei 8, Taiwan, and the reference is included under Attachment 2.
6. Several letters reflecting a sharp difference of opinion between Pawley and Chennault regarding aircraft maintenance at Loiwing may be found in the Chennault Papers at the Hoover Library, Stanford University.
7. Interview with David B. ("Tex") Hill, San Antonio, TX March 1977.
8. Taped interview with Edward Rector, Washington D.C. May 1983.
9. *American Volunteer Group,* Attachment 4.
10. Chennault Papers, Hoover Library.
11. Interview with Spencer Moosa in Taiwan, August 1976.
12. *China At War,* (Chinese Ministry of Information) Chungking, March 1942.

Chapter XVI.

1. *Way of a Fighter,* pp. 171-172.

Chapter XVII.

1. Transcript of trial is held at National Records Center, Suitland, Md.
2. *Ibid.*
3. *Ibid.*
4. Telephone conversation with John Hendren from Jefferson City, MO, 1978.
5. The author drove a jeep north from Shangjao, Kiangsi, in the spring of 1945 where a mountain road skirted three large villages, all totally destroyed by Japanese forces three years earlier. The villages were still deserted in 1945.
6. *United States Strategic Bombing Survey.*

7. Lowell Thomas and Edward Jablonski use the figure of "some 250,000 Chinese civilians and soldiers" who died in a three-month "rampage" by Japanese forces. There were few Chinese troops in Kiangsi at this time and the figures would refer largely to civilians. Figures quoted from p. 198 of book on Doolittle.

Chapter XVIII.

1. Copy of Order No. 1 in author's collection, courtesy of Robert J. Gillespie, 14th Headquarters Squadron.
2. The author replaced Birch in late December 1943 at Changsha while Birch was to be moved to a Chinese war zone north of the Yangtze River. He was delayed several weeks by illness in Chungking.
3. Original order in author's collection.
4. *Ibid.*
5. *Time,* 14 November 1947.
6. *Associated Press* from Shanghai 16 November 1945.
7. The author at this time served as assistant liaison officer under Neal Ausman, USA, code name "Sandy" at the 6th War Area Headquarters in Enshih.
8. Original message in author's collection.
9. Air-ground liaison station DS8 was commanded by the author from 1 January 1944 to June 1944 when code designation was changed to LZ5 after retreat from Changsha.
10. Original message in author's collection.

Chapter XIX.

1. Copy of mission report in author's collection made from original mission report held at Air University, Maxwell Field.
2. John Stewart to the author, December 1983.

Chapter XX.

1. Interview with Eugene L. Strickland, Phoenix, AZ, April 1982.
2. Interview, Malair Air Base (Karachi), November 1943.
3. Copy in author's collection.
4. J.J. McCann to the author 7 May 1983.

Chapter XXI.

1. Eugene H. Beebe to the author July 1983.
2. Copy of squadron journal in author's collection.
3. William L. Dwyer to the author, September 1983.
4. In author's collection.
5. Edward P. Zeidler to the autahor, June 1983.
6. Leonard E. Gelhaus to the author November 1983.

Chapter XXII.

1. Quoted by Thomas W. Coffee in *Hap,* p. 343.
2. Interview with Robert Wix, Merrill, WI June 1982.
3. Telephone conversation with Denny D. Pidhayny from Los Angeles, June 1983.
4. Irving Newman to the author, December 1983; also telephone conversation with Newman from Boston.
5. Telephone conversation with James V. Edmundson from Longboat Key, FL, September 1983.
6. Telephone conversation with Pidhayny, June 1983.
7. Quoted from diary of Harry M. Changnon, Cupertino, CA.
8. Quoted in *Mission with LeMay,* p. 332.
9. In collection of the author, courtesy Mrs. William D'Ettore.
10. Interview with T. Alan Bennett, Kansas City, MO, August 1983.
11. Letter, "The History and Status of the First American Volunteer Group," Toungoo, Burma under date of 19 October 1941, p. 11, in author's collection.
12. Event attended by the author and Mrs. Rosholt.
13. From diary of Harry Changnon.
14. Copy of letter, undated, from "Blondie" Saunders to members of "the old 40th Bomb Group" in collection of the author; also, Saunders to the author, August 1983; also telephone conversation with Saunders from Aberdeen, SD.
15. United States vs Major General Masataka Kaburagi, *et al,* in National Records Center, Suitland; portions of transcript in author's collection.
16. Telephone conversation with Vernon Schaefer from Kimball, MN 23 April 1984.

Chapter XXIII.

1. Quoted from *United States Strategic Bombing Survey,* Monograph No. 76.
2. *Shanghai Mainichi,* 6 July 1937.
3. *Japanese United Press* (Shanghai) 23 October 1937.
4. *Domei* from Shanghai, 8 October 1937.
5. *United States Strategic Bombing Survey,* Monographs 76.
6. Since no time limit is set on the number of months of imports, these figures are probably exaggerated. (See Chapter XI on "Soviet Aid to China").
7. Order cut 28 July 1943, APO 430 (Kueilin) in author's collection.

PICTURE CREDITS

Illustrations and maps in this book are given credit on page numbers from left to right and down. A first citation of a donor's name is given in full: thereafter only person's surname. Painting by Susan M.P. Groshek, 8-9; Glenn H. Curtiss Museum of Local History, 7, 10; National Air and Space Museum, Smithsonian Institution, 10-11; Curtiss Museum of Local History, 13-14; John Underwood 16-17; Alva L. Harvey, 18; Lincoln Reynolds, 19; Ralph F. Mitchell, 20; Library of Congress, 20-21; W.A. Spencer, 22-23; Lawrence Wilson, 24; American Hermitage Center, University of Wyoming, 25; Spencer and Malcolm Rosholt, 26; Spencer 27; Rosholt 28-29; Reynolds, 31; Spencer, 32; Reed Knight, 33-34; Spencer, 35; Knight, 36; John Schweitzer, 37; Knight and Spencer, 38; Spencer, 39; Schweitzer, 40-41; Schweitzer and Knight, 42-43; Barry Goldwater Collection, Arizona State University, 44; Schweitzer, Knight, Wilson, 45; Spencer, Wilson, Knight, 47; San Diego Aero-Space Museum, 48; Spencer and Knight, 49; Knight and Sebie Biggs Smith, 50; Schweitzer, 51; Edna Pawley, 52-53; San Diego Aero-Space Museum, 53; Ken Fuglien, 54; Pawley, 55; Ivaloo Watson and Robert Fausel, 56; Watson, 57; Rosholt, 59; National Archives and Mitchell, 60; Pan American World Airways, 61; Mitchell, 62; Mitchell and Spencer, 63; Mitchell, 64; Pan Am, 65; Nancy Allison Wright, 66; author collection, 67-68; Rosholt, 69; *The Far Eastern Review,* 70; CAT *Bulletin,* 71; author collection, 72; Harry Kjus, 74; Spencer, 75; Kjus, 76; Wilson, 78; author collection, 79-80; Rosholt, 81; Wilson, 82; National Air & Space Museum and San Diego Aero-Space Museum, 84; National Archives, 85; U.S. Air Force, 86; Wilson, 88-89-90-91; author collection, 92; Rosholt, 93, 95, 97; author collection, 100; National Archives, 101; author collection, 102; Sebie Smith, 103-104; San Diego Aero-Space collection, 104, 106; Wayne Ricks, 107; Walter A. Pentecost Collection, National Air & Space Museum, 106, 109, 111-112-113; National Air & Space Museum, 114; Ricks and Pentecost, 116; U.S. Air Force, 117; author collection, 118; U.S. Air Force and Rosholt, 119; U.S. Air Force and Douglas Erickson, 120; author collection, 121; U.S. Air Force, 122; National Air and Space Museum, 123; U.S. Air Force and Leo Braun, 124; Braun, 125, 126, 127, 128, 130-133; W.H. Van Vleck, 134; National Archives, 136; U.S. Air Force 137; Charles R. Fox, 138; E. Jay Curtis, 139; author collection, 140; Park Morrow, 141; Curtis and Robert Sheldon, 142; U.S. Air Force, 144; Amil C. Stafford, 145; Frank W. Schuster, 146; author collection, 147; U.S. Air Force, 148; Oscar Harper, 149; author collection and Pan Am, 150; U.S. Air Force, 151; painting by Hal Rozema, 152-153; National Archives, Harper and Barbara Felt, 154; U.S. Air Force and Charles A. Breingen, 155; author collection, 156; U.S. Air Force, 157; author collection, 158-159; map by Wayne Johnson, 160; Eugene H. Beebe, 161; map by Randell E. Rohe, 162; Sam Palmer, 163-164; John S. Stewart, 165; National Archives 168-169; U.S. Air Force, 170; U.S. Air Force and J.J. McCann, 171; U.S. Air Force and author collection, 172; William F. Carroll, 173; Robert L. Cowan, 174-175; Elmer Carpenter, Sr., 176; Rosholt, 177; William J. Peck and National Archives, 178-179; Morrow, 180; Beebe, William D. Hopson and Schuster, 181; Joseph Del Masso, 182; Burlingham, 183; Morrow, 184; Morrow and Schuster, 185; Morrow, 186; U.S. Air Force, 187-189; Flo Mallory, 190; James J. O'Keefe, 191; James V. Edmundson, Alva L. Harvey and Stanley Andrews, 192; Edmundson and Harvey, 193; William R. Rooney, 194; Robert Wix and Denny D. Pidhayny, 195; O'Keefe, 196; Harry Changnon and U.S. Air Force, 197; Glenn E. McClure, 198; Wix and William Klapper, 199; Wix, 200-201; Curtis, 202; Kohichi Minobe, 203; Wilson and Rosholt, 204; Minobe, 205-206; National Archives, 207; author collection and Eugene L. Strickland, 208; Wilson and National Archives, 209; author collection, 210; Rosholt, 211; Winthrop S. Shaw from National Air & Space Museum, 212; National Air & Space Museum, 213; author collection 214-215; U.S. Air Force, 216; Henry L. Mazurie, A.L. Patterson, William Hayes, 217; author collection, 219; Robert J. Kennedy, 219; Wix, 220-221; Edward P. Zeidler, 222; O'Keefe, 223; author collection, 224; author collection and C.H. Laughlin, 225; author collection, 226-229.

INDEX

A

"Adam and Eves," 111
Adams, *Dr.* Eldrige, 36
Aero Commission, 26
Air bases (aerial views)
—Chabua, 119; Kunming, 123; Yang Tong, 125, 126, 127; Erh Tong, 128; Lichiachen, 129; Liuchow, 130; Lingling, 132; Kanchow, 132; Suichuan, 131; Mukden, 133.
Air operations chart, 160
Air raid alarm system, 99, 108
Air raid precautions, 99
Air Transport Command, 143, 218
Aircraft in China
A-29, 97
A5M2b Model 2-2, 80
Aichi AB-3, 206
Armstrong-Whiteworth, 32
AT-6, 166
Atlas, 32
Avian, 32
Avro, 31, 32
B-17 Fortress, 187
B-24, 98, 143, 173, 174, 176, 177, 179, 181
B-24L, 179
B-25, 98, 143, 147, 149, 168, 171, 172, 188
B-25 C or D, 173
B-25H, 171, 172, 173
B-25J, 173, 180
B-26, 171
B-29, 147
Beech, 18, 199
Bellanca, 47, 48, 53
Blackburn Lincock, 36, 111
Bleriot, 7
Boeing, 187, 188
Breda, 48
Breguet XIX, 18
Brewster Buffalo, 110
C-46, 67, 72, 98, 121, 139, 197
C-47, 67
C-54, 197
C-87, 197
C-109, 197
Caudron, 15, 16, 17
Cessna, 72, 195
Consolidated Commodore, 62
Curtiss Condor, 42, 51, 62, 83
Curtiss Falcon, 45
Curtiss Hawk II, 21, 37, 51, 82
Curtiss Hawk CW-21, 56, 106
Curtiss Hawk III, 53
Curtiss Hawk H-75M, 84
Curtiss Hawk 75H, 48
Curtiss Hawk, 75Q, 56, 106
Curtiss JN, 25
Curtiss 19R (trainer), 49
Curtiss pusher, 13, 14
Curtiss Shrike, 40
Curtiss SPC-2s "Seagull," 80
DeHaviland, 24, 46
Dewoitine, 31, 217
Douglas cruiser, 18, 19
Douglas Dolphin, 61
Douglas 02MC-4, 38, 82
DC-2, 55, 61, 62
DC-3, 62
Etrich, 6, 21
Fairchild, 61
Farman, 12, 21
Fiat BR3, 37, 46
Fleet trainer, 21, 38
Fokker D-7, 31
Ford Trimotor, 62, 74
French Farman, 48
Gloster Gladiator, 96
Gypsy Moth, 19
Handley-Page, 30
Haviland Dragon Rapide, 62
Hayabusa, 22, 207
Hayate (Gale), 211
Italian Caproni CA-III, 101
Junkers A-35, 31
Junkers D-ANYK, 64
Junkers JU-52, 68
Junkers K-47, 26, 27, 45
Junkers W-33, 32, 69
Junkers W-39, 67
Kawasaki C-5, 20
Kawasaki Ki-45, 190
Kawasaki Ki-61, 213
Laird Looper, 17
Liberator, 174, 186
Lockheed Hudson 3-A-29, 67
Lockheed Sirius, 20
Loening, 59, 66
Martin, 61, 85
Mitsubishi A5M III, 204
Mitsubishi BIM, 22
Mitsubishi G3M2, 93
Mitsubishi Ki-57-11, 214
Mitsubishi Type O, 205
Mitsubishi Type 3, 205
Mitsubishi Type 96, 80, 205
Moraine Saulnier, 18
Nakajima 97, 191
Nakajima A2N1, 80
Nakajima A4N1, 80
Nakajima Army Type 90, 205
Nakajima Army Type 97, 208, 209
Nakajima Army Type 99, 209
Nakajima B5N, 204
Nakajima E8Ni, 204
Nakajima Ki-43, 207
Nakajima Ki-44, 206
Nakajima Ki-84, 206, 211
Newport-Delage, 31
Northrop, 53, 78, 83
P-40E, 98, 212
P-43, 97, 98, 117
P-47Ds, 97, 149
P-51, 139, 145, 149
P-51A and B, 157, 163, 165
P-51D, 155
P-61A, 157
P-66, 97, 98
Partridge-Keller, 17
Polikarpov E-15, 87, 88
Polikarpov E-16, 87, 88, 91, 92
Potez 25A, 2, 31
Ryan STC-4, 58
Savoia-Marchetti, 46, 95
SB-2 bomber, 88
Sikorsky S-38, 26, 62
Sommer, 11
Soviet Katiuska, 28
Stinson, 17, 61, 105, 116
TBF "Avengers," 195
Tupolev bomber, 89
Vimy (trainer), 30
Vought-Corsair, 31, 47
Vultee, 52, 53
Aircraft salesmen, 58
Alexander, A.M., 76
Alison, James W.M., 103, 105
Allen, Tommy, 103
Allison, E.M., 59, 60, 65
Alspaugh, Walter N., 193
Altamire, Phyllis, 71
AVG *see* First American Volunteer Group
Anderson, Albert, 55
Andrews, Stanley S., 192
Armstrong, John G., 174, 177, 178
Army Advisory Group (U.S.), 98
Arnegard, Howard M., 139
Arnold, *General* Henry H. ("Hap"), 143, 188
Arnold, Julian, 35
Arnold, Leslie, 18
Atack, George J., 139
Athey, Durdan S., *jr.*, 195
U.S.S. *Augusta*, 70
"Aviation Explorations," 59
Aviation Hall of Fame, 196
Aviation Parts & Equipment, Inc., 173
Aviation Schools, 15, 25, 32, 33
"Awakening Village," 35
"Axis Nitemare," 181

B

Baldwin, Thomas Scott, 8, 12
Ballooning, 8, 9
Bane, LeRoy E., 195
Barr, George, 136
Barr, Julius, 76, 102, 103, 105
Beaucrest (no first name given, hereafter as NFN), 163
Beckham, William A., 193
Beder, Robert J., 136
Beebe, Eugene H., 161, 174, 175
Bell, Dale, 163, 164
Bell, Ernest J., 163
Bemis, Sterling, 72
Bennett, T. Alan, 171, 196
Bingham, Cleve L., 163
Birch, John, 3, 139, 217
Bissell, Clayton, 117
Bixby, Harold M., 62
Blagoveschenskii, *General* A., 89, 91, 92
Blanchard, William R. ("Butch"), 188
Blom, Benjamin, 190
Bombing of Chungking, 100
Bonus payments to A.V.G., 112, 113
Booth, John L., 163
Boulingre, A., 103, 105
Boyce, Robert, 179
Boycott of Japanese goods, 33
Branch, Irving L. ("Twig"), 163, 168
Braun, Leo, 3
Brennan, James, 73
Brenner, William A., 163
Bridges bombed, 146-148
Brissenden, Harry G., 193
Brooksband, James M., 139
Brundage, Ernest J., 193
"Bugs Bunny," 187, 195
Burch, *1st Lt.* Donald J., 168
Burial at sea, 176
"Buzz" Felio, 55

C

Cain, Edmund, 217
Cannon, Carl, 163
Cantonese Air Force, 15, 16
Carmichael, Richard, 188
Carroll, William F., 173
Carson, W.P., 163
Carswell Air Force base, 180
Carswell, Horace S., 180
Castell-Rudenhausen, *Count* Wulf-Diether Graf zu, 69
Caudron Brothers, 6
Central Air Transport Corporation, 67
Central Aircraft Manufacturing Company, 52, 53, 55
Central Aviation Manufacturing Company (CAMCO), 106
Central Aviation School, 33, 34, 35, 36, 40
Central Intelligence Agency, 73
Champion, Frank, 17, 18
Chang Chih-kao, 32
Chang Ching-yu, 59, 61
Chang, E.C., 163
Chang, *1st, Lt.* L.M., 170
Chang Hing-wan, 32
Chang Hsieh-chang, 26
Chang Hsueh-liang
—made air marshal, 31; reviews air cadets, 39; buys Ford airplane, 74; taken ill, 75; kidnaps Chiang Kai-shek, 77; placed under house arrest, 77.
Chang Hsuei-chang, 32
Chang Hsun, 22
Chang Kwang-ming, 94
Chang Ming-chi, 11
Chang Ming-sheng, 94
Chang (NFN), 163
Chang, S.C., 163
Chang Shang-kan, 39
Chang, *Sub Lt.* S.C., 168
Chang Tso-lin, 30, 31, 32, 102
Chang Yuk-hang, 32
Changnon, Harry, 191, 197
Changte campaign, 150-151
Chao Keng-yun, 31
Chao Liu, 36
Chao, *Sub Lt.*, L.P., 170
Chen, Art, 96, 103
Chen, C.S., 163
Chen, Eugene, 29
Chen, K.C., 163
Chen, K.Y., 30
Cheng, H.H., 163
Cheng Siao-yu, 96
Chennault, *General* Claire L.
—recommends instructors, 39; arrives in China, 41; meets Generalissimo, 42; enters civilian aviation, 70; sells airline, 73; recalls Soviet air battle, 87, 91; assigned as advisor to volunteer squadron, 102; describes air battles, 103; organizes Flying Tigers, 107, 108; activates 14th U.S. Air Force, 138; cites enemy losses, 139; organizes composite wing, 166; decorates Eugene Beebe, 181; urges use of incendiary bombs, 196.
Chiang Kai-shek, *General*
—establishes Nanking government, 25; fights Fukien rebels, 29; inspects air force cadets, 39; meets foreign advisors, 40; sample of his calligraphy, 40; consolidates military power, 59; kidnapped, 77; refuses Japanese peace offer, 101; attempt by Japanese to influence bombing raids, 201
Chiang Kai-shek, *Mme.* (*nee* Soong Mei-ling)
—instructs foreign advisors, 40; hears report on air force strength, 40; attends funeral at Arlington National Cemetery for Chennault, 42; tours bomb damage in Chungking, 101.
Chichester, Francis C., 19
Chihkiang campaign, 170
Chin, H.H., 163
Chin Tsch-liu, 94
China Air Task Force, 117
China Airways, 59, 61
"China Clipper," 67
China National Aviation Corporation, 59
Chinese Air Force
—founding of, 24; fights rebels in north China, 30, 31; attacks Japanese fleet, 78; air battles over Nanking, 82, 83; logo adopted, 82; Soviet aid, 90, 91; combat over Wu-Han, 93; military operations 1942-45, 97, 98; recruits foreign volunteers, 103.
Chinese "Air Force Day," 78
Chinese-American Composite Wing, 42, 166
Chinese Flying Club, 29
Chinese National Relief & Rehabilitation Administration *see* United Nations
Chinese Revolution Beginning 1911, 6
Chiu Kow, 94
"Choppers," 155
Chou Chih-jou, 27, 39, 98
Chou Tsing-feng, 39
Chu Kia-shun, 94
Chu Teh, 90, 91, 92
Chung De Sean, 7
Churchill, S., 163
Civil Air Transport Inc., 70
Clark, G.B., 36, 75
Clinkscales, Robert S., 190, 192
"Cocky Bobby," 179
Colbert, Robert T., 163, 165
Cole, Frank, 75, 76
Colvin, Howard, 175
Commercial aviation
—China National Aviation Corporation, 59, 61; China Airways, 59, 61; Eurasia Aviation Corporation, 62; Civil Air Transport Inc., 70; Southwest Aviation Corporation, 67; Central Air Transport Corporation, 67.
Congressional Medal of Honor, 180
Conte Verde (vessel), 178
Cooper, Marian, 117
Corcoran, Thomas G., 73
Costello, John, 163
Cotton, Thomas, 145
Coughlin, James A., 36
Court martial trial, 150
Cowan, Robert L., 175
Crane, Frank, 55
Crow, Carl, 99
Curtiss, Glenn H., 11, 12

D

Dai Enki, 61
Daniels, L., 163
Daniels, Walter, 145
De Bemardi, Mario, 52
Deshiega, Jacob D., 136
d'Ettore, William, 140, 193
Dickmann, Paul J., 163
Dillon, Clarence, 59
Doisy, Georges Pelletier, 18, 31
Donald, W.H., 40, 41, 50
Donovan, William, 73
Doolittle, J.H. ("Jimmie"), 21, 23, 63, 134, 135
Doolittle raid on Tokyo, 134, 135; trial of fliers, 136.
Dorsey, E.D., 49
"Double Tenth," Chinese Independence Day, 12
Douglas Aircraft Corporation, 19
Dowis, Kendal B., 163, 164
"Dragonflies," 196
Dunn, Irene, 44
Dunning, John A., 170
Durdin, F. Tillman, 87, 190
Dwyer, William L., 174
Dykstra, George E., 163

E

"Eddie Allen," 192
Edmundson, James V., 3, 190, 192, 193
Eickenberry, Arthur, 217
Eiken, Alfred F., 195
Elder, Jimmy *jr.*, 76
Elker, Oswin H., 145
Elrick, *Mrs.* David (*nee* Mei-fei Rosholt), 3
Engler, Howard P., 188
Enola Gay, 199
Enshih air base (photo), 97
Eurasia Aviation Corporation, 19, 64
Evacuation of Liuchow, 159
Evacuation of Suichuan, 158

F

Factory No. 10, 145
Farnsworth, Clyde D., 71
Farrow, William G., 136
Faulkner, Theodore S., 190
Fausel, Robert W., 56, 106
Felt, Eugene 154
Feng Ju-ho, 94
Feng, *Dr.* Nelson, 36, 51
Feng Yu-hsiang, 25
"Fifi," 187
58th Bomb Wing Association, 195
First American Volunteer Group (AVG)
—staff named, 35; maintenance problems, 52; organized, 107, 108; record of air victories, 112, 113; origin of Flying Tiger logo, 114; deactivated, 117
Fisher, William P., 174, 178, 181
Fleming, Robert E., 139
Flores Ruiz, Santiago, 3
Floueir, Marcel, 103
"Flying Dragon," 15
Flying Tiger Line, 70
"Flying Tigers" *see* First American Volunteer Group
Folmar, Cecil, 36, 51
Fong, C.F., 36
Forbes, James, 201
14th U.S. Air Force
—activated, 138; losses, 139; damage assessment to enemy, 140; 23rd Fighter Group, 147; 341 Bombardment Group, 147; air operations chart, 160; attack on Shinchiku, 163;

Chinese-American Composite Wing, 166; 308th Bombardment Group, 173
Fox, Charles, 163
Fradsham, G.W., 30
Francillon, Rene J., 80
Frees, Henry *jr.*, 195
Fu, W.S., 163
Fu-chi, 12
Fukien rebellion, 29
Fung Joe Guey, 6
Fung U *see* Fung Joe Guey

G

Gandhi, Mohatma, 188
Garrold, Mervyn ("Murph"), 55
Gay, Frank, 36, 41
Gelhaus, Leonard, 178
Gibbon, Elwyn, 48, 87, 103, 104, 106
Gibson, Frank H., 163
Gilbert, Rodney, 30
Gilkson, Mark T., 139
Gillespie, Robert J., 139
Glass, Edwin, 194
Glenn H. Curtiss Museum, 12, 15
"Glip" bombing, 147
"Gold Tigers," 196
Goldwater, Barry, 44
Goss, *Major* (NFN), 174
"Greater East Asia Co-Prosperity Sphere," 101
Green, Nathan E., 145
Greenlaw, Harvey, 35
Greenlaw, *Mrs.* Harvey (nee Olga Sower), 35, 96
Groshek, *Mrs.* Norman (*nee* Susan Peske), 8-9
Grosvenor, William J., 145
Grottle, George T., 163
Grover, F., 163
Gunn, Tom, 12, 15
Gustavson, Walter R., 195

H

Hachimaki, 210
Hagenberger, Alfred, 117, 139
Hall, Weston Burt, 21
Hallmark, Dean, 136
Hamilton, Edward, 74
Hamilton, Robert T., 163
Han Sheng, 96
Hancock, *1st Lt.*, J.T., 168
Harding, John, 18
Harvey, Alva L., 3, 18, 19, 193
Hasegawa, Kiyoshi, 203
Hawks, Frank M., 42, 51, 77
Haynes, Caleb, 117
Hayes, William B., 217
Hayward, R.O., 59
Hearn, Russell, 102
Hedreck, Donald, 163
Hofferman, Rollin B., 193
Heise, Paul, 76
Helliwell, Paul G., 73
"Hellbirds," 196
"Hells Angels," 111, 115
Hendren, John, 136, 137
Henrickson, James, 103, 105
Herbst, John ("Pappy"), 213
H.M.S. *Hermes*, 20
Hill, David ("Tex"), 117, 163, 164
Hill, John R., 163
Hilliard, John K., 195
Hindustani Aircraft Ltd., 55
Hirohito, *Emperor of Japan*, 199, 214, 215
Hirota, Koki, 102
Hite, Robert L., 136
Ho, *2nd Lt.*, C.S. 163
Ho, T.J., 163
"Hoboken Barber," 81
Holbrook, L.R. (or as Roy), 35, 40, 45
Holland, John, 39, 40, 50
Holloway, Bruce, 117, 138
Holt, F.V., 30
Hopson, William D., 98, 174, 177, 178, 181
Howard, Edward P., 31, 35
Howard, Sanborn V., 163
Hsieh Kung-chien, 32
Hsieh Mang, 25, 98
Hsin Shou-shun, 94
Hsu, *Capt.* J.S., 168
Hsu Huang-sheng, 85
Hsu, *Sub Lt.* S.Y., 170
Hu, *2nd Lt.* H.K., 170
Hu Tsung, 6
Huang, Kwang-han, 29
"Huitang Corporation," 70
Hull, Cordell, 102
Hump *map*, 162
Hungjao air base, 25
Hunter, Charles, 55, 106
Hutton, Letha, 76
Hutton, Perry, 74, 75, 76
"Hydroaeroplane," 14

I

Ichi-Go campaign, 170, 210
H.I.J.M.S. *Idzumo*, 78
Irving, Newman, 190
Ivy, William, 6

J

Japanese Air Force
—attacks China, 79; establishes air superiority, 80; attacks Soviet air unit, 89; losses to Soviet airmen, 90; assessment of Japanese losses, 210; air victory claims, 212
Japanese surrender at Chihkiang, 215
Jen Yung-hou, 83
Johnson, Berger, 61
Johnson, Eldon E., 139
Johnson, Stanley, 163
Johnson, Wayne G., 160
Johnson, William, 163
Joint Board of Air Raid Relief, 99
Jones, William G., 163
Jouett, John H., 35, 36, 38
Jouett Mission
—organized, 35; training problems, 36; termination of contract, 40
Jung Wai-lung, 15

K

Kaburagi, Masataka, 201
H.I.J.M.S. *Kaga*, 203
Kalberer, Alfred P., 190, 196
Kao Chia-hsing, 36
Kay, Kenneth E., 167
Keh, *Sub Lt.* H.S., 170
Kent, W.C. ("Foxy"), 35, 39, 42
Kessler, Robert E., 139
Keyes, Clement M., 59, 62
Khryukin, Timofey, 90
Kiangwan race course, 11
King Kia-sze, 103
King, Woods, 140
Kjus, Carl, 3, 74, 75, 76, 77
Kjus, Harry, 3
Knight, Morris B. *see* Knight, Reed
Knight Reed, 3, 34, 35, 37, 41, 42, 51
Koester, Hans, 64
Kouzminski, Russian pilot, 9
Kung, *Dr.* H.H., 41, 42
Kung, *1st Lt.*, S.H., 170
Kung Yeh-ti, 94

L

L.E. Gale & Co., 22
LaBussiere, William, 103
L'Aero Club de France, 12
Lackell, Frank, 55
The Lady and the Tigers, 36, 96
Lam Wei-hsing, 32
LaRose, W., 103
Lassiter, Jerome H., 217
LaValle, Carl J., 163
Leaflet raid on Japan, 85
Leary, William M. *jr.*, 3
Lee, *1st Lt.* C.K., 168
Lee, H.P., 163
Lee, Le-ong, 103
Lee, *Major*, 36
Lee, Y.C. *see* Li Ju-jen
Lee, Y.L., 163
Leighton, Bruce, 106
Leiser, Edward, 3
LeMay, *General* Curtis E., 187, 188, 190, 191, 193, 196, 197, 199
Leonard, Royal, 51, 102, 106
Leuba, Dorothy *see* Yuen, Dorothy (*Mrs.* Harold Lueba), 71
Levan, Jay E., 181
Lewis, *1st Lt.* Armit W., 168, 170
Li Ju-jen, 6
Li Keh-yuan, 95
Li Kwei-tan, 93, 94
Li Hwai-ming, 95
Li Pei-li, 36
Li Peng-chiang, 94
Li, *2nd Lt.* H., 170
Li. *Sub Lt.* T.T., 168
Li Wen-chiang, 94
Li Yuan-hung, 6, 17, 22
Lian Hong-yuan, 39
Liang Hung-yung, 83
Liang, T.H., 163
Lin, *2nd Lt.* S.C., 168
Lin Sen, 40
Liquor ration, 158
Lisowski, Benjamin, 199
Liu Chieh-sheng, 93, 94, 96
Liu Chung-wu, 94
Liu Hsiang, 32
Liu Mu-chuen, 98
Liu Tse-han, 94
Liu Tsui-kang, 83
Lo, S.Y., 163
Lo Yi-chien, 79, 83
Lo Ying-teh, 96
Loiwing, 52, 62

Long, Dudley, 103
Loofbourrow, Philip C., 145
Louie, Clifford, 96
Lu Tsi-shun, 94
Lufthansa, 19
Lundy, Glen E., 145
Lym, Art, 13, 14, 15, 16
Lym Fu-yuan, *see* Lym, Art

M

Ma Eng-lin, 137
McAllister, Gerald J., 139
MacArthur, *General* Douglas, 204
McBride, James C. *see* Mars, James C.
McCann, J.E., 171
McConnel, John Paul, 98
McDonald, Franklin J., 193
MacDonald, William C. ("Billy"), 39, 40, 41, 50, 103, 107
McEwen, Arch, 106
McGregor, Robert W., 139
McHahan, William, 199
McMillin, George, 163, 164
McNair, William, 194
McNamara, John, 15
Madeline, John J., 194
Mainichi (Shanghai), 203
Malir air base, 166
Manbeck, Les, 163
"Manchuria Aviation Company," 70
Mann, Harold, 190
Mao P'an-tsu *see* Mow, Peter
Mao Tso-tung, 77, 91, 193
Mao Ying-chu, 29
Mao Ying-tzu, 94
Marco Polo Bridge, 41
Mars, James C. ("Bud"), 11
Mars Task Force, 112
Marshall, *General* George C., 98
Martin, Frederick C., 18, 19
Matsui, Iwane, 203, 204
Matterhorn (code name), 188
Matthews, Ira V., 192
Mathewson, Christopher, 35, 36
Mauritson, Richard A., 145
May, Charles, 67
May, John, 36, 50
May, *Sir* Francis, 15
Mazanowski, Raymond J., 163
Mazurie, Henry L., 217
Meehan, James P., 193
Meeus, Charles L., 137
Merrill, Henry, 112
"Merrill's Marauders," 112
Meyer, Fred P., 145
Meyers, C.C., 190
Miller, Everett, 195
Mimmack, Russell F., 145
"Miss Beryl," 180
Mitchell, Hewitt, 59, 63
Mitchell, Ralph, 3
Mo Teh-yen, 94
"Monsoon Maiden," 174, 175, 178
Moon, Ryan, 163, 164
Mooney, R.M., 120
Moosa, Spencer, 114
Morgan, Herbert ("Butch"), 117, 138
Morrow, Dwight W., 20
Mountbatten, *Admiral* Lord Louis, 199
Mow, Peter, 39, 41, 42, 75, 91
Mukden air base, 133
Murrath, Herbert, 195
Musick, Edwin C., 62
Mussolini, Benito, 41, 42

N

Nahmacker, Carl, 23
Nakazone, Moritaka, 208
Nanking government established, 25
Nanyuan flight school, 16, 17, 30, 59
Nelson, Erick, 18, 19
Newman, Mike, 15
Neutrality Act, 110
Ng Yung-chung, 16
Nickols, Loren, 163
Nielsen, Chase Jay, 136
Norstad, Lauris, 197
Northeast Air Force, 31
Northern Punitive Expedition, 24, 25
Northrop, John K., 20
Norton, Frank, 163
H.I.J.M.S. *Notoro,* 206

O

Oblender, Harold T., 197
O'Brien, Donald G., 193
O'Connel, Joseph, 163
O'Connor, Morgan E., 139
O'Day, Gerald, 163
O'Donnell, Emmett *jr.* ("Rosie"), 199
O'Keefe, J.J., 223
"O'Reilly's Daughter," 193
Ogden, Henry H., 18
Olavson, Charles, 194
Olney, Richard O., 163, 164, 165
Olson, Arvid, 107
Operation Rooster, 140
Operational Training Unit (OTU), 166
Orakhov, Sasha, 90
Otero, Miguel A., 18
Otte, Frank R., 139

P

Pa Tsing-chen, 94
Palmer, Sam, 163, 164
Pan American Airways, 62
"Panda Bears," 111
Pao *Lieutenant,* 16
Parker, Lewis R, 188
Patterson, A.L. ("Pat"), 217
Pawley, William D.
—entertains foreign advisors, 40; as salesman for Curtiss aircraft, 51, 58; builds factory at Shien Chiao, 52; builds factory at Loiwing, 54; builds factory in India, 55; made U.S. envoy, 55; passenger to Chungking, 106; helps recruit volunteers, 110
Pennington, W.H., 36
Penny (NFN), 163
Pentecost, Walter, 110
Pickett, Ernest, 190
Pidhayne, Denny, 3, 188
Pierce, James O., 139
Pike, Harry, 229
Pilot's Handbook, 167
Polynin, Fedor, 90
Poncetti, H. ("Nemo"), 36, 50
Potapov, Ivan, 90
Prescott, Louis, 70
President Hoover (vessel), 78
Price, Frank B., 59

R

Radford, John, 197
Ramey, Roger, 197
Remy (NFN), 24
Randolph, Archie, 103, 105
Rector, Edward, 111, 113, 117
Red Guards, 89, 90
Reed, William N., 168
Reinburg, George E.A., 39, 50
Renieri, Deno, 174
Rescue patch, 106
Richardson, Elmer W., 145
Rickenbacker, Eddie, 174
Riggs, Jack, 195
"Ringer Squadron," 154
Risztics, Johann, 68
Robertson, *Mrs.* Renee, 16
Robinson, Herbert I., 163
Roby, William, 179
Rochelle, Harold, 163
Roeffear, Jan, 103
Rohe, Randall E., 162
Role of Defensive Pursuit, 108
Rooney, William, 194
Roosevelt, *President* Franklin D., 110, 159
Rowland, Harry T., 35
Rose, Walter L., 163, 164
Rosholt, Malcolm, 68, 217, 229
Rozema, Hal, 149
Rucker, Albert E., 139
H.I.J.M.S. *Ryujo,* 203

S

Sakai, Saburo, 92
Sansbury, Roland L., 35
Sargeant, Andrew, 55
Sato, Susie, 3
Saunders, LaVerne ("Blondie"), 187, 188, 192, 197
Schaefer, Vernon, 201
Schiel, Frank, 117
Schmidt, John, 163
Schmidt, Vincent, 103, 105
Schnepf, Carlton S., 180
Schriver, Todd, 11, 12
Schultz, Robert B., 163
Schultz, William, 55
Schwartz, Richard L., 163, 164
Schweitzer, Jack, 3, 35, 37, 45, 51
Scott, Robert L., 117, 139
Seagrave, *Dr.* Gordon, 56
Searle, Harold K., 163
The "Second Revolution," 24
The "Seventy-two Martyrs," 12
"Shanghai Incident," 33
Shannon, Ellis D., 35, 38, 39
Sharp, Charles L., 63
Shaw, Bacon, 55
Sheldon, (NFN), 163
Shen Tieh-hsieh, 42
Shien Chiao air base, 35, 36, 38, 41, 42, 75
Shih Kuang-fang, 98
Shinchiku air base, 152-153
Short, Elizabeth, 23
Short, Robert, 22, 23, 226

Shu, P.Y., 36, 41, 107
Shytle, John D. *jr.,* 181
Skaer, William K., 188
Smirnov, Sergei, 90
Smith, E.M., 54
Smith, Gail, 173
Smith, Harry G., 59, 66, 76
Smith, Lawrence W., 145
Smith, Lowell, 18, 19
Smith, Rose M., 144
Smith, Sebie Biggs, 3, 39, 40, 41, 50, 54, 55, 103, 105
Smith, Wilfred J., 139
Sobel, Andrew, 55
Sobel, Walter, 44
Sommer, Frank, 199
Sommers, *Lt. Col.* (NFN), 168
Soong, T.V., 35, 41, 44, 110
H.I.J.M.S., *Soryu,* 203, 216
Southeast Aviation Corporation, 67
Soviet aid to China
—air mission, 87; Soviet fighter tactics, 89; casualties, 90; number of aircraft sent to China, 90; Soviet raid on Formosa, 92
Spatz, Harold, 136
Spencer, W.A. ("Al"), 36, 39, 41, 42, 50
Stanovich, Joseph J., 139
Stelmach, Casimer, 193
Stevenson, Donald B., 36
Stewart, John S., 163, 165
Stickler, Merrill, 12
Stilwell, *General* Joseph W., 12, 117
Stinson Aircraft Corporation, 17, 59
Stinson, Eddie, 17, 59
Stinson, Katharine, 16, 17, 18
Stocker, William, 55
Stokes, Harry, 103, 105
Stone, Charles B., 147, 148
Strickland, Eugene L, 166, 167, 168
Strickland, Henry F., 138
Stuaffer, Ornell H., 190
Sullivan, Henry R., 188
Sun Fe, 59, 61
Sun, J.C., 163
Sun (NFN), 163
Sun Tung-kang, 27, 29
Sun Yat-sen
—political leader, 6; sends Tom Gunn to U.S., 15; organizes new government, 15; resigns presidency, 24; heads "Second Revolution," 24; accepts Soviet aid, 24-25; airplane named for him, 25.
Sutter, Harry, 50

T

Takasu, Shiro, 203
Tomaczak, Anthony, 199
Tan, *1st Lt.* K., 168
Tang, P.L., 163
Tang Shih-yuan, 30
Tanner, Vern A., 145
Tatum, Sterling, 39, 40, 50
Taylor, Richard V., 139
Taylor, T.L., 36
Teng, C.K., 163
Tenny, Parker C., 31
10th Air Force, 140
"Three Men on a Flying Trapeze," 41
Tiger Wang *see* Wang Hsu-ming
Ting Chi-hou, *Colonel,* 26
Ting Chin *see* Tinn, Mouhan K.
Ting Shih-yuan, 16
Tinian, *map,* 198
Tinn, Mouhan K., 30
Tong, Hollington, K., 59, 101
Treiber, Richard, 155
Tsai Ting-kai, 29
Tsai Tso, *Prince,* 16
Tsao Kun, 16
Tsao, W.P., 36
Tsing, *Colonel see* Ting Shih-yuan
Tsing Shao-yu, 94
Tu, P.S., 163
Tu, Ruthy, 18
Tuan Chi-jui, 22
Tung Ching-hsiang, 83
Tung King-chang, 95
Tung Ming-teh, 93, 94, 95
Tung Yen-po, 85
Turner, Arthur, 18
U.S.S. *Tutuila,* 196
"20th Century Unlimited," 197

U

Unit Citations, 170, 177
United Nations, 71, 73
United Press (Shanghai), 203
U.S. code names for Japanese aircraft
—Betty, 165; Claude, 93, 204; Dave, 204; Frank, 206; Lilly, 209; Nate, 164; Nell, 93, 165; Nick, 190; Oscar, 165, 207; Sally, 165; Tojo, 206, 207; Tony, 213; Topsy, 214; Zeke, 164, 208
United Nations Strategic Bombing Survey, 203, 213
Underwood, John, 3

V

Van Ausdall, Robert L., 170
Van den Born, Charles, 12, 15
Van Vleck, W.H., 134
Vargo, Anton M., 139
Vaughn, Charles S. ("Chillie"), 228
Vidovitch, Harland S., 145
Villon, Louis, 7, 11
Vincent, Clinton D. ("Casey"), 42, 117, 138, 139, 163, 212, 213
Vurgaropulas, James C., 145
Vultee, Gerald, 20

W

Wade, Alma, 35
Wade, Leigh, 18, 19
Wagner, Ray, 3
Waidner, Robert A., 139
Walker, Herbert, 103
Walsh, M.D. ("Doc"), 55, 56
Wang *Captain* (NFN), 166
Wang I, 94
Wang, *1st Lt.* K.F., 168
Wang Ching-wei, 207
Wang Hsu-ming, 16, 84, 107
Wang Ken, 30
Wang Samg-choi, 12
Wang, *2nd Lt.* M.T., 170
Wang Teh-chien, 94
Wang Tien-hsing, 36
Wang Yuan-po, 94, 96
Wang Yu-kun, 94
War crimes trials, 137
Watson, Harold F., 137
Watson, *Mrs.* Rolfe (Ivaloo), 3, 52-55, 57
Watson, Rolfe, 39, 40, 49, 50, 103, 106
Weigle, George H., 103. 106
Welch, Harold, 49, 103
Wells, Joseph B. ("Preacher"), 163
Wen, K.C., 163
Whampoa Military Academy, 15, 24
Whang Ping-han, 25, 26
Wheaton, Henry, 201
White, Lester W., 201
White, Theodore H., ("Teddy"), 163
"White Wolf" operations, 16
Whitehead, R.G., 103
The *Wilhelm Schmidt* (commercial plane), 68
Willard, PFC Clorada, 176
Willauer, Whiting, 70, 71, 73
Williams, Jesse C., 139
Williams, James M., 163
Williams, John M., 71, 108, 139
Williamson, John ("Luke") 39, 41, 50
Wine, James, 190
Wingate, Orde, 112
Wingerter, Edward, 36
Winkler, Eddie E., 193
Wise, Richard, 71
Wix, Robert, 187, 195, 199
Woelpel, Lyman, 103
Wolfe, Kenneth B. (or as K.B.), 187
Wong, Albert, 92
Wong H.S. ("Newsreel"), 79
Wong Un-ki, 15
Wright Brothers, 7, 11
Wu (NFN), 163
Wu Te-chen, 29
Wu Ting-cheng, 94

XYZ

Yang, C.C., 163
Yang, *1st Lt.* Y.K., 168
Yang Ku-fan, 94
Yang Shang-hsien, 94
Yang *Sub Lt.* C.T., 170
Yao, *Lieutenant,* 16
Yawata raid, 188, 190, 191
"Yellow Scorpions," 196
Yen Hsi-shan, 25
Yen Yu-chung, 98
Yi Chou-ma, 36
Yieh, *Capt.* W.F., 169
Yonai, Mitsumasa, 79
Yorston, Alfred *jr.,* 163, 164
Yoshiro, Sakemego, 23
The "Young Marshal" *see* Chang Hsueh-liang
Yuan, *Major* Ching-han, 171
Yuan Po-kang, 83
Yuan Shih-kai, 6, 15, 16
Yuen, Dorothy, 71
Yui, O.K., 78
Zaboluev, Kostyn, 90
Zeidler, Edward P., 178, 222
Zimmerman, Paul G., 36